T0215456

Computational Algebra

Course and Exercises
with Solutions

Computational Algebra
Course and Exercises
with Solutions

Ihsen Yengui
Université de Sfax, Tunisia

 World Scientific

NEW JERSEY · LONDON · SINGAPORE · BEIJING · SHANGHAI · HONG KONG · TAIPEI · CHENNAI · TOKYO

Published by

World Scientific Publishing Co. Pte. Ltd.

5 Toh Tuck Link, Singapore 596224

USA office: 27 Warren Street, Suite 401-402, Hackensack, NJ 07601

UK office: 57 Shelton Street, Covent Garden, London WC2H 9HE

Library of Congress Control Number: 2021017843

British Library Cataloguing-in-Publication Data
A catalogue record for this book is available from the British Library.

COMPUTATIONAL ALGEBRA
Course and Exercises with Solutions

ISBN 978-981-123-824-6 (hardcover)
ISBN 978-981-123-930-4 (paperback)
ISBN 978-981-123-825-3 (ebook for institutions)
ISBN 978-981-123-826-0 (ebook for individuals)

For any available supplementary material, please visit
https://www.worldscientific.com/worldscibooks/10.1142/12313#t=suppl

Desk Editor: Soh Jing Wen

Printed in Singapore

Contents

Introduction

This book is intended to provide material for a graduate course of one or two semesters on computational commutative algebra and algebraic geometry spotlighting potential applications in cryptography. Also, the topics in this book could form the basis of a graduate course that acts as a segue between an introductory algebra course and the more technical topics of commutative algebra and algebraic geometry. It contains a total of 124 exercises with detailed solutions as well as an important number of examples that illustrate definitions, theorems, and methods. This is very important for students or researchers who are not familiar with the topics discussed. Experience has shown that beginners who want to take their first steps in algebraic geometry are usually discouraged by the difficulty of the proposed exercises and the absence of detailed answers. So, exercises (and their solutions) as well as examples occupy a prominent place in this course. This book is not designed as a comprehensive reference work, but rather as a selective textbook. The many exercises with detailed answers make it suitable for use both in a math or computer science course.

Polynomial systems play a decisive role in many branches of computer and engineering science, e.g., in cryptography, robotics, and signal processing. Gröbner basis computation is one of the main practical tools for solving systems of polynomial equations. They were introduced in 1965, together with an algorithm to compute them (Buchberger's algorithm), by Bruno Buchberger in his Ph.D thesis. He named them after his advisor Wolfgang Gröbner. Buchberger's algorithm can be seen as both multivariate and nonlinear generalization of the Euclidean algorithm for computing polynomial gcd (greatest common divisor) and Gaussian elimination for solving linear systems.

This course focuses on the concept of Gröbner bases as a powerful tool for problem solving in algebraic geometry via the algebra-geometry dictionary, and also on elliptic curves and their applications in cryptography. Chapters 3 and 4 can be considered as "background" while the other chapters are the main content.

To differentiate from other books on Gröbner bases, we give in Chapter 1 a theory of Gröbner bases in the broader framework of Bézout domains of Krull dimension ≤ 1 or coherent Bézout rings (possibly with nonzero zero-divisors) of Krull dimension 0. This grants Gröbner bases a foothold in the non-Noetherian world and this is also important in concrete applications. As a matter of fact, besides the undeniable importance of Gröbner bases over fields, recently Gröbner bases over rings which are not fields (in particular $\mathbb{Z}/2^\alpha \mathbb{Z}$ and $(\mathbb{Z}/2^\alpha \mathbb{Z}) \times (\mathbb{Z}/2^\alpha \mathbb{Z})$) have attracted some attention due to their potential applications in formal verification of data paths, cryptography, error correcting codes, and coding theory. We present a method for the construction of Gröbner bases over valuation rings (possibly with zero-divisors) inspired from Buchberger's original method over fields. Recall that, a valuation ring is a ring in which every two elements are comparable under division. A typical example of a valuation domain which is not a field is $\mathbb{Z}_{p\mathbb{Z}} := \{ \frac{a}{b} \in \mathbb{Q} \mid a \in \mathbb{Z} \text{ and } b \in \mathbb{Z} \setminus p\mathbb{Z} \}$, where p is a prime number. A typical example of a valuation ring with nonzero zero-divisors is $\mathbb{Z}/p^\alpha \mathbb{Z}$, where p is a prime number and $\alpha \geq 2$. The dynamical method (see Remark 403) will be used to extend this Gröbner bases construction to arithmetical rings (possibly with zero-divisors). Arithmetical rings are rings which are locally valuation rings ($\mathbb{Z}$ and $\mathbb{Z}/m\mathbb{Z}$ are typical examples). This gave birth to the notion of "dynamical Gröbner bases" [32, 64, 66]. It is a new alternative for computation with multivariate polynomials over rings. Contrary to the methods that have been proposed, which suggest that for (Noetherian) rings the analog of Gröbner bases over field should be computed, a dynamical substitute is proposed. Instead of a Gröbner basis describing the situation globally, use a finite number of Gröbner bases, not over the base ring, but over comaximal localizations of this ring. At each localization, the computation behaves as if a valuation ring were present. In a nutshell, it is somewhat like Serre's method in "Corps locaux" [58] but follows the lazy fashion of computer algebra [13, 16]. In fact, the rings $\mathbb{Z}$ and $\mathbb{Z}/N\mathbb{Z}$ are more than arithmetical rings, they are Bézout rings (arithmetical rings with a Bézout identity between any two elements whose gcd is 1). We will explain that, over such rings, when computing dynamical Gröbner bases, one can avoid branching. This fact allows, for the first time, to give constructive versions of the celebrated Hilbert's syzygy theorem over $\mathbb{Z}$, $\mathbb{Z}_{p\mathbb{Z}}$ and $\mathbb{Z}/N\mathbb{Z}$ following Schreyer's method which uses Gröbner bases.

The first part of Chapter 2 is devoted to Noether normalization lemma: a cornerstone for the Hilbert's Nullstellensatz and other fundamental theorems. The little extra that is proposed for the first time is a wink to a recent version of Noether normalization lemma over the integers. Here, it is worth pointing out that as this result is new, more work needs to be done in order to interpret it geometrically and to find concrete applications. One main goal of Chapter 2 is to explain the algebra-geometry dictionary and the relevance of the use of Gröbner bases when doing effective computations. Particular emphasis will be given to the zero-dimensional polynomials systems (those having a finite number of solutions) which are for obvious reasons pertinent in concrete applications. Following Faugère, we will present a method for parametrizing such systems using only linear algebra (computation of ranks of matrices).

As finite fields are widely used in concrete applications in cryptography (this is one main possible field of application of the algorithms presented in this book), error correcting codes, biology, statistics, Chapters 3 and 4 will be devoted to a quick overview of finite fields and some basic notions of cryptography. Of course, these two chapters are optional and rather serve only as rapid overviews. So, students who already studied finite fields can skip Chapter 3 (for example, students with good background in mathematics), while students who already studied basic notions of cryptography can skip Chapter 4 (for example, students with good background in computer science).

Having an objective in Chapter 6 to introduce elliptic curves and elliptic cryptosystems over finite fields, we will give in Chapter 5 an overview of algebraic plane curves, and, particularly, we will introduce Bézout's theorem, Riemann-Roch theorem and genus computation via quadratic transformations.

This course essentially follows [14, 22, 56, 66] and five courses held by François Brunault (Courbes elliptiques, ENS Lyon, 2008–2009), Olivier Debarre (Introduction à la géométrie algébrique), Christophe Delaunay (Courbes elliptiques), Alain Kraus (Cours de cryptographie, 2012/2013), and Franz Lemmermeyer (Algebraic Geometry, 2005).

The present notes are based on lectures on Computational Commutative Algebra and Algebraic Geometry the author gave at Verona (2017) and Trento (2015) universities.

The undefined terminology is standard as in [14, 19, 24, 33, 59, 61] for algebraic geometry, [2, 5, 8, 14, 19, 41, 42, 43, 69] for commutative algebra, [28, 30] for computer algebra, and [39] for cryptography. All the considered rings are commutative and unitary.

Prerequisites

Readers should have taken courses in elementary number theory, elementary group theory, linear algebra, abstract algebra, and topology (basic notions). Everything that is assumed is contained in the books [5, 43]. In particular, readers should have an understanding of the following notions:

- definition of a commutative ring,
- ideals, prime ideals, maximal ideals,
- zero-divisors,
- quotient rings,
- localization of a ring,
- Krull dimension of a ring,
- principal ideal rings, unique factorial domains.

Chapter 1

Gröbner bases over arithmetical rings

The method of Gröbner bases is a powerful effective technique for solving problems in commutative algebra and algebraic geometry that was coined by Bruno Buchberger in his PhD thesis [10]. Roughly speaking, a Gröbner basis G for a system S of polynomials is an equivalent system (i.e., generating the same ideal) that possesses useful properties, for example, that another polynomial f is a combination of those in S iff the remainder of f on division by G is 0 (a monomial order being chosen). In the univariate case, a Gröbner for S is nothing but the gcd of the elements of S. Buchberger's algorithm for computing Gröbner bases can be seen as both multivariate and nonlinear generalization of the Euclidean algorithm for computing polynomial gcd and Gaussian elimination for solving linear systems. Gröbner bases provide a uniform approach for solving problems that can be expressed in terms of systems of multivariate polynomial equations. It happens that many practical problems, e.g. in signal processing and cryptanalysis, can be transformed into sets of polynomials, thus solved using Gröbner bases technique.

In addition to the growing interest in using Gröbner basis over fields, Gröbner basis techniques in polynomial rings over $\mathbb{Z}/m\mathbb{Z}$ and $(\mathbb{Z}/p^\alpha\mathbb{Z}) \times (\mathbb{Z}/p^\alpha\mathbb{Z})$ (in particular $\mathbb{Z}/2^\alpha\mathbb{Z}$ and $(\mathbb{Z}/2^\alpha\mathbb{Z}) \times (\mathbb{Z}/2^\alpha\mathbb{Z})$) have attracted some attention due to their potential applications in formal verification of data paths [7, 31, 60], and coding theory [11, 50, 51, 52, 53] (see also the recent Ph.D thesis of Wienand [67]). Also, some authors [4, 18, 29, 54, 55, 68] have been interested in computing Gröbner bases over $\mathbb{Z}/p^\alpha\mathbb{Z}$ (where p is a "lucky" prime number), because modular methods give a satisfactory way to avoid intermediate coefficient swell with Buchberger's algorithm for computing Gröbner bases over the rational numbers.

In Section 1.2, we will cast light on the approach of Gröbner bases via valuation rings used in the papers [32, 64, 65]. This is still not very well-known to people working on this subject despite its simplicity and the fact that it enables one to easily resolve the delicate problem caused by the appearance of zero-divisors as leading coefficients (see for example [12], and [36] in which Boolean rings are used to model propositional calculus). We will specify the presented method to the rings $\mathbb{Z}/p^{\alpha}\mathbb{Z}$ (where p is a prime number), and to the ring $\mathbb{F}_2[Y]/\langle Y^2 \rangle$ (a useful ring in coding theory, where $\mathbb{F}_2 = \mathbb{Z}/2\mathbb{Z}$).

In Section 1.3, we will explain how to compute "dynamical Gröbner bases" [32, 64, 65] over arithmetical rings, i.e., rings which are locally valuation rings, with $\mathbb{Z}/m\mathbb{Z}$ and $(\mathbb{Z}/p^{\alpha}\mathbb{Z}) \times (\mathbb{Z}/p^{\alpha}\mathbb{Z})$ as main examples. We will also explain that when computing dynamical Gröbner bases over $\mathbb{Z}$ (and more generally, over a Bézout domain of Krull dimension ≤ 1) one can avoid branching. The theory of Gröbner bases being developed, we will show in Section 1.5 how to compute free resolutions using Gröbner bases.

1.1 Dickson's lemma and Gröbner bases over fields

The main thread of this section is the following central problem:

Ideal Membership Problem (IMP, in short):

For $f \in \mathbf{K}[X_1,\ldots,X_n]$ ($\mathbf{K}$ a field) and $I = \langle f_1,\ldots,f_s \rangle \subseteq \mathbf{K}[X_1,\ldots,X_n]$, can we decide whether $f \in I$ and, in case of positive answer, can we compute $u_1,\ldots,u_s \in \mathbf{K}[X_1,\ldots,X_n]$ such that $f = u_1 f_1 + \cdots + u_s f_s$?

For $n = 1$, it suffices to calculate the greatest common divisor $\Delta = \gcd(f_1,\ldots,f_s)$. Then one has:

$$f \in \langle f_1,\ldots,f_s \rangle = \langle \Delta \rangle \Leftrightarrow \Delta/f.$$

In case of positive answer, say $f = g\Delta$ with $g \in \mathbf{K}[X_1]$, to find $u_1,\ldots,u_s \in \mathbf{K}[X_1]$ such that $f = u_1 f_1 + \cdots + u_s f_s$ it suffices to write $\Delta = v_1 f_1 + \cdots + v_s f_s$ by the Extended Bézout algorithm with $v_i \in \mathbf{K}[X_1]$, and to take $u_i = g v_i$.

However, it is much more difficult to address the IMP when $n \geq 2$ since $\mathbf{K}[X_1,\ldots,X_n]$ cannot be equipped with an Euclidean division algorithm (it is no longer an Euclidean ring). Our goal is to present a substitute to this algorithm in $\mathbf{K}[X_1,\ldots,X_n]$ when $n \geq 2$.

The Bézout algorithm for computing the gcd is nothing but a finite number of calls of the Euclidean division algorithm. This latter plays a decisive and unavoidable role when dealing with univariate polynomials with coefficients in a field $\mathbf{K}$. It works as follows: if one wants to divide a $f(X) = a_n X^n + a_{n-1} X^{n-1} + \cdots$ ($a_i \in \mathbf{K}$, $a_n \neq 0$) by a polynomial $g(X) = b_{n-r} X^{n-r} + b_{n-r-1} X^{n-r-1} + \cdots$ ($b_i \in \mathbf{K}$, $r \geq 0$, $b_{n-r} \neq 0$), one should subtract $a_n b_{n-r}^{-1} X^r g$ from f to cancel the leading term $a_n X^n$ of f, and then to resume the same process with $f - a_n b_{n-r}^{-1} X^r g$ until one obtains a polynomial of degree less than $n - r$. Note that the monomials of the polynomials above are written in decreasing order by degree in X:

$$\cdots > X^{k+1} > X^k > \cdots > X^3 > X^2 > X > 1.$$

In order to generalize the division algorithm to the multivariate case, as a first step, one has to define a total order on monomials which is compatible with multiplication and which is a well-ordering, in the sense that any nonincreasing sequence of monomials pauses (Corollary 13).

Definition and notation 1. Let **R** be a ring, $n \geq 1$, and consider n independent variables $X_1, \ldots, X_n$ on **R**.

(1) For $\alpha = (\alpha_1, \ldots, \alpha_n) \in \mathbb{N}^n$, we denote by $X^\alpha := X_1^{\alpha_1} \cdots X_n^{\alpha_n}$.

(2) We denote by $\mathbb{M}_n := \{X^\alpha \mid \alpha \in \mathbb{N}^n\}$ the set of monomials at $X_1, \ldots, X_n$. Of course, there is a one-to-one correspondence between $\mathbb{M}_n$ and $\mathbb{N}^n$ given by $X^\alpha \leftrightarrow \alpha$, with $1 \leftrightarrow (0, \ldots, 0)$.

(3) A monomial order on $\mathbf{R}[X_1, \ldots, X_n]$ is a relation $>$ on $\mathbb{M}_n$ satisfying:

   (i) $>$ is a total order on $\mathbb{M}_n$.

   (ii) For $\alpha, \beta, \gamma \in \mathbb{N}^n$, if $X^\alpha > X^\beta$ then $X^\alpha X^\gamma > X^\beta X^\gamma$.

   (iii) $X^\alpha \geq 1$ for all $\alpha \in \mathbb{N}^n$.

Example 2. Let $\alpha, \beta \in \mathbb{N}^n$.

(1) *Lexicographic order* with $X_1 > X_2 > \cdots > X_n$: $X^\alpha >_{\text{lex}} X^\beta$ if the left-most nonzero entry of $\alpha - \beta$ is positive. For example, $X_1^2 X_2 X_3^2 >_{\text{lex}} X_1^2 >_{\text{lex}} X_1 X_2^2 X_3^3 >_{\text{lex}} X_1 X_2 X_3^3 >_{\text{lex}} X_3^7$.

(2) *Graded lexicographic order* with $X_1 > X_2 > \cdots > X_n$: $X^\alpha >_{\text{grlex}} X^\beta$ if $\sum_{i=1}^n \alpha_i > \sum_{i=1}^n \beta_i$ or ($\sum_{i=1}^n \alpha_i = \sum_{i=1}^n \beta_i$ and $X^\alpha >_{\text{lex}} X^\beta$). For example, $X_3^7 >_{\text{grlex}} X_1^2 X_2 X_3^2 >_{\text{grlex}} X_1 X_2^2 X_3^2 >_{\text{grlex}} X_1 X_2 X_3^3 >_{\text{grlex}} X_1^2$.

(3) *Graded reverse lexicographic order* with $X_1 > X_2 > \cdots > X_n$: $X^\alpha >_{\text{grevlex}} X^\beta$ if $\sum_{i=1}^n \alpha_i > \sum_{i=1}^n \beta_i$ or ($\sum_{i=1}^n \alpha_i = \sum_{i=1}^n \beta_i$ and the right-most nonzero entry of $\alpha - \beta$ is negative). For example, $X_3^7 >_{\text{grevlex}} X_1^2 X_2 X_3^2 >_{\text{grevlex}} X_1 X_2^2 X_3^2 >_{\text{grevlex}} X_1 X_2 X_3^3 >_{\text{grevlex}} X_1^2$.

Of course, for $n = 1$, all monomial orders coincide.

Definition 3. (Monomial orders on finite-rank free $\mathbf{R}[\underline{X}]$-modules)

Let **R** be a ring, $n, m \geq 1$. Consider n indeterminates $X_1, \ldots, X_n$ and $\mathbf{R}[\underline{X}] = \mathbf{R}[X_1, \ldots, X_n]$. Let $\mathbf{H}_m \simeq \mathbb{A}^m(\mathbf{R}[\underline{X}])$ be a free $\mathbf{R}[\underline{X}]$-module with basis $(e_1, \ldots, e_m)$.

1. A *monomial* in $\mathbf{H}_m$ is a vector of type $\underline{X}^\alpha e_i$ $(1 \leq i \leq m)$, where $\underline{X}^\alpha = X_1^{\alpha_1} \cdots X_n^{\alpha_n}$ is a monomial in $\mathbf{R}[\underline{X}]$; the index i is the *position* of the monomial. The set of monomials in $\mathbf{H}_m$ is denoted by $\mathbb{M}_n^m$, with $\mathbb{M}_n^1 \cong \mathbb{M}_n$ (the set of monomials in $\mathbf{R}[\underline{X}]$). For example, $X_1 X_2^3 e_2$ is a monomial in $\mathbf{H}_m$, but $2X_1 e_3$, $(X_1 + X_2^3)e_2$ and $X_1 e_2 + X_2^3 e_3$ are not.

   If $M = \underline{X}^\alpha e_i$ and $N = \underline{X}^\beta e_j$, we say that M *divides* N if $i = j$ and $\underline{X}^\alpha$ divides $\underline{X}^\beta$. For example, $X_1 e_1$ divides $X_1 X_2 e_1$, but does not divide $X_1 X_2 e_2$. Note that in the case that M divides N, there exists a monomial $\underline{X}^\gamma$ in $\mathbf{R}[\underline{X}]$ such that $N = \underline{X}^\gamma M$. In this case, we define $N/M := \underline{X}^\gamma$. For example, $(X_1 X_2 e_1)/(X_1 e_1) = X_2$.

   A *term* in $\mathbf{H}_m$ is a vector of type cM, where $c \in \mathbf{R} \setminus \{0\}$ and $M \in \mathbb{M}_n^m$. We say that a term cM *divides* a term $c'M'$, with $c, c' \in \mathbf{R} \setminus \{0\}$ and $M, M' \in \mathbb{M}_n^m$, if c divides c' and M divides M'.

2. A *monomial order* on $\mathbf{H}_m$ is a relation $>$ on $\mathbb{M}_n^m$ such that

   (a) $>$ is a total order on $\mathbb{M}_n^m$,

   (b) $\underline{X}^\alpha M > M$ for all $M \in \mathbb{M}_n^m$ and $\underline{X}^\alpha \in \mathbb{M}_n \setminus \{1\}$,

   (c) $M > N \Rightarrow \underline{X}^\alpha M > \underline{X}^\alpha N$ for all $M, N \in \mathbb{M}_n^m$ and $\underline{X}^\alpha \in \mathbb{M}_n$.

   Note that, when specialised to the case $m = 1$, this definition coincides with the definition of a monomial order on $\mathbf{R}[\underline{X}]$. Any *nonzero* vector $h \in \mathbf{H}_m$ can be written as a sum of terms
   $$h = \sum_{i=1}^t c_i M_i,$$
   with $c_i \in \mathbf{R} \setminus \{0\}$, $M_i \in \mathbb{M}_n^m$, and $M_t > M_{t-1} > \cdots > M_1$. We define the *leading coefficient, leading monomial,* and *leading term* of h as in the ring case (see Definition 4 below): $\mathrm{LC}(h) = c_t$, $\mathrm{LM}(h) = M_t$, $\mathrm{LT}(h) = c_t M_t$. Letting $M_t = \underline{X}^\alpha e_\ell$ with $\underline{X}^\alpha \in \mathbb{M}_n^m$ and $1 \leq \ell \leq m$, we say that α is the *multidegree of h* and write $\mathrm{mdeg}(h) = \alpha$, and that the index ℓ is the *leading position* of h and write $\mathrm{LPos}(h) = \ell$.

   We stipulate that $\mathrm{LT}(0) = 0$ and $\mathrm{mdeg}(0) = -\infty$, but we do not define $\mathrm{LPos}(0)$.

3. A monomial order on $\mathbf{R}[\underline{X}]$ gives rise to the following canonical monomial order on $\mathbf{H}_m$: for monomials $M = \underline{X}^\alpha e_i$ and $N = \underline{X}^\beta e_j \in \mathbb{M}_n^m$, let us define that

$$M > N \quad \text{if} \quad \left| \begin{array}{l} \text{either } \underline{X}^\alpha > \underline{X}^\beta \\ \text{or both } \underline{X}^\alpha = \underline{X}^\beta \text{ and } i < j. \end{array} \right.$$

   This monomial order is called the *term over position* (TOP) order because it gives more importance to the monomial order on $\mathbf{R}[\underline{X}]$ than to the vector position. For example, when $X_1 > X_2$, we have
   $$X_1 e_1 > X_1 e_2 > X_2 e_1 > X_2 e_2.$$

4. Most of the results given in this book related to Gröbner bases for finitely-generated ideals in $\mathbf{R}[X_1,\ldots,X_n]$ can fairly be generalized to finitely-generated submodules of $\mathbf{R}[X_1,\ldots,X_n]^m$ using monomials order on $\mathbf{R}[X_1,\ldots,X_n]^m$, namely, Dickson' lemma (Theorem 12), the division algorithm, Buchberger's algorithm, the ideal membership test, and so on. We sometimes chose to limit ourselves to ideals simply to make the text easier to read.

Definition 4. Let $\mathbf{R}$ be a ring, $f = \sum_\alpha a_\alpha X^\alpha$ a nonzero polynomial in $\mathbf{R}[X_1,\ldots,X_n]$, and $>$ a monomial order on $\mathbf{R}[X_1,\ldots,X_n]$.

(1) The X^α (resp. the $a_\alpha X^\alpha$) are called the monomials (resp. the terms) of f.

(2) The *multidegree* of f is $\mathrm{mdeg}(f) := \max\{\alpha \in \mathbb{N}^n : a_\alpha \neq 0\}$.

(3) The *leading coefficient* of f is $\mathrm{LC}(f) := a_{\mathrm{mdeg}(f)} \in \mathbf{R}$.

(4) The *leading monomial* of f is $\mathrm{LM}(f) := X^{\mathrm{mdeg}(f)}$.

(5) The *leading term* of f is $\mathrm{LT}(f) := \mathrm{LC}(f)\mathrm{LM}(f)$.

(6) For $g,h \in \mathbf{R}[X_1,\ldots,X_n] \setminus \{0\}$, we say that $\mathrm{LT}(g)$ divides $\mathrm{LT}(h)$ if $\mathrm{LM}(g)$ divides $\mathrm{LM}(h)$ and $\mathrm{LC}(g)$ divides $\mathrm{LC}(h)$.

Example 5. Let $f = 1 - 3XY^2 + 2X^2Y \in \mathbb{Z}[X,Y]$ equipped with the lexicographic monomial order with $X > Y$. Then: $\mathrm{LM}(f) = X^2Y$, $\mathrm{LC}(f) = 2$, and $\mathrm{LT}(f) = 2X^2Y$.

In order to give a constructive proof of Dickson's lemma following [47], we give the following definition.

Definition 6. A partially ordered set $(E, \leq)$ is said to satisfy the *Descending Chain Condition* (in short, DCC) if for every nonincreasing sequence $(u_n)_{n\in\mathbb{N}}$ in E, there exists $n \in \mathbb{N}$ such that $u_n = u_{n+1}$. A partially ordered set $(E, \leq)$ is said to satisfy the *Ascending Chain Condition* (in short, ACC) if for every nondecreasing sequence $(u_n)_{n\in\mathbb{N}}$ in E, there exists $n \in \mathbb{N}$ such that $u_n = u_{n+1}$.

Example 7. $\mathbb{N}$ with the usual order satisfies DCC.

Let $(E, \leq)$ be a partially ordered set. We will denote by $\leq_d$ the order on E^d defined by $(x_1,\ldots,x_d) \leq_d (y_1,\ldots,y_d)$ if and only if $x_i \leq y_i$ for all $1 \leq i \leq d$. We shall write $\leq$ instead of $\leq_d$ when there is no risk of confusion.

Lemma 8. If a partially ordered set $(E, \leq)$ satisfies DCC (resp., ACC), then so does $(E^d, \leq_d)$ (resp., ACC).

Proof. It suffices to prove the result in the case $d = 2$. The same reasoning can be used to prove the general case by induction. Let $(u_n, v_n)_{n \in \mathbb{N}}$ be a nonincreasing sequence of elements of E^2. It is easy to see that, since the sequence $(u_n)_{n \in \mathbb{N}}$ is nonincreasing, one can find $n_1 < n_2 < \cdots$ such that $u_{n_i} = u_{n_i+1}$ for all $i \in \mathbb{N}$. The sequence $(v_{n_i})_{i \in \mathbb{N}}$ being nonincreasing, there exists $j \in \mathbb{N}$ such that $v_{n_j} = v_{n_j+1}$. But, as $v_{n_j} \geq v_{n_j+1} \geq v_{n_j+1}$, we have $v_{n_j} = v_{n_j+1}$, and, thus, $(u_{n_j}, v_{n_j}) = (u_{n_j+1}, v_{n_j+1})$.

For the ACC case, consider the reverse order.

Definition and notation 9. Let $(E, \leq)$ be a partially ordered set.

(1) For $Y \in E$, we define

$$Y^{\uparrow} := \{Z \in E \mid Z \geq Y\},$$

and for $Y_1, \ldots, Y_m \in E$, we define

$$\mathcal{M}_E^+(Y_1, \ldots, Y_m) := \cup_{i=1}^m Y_i^{\uparrow} = \{Z \in E \mid Z \geq Y_1 \vee \cdots \vee Z \geq Y_m\}.$$

$\mathcal{M}_E^+(Y_1, \ldots, Y_m)$ is called a *final subset of finite type* of E (generated by $Y_1, \ldots, Y_m$).

The set of final subsets of finite type of E, including the empty subset considered as generated by the empty family, will be denoted by $\mathcal{F}(E)$.

(2) In the particular case $E = \mathbb{N}^d$, for $Y = (y_1, \ldots, y_d) \in \mathbb{N}^d$, we have

$$
\begin{aligned}
Y^{\uparrow} &:= \{Z = (z_1, \ldots, z_d) \in \mathbb{N}^d \mid z_i \geq y_i \forall 1 \leq i \leq d\} \\
&= (y_1, \ldots, y_d) + \mathbb{N}^d.
\end{aligned}
$$

The set $\mathcal{F}(\mathbb{N}^d) \setminus \{\emptyset\}$ will be denoted by $\mathcal{M}_d$. So, $\mathcal{F}(\mathbb{N}^d)$ is isomorphic to $\mathcal{M}_d \cup \{-\infty\}$.

Proposition 10. [47]

(1) *Every $A \in \mathcal{M}_d$ is generated by a unique minimal family (for $\subseteq$). This family can be obtained by taking the minimal elements (for $\leq_d$) of any family of generators of A.*

(2) *Given $A, B \in \mathcal{M}_d$, one can decide whether $A \subseteq B$ or not.*

(3) *The ordered set $(\mathcal{M}_d, \subseteq)$ satisfies ACC.*

Proof. (1) It is clear that for any $Y, Y_1, \ldots, Y_n \in \mathbb{N}^d$, we have

$$Y^\uparrow \subseteq Y_1^\uparrow \cup \cdots \cup Y_n^\uparrow \;\Leftrightarrow\; Y \in Y_1^\uparrow \cup \cdots \cup Y_n^\uparrow \;\Leftrightarrow\; Y_1 \leq_d Y \text{ or } \cdots \text{ or } Y_n \leq_d Y.$$

So, starting from a finite family of generators of A, to obtain a minimal family of generators of A, one has only to keep the minimal elements (for $\leq_d$) of the considered family. This proves the existence part of (1).

If $Y_1, \ldots, Y_n$ and $Z_1, \ldots, Z_m$ are minimal families of generators of A, then for each $1 \leq i \leq n$, there exists $1 \leq j \leq m$ such that $Z_j \leq_d Y_i$ and vice versa. By minimality of the two families, we deduce that $\{Y_1, \ldots, Y_n\} = \{Z_1, \ldots, Z_m\}$.

(2) This is straightforward.

(3) We will induct on d. The case $d = 1$ is clear. Suppose that $d \geq 2$ and consider a nondecreasing sequence $(A_m)_{m \in \mathbb{N}}$ in $\mathcal{M}_d$. Let $a = (a_1, \ldots, a_d) \in A_0$. For all $1 \leq i \leq d$ and $r \in \mathbb{N}$, let

$$H_{i,d}^r := \{(x_1, \ldots, x_d) \in \mathbb{N}^d \mid x_i = r\}.$$

There is an obvious order isomorphism between $(H_{i,d}^r, \leq_d)$ and $(\mathbb{N}^{d-1}, \leq_{d-1})$. So $(\mathscr{F}(H_{i,d}^r), \subseteq)$ satisfies ACC by induction hypothesis (it is isomorphic to $\mathcal{M}_{d-1} \cup \{-\infty\}$). The crucial point in the proof is the following observation:

$$\mathbb{N}^d \setminus a^\uparrow = \cup_{i=1}^d \cup_{r < a_i} H_{i,d}^r \text{ (a finite union)}.$$

It follows that for all $m \in \mathbb{N}$, we have

$$A_m = a^\uparrow \bigcup \cup_{i=1}^d \cup_{r < a_i} (A_m \cap H_{i,d}^r).$$

The desired result follows since all the nondecreasing sequences $(A_m \cap H_{i,d}^r)_{m \in \mathbb{N}}$ pause by induction hypothesis.

Definition 11. Consider d variables $X_1, \ldots, X_d$ over a field $\mathbf{K}$. As usual, for $\alpha = (\alpha_1, \ldots, \alpha_d) \in \mathbb{N}^d$, X^α denotes the monomial $X_1^{\alpha_1} \cdots X_d^{\alpha_d}$.

A *monomial ideal* of $\mathbf{K}[X_1, \ldots, X_d]$ is an ideal generated by a family of monomials at $X_1, \ldots, X_d$. Clearly, two monomial ideals are equal if and only if they contain the same monomials, and the set of finitely-generated monomial ideals is in one-to-one correspondence with $\mathcal{M}_d$.

The third assertion of Proposition 10 is equivalent to Dickson's lemma.

Theorem 12. (Dickson's lemma)

The set of finitely-generated monomial ideals of $\mathbf{K}[X_1,\ldots,X_d]$, *ordered with* $\subseteq$, *satisfies ACC.*

Corollary 13. *Let* $\mathbf{R}$ *be a ring,* $>$ *be a monomial order on* $\mathbf{R}[X_1,\ldots,X_d]$, *and denote by* $\mathbb{M}_d := \{X^\alpha = X_1^{\alpha_1}\cdots X_d^{\alpha_d} \mid \alpha \in \mathbb{N}^d\}$. *Then* $(\mathbb{M}_d,\leq)$ *satisfies DCC. In other words, any monomial order is a well-ordering.*

Proof. There is a one-to-one correspondence between $\mathbb{M}_d$ and $\mathbb{N}^d$ given by $X^\alpha \leftrightarrow \alpha$. A nonincreasing sequence $(u_n)_{n\in\mathbb{N}}$ in $(\mathbb{N}^d,>)$ pauses at step n if and only if the nondecreasing sequence $(\cup_{i=0}^n u_i^\uparrow)_{n\in\mathbb{N}}$ in $(\mathcal{M}_d,\subseteq)$ pauses at step n. The desired result follows from Proposition 10.

Now we have all the necessary tools to give a generalization of the Euclidean division algorithm to multivariate polynomials.

Algorithm 14. (Division Algorithm in $\mathbf{K}[X_1,\ldots,X_d]$)

Input: $f_1,\ldots,f_s,f \in \mathbf{K}[X_1,\ldots,X_d]$, and $>$ a monomial order on $\mathbf{K}[X_1,\ldots,X_n]$, where $\mathbf{K}$ is a field.

Output: $q_1,\ldots,q_s,r \in \mathbf{K}[X_1,\ldots,X_d]$ such that

$$f = q_1 f_1 + \cdots + q_s f_s + r,$$

$\mathrm{mdeg}(q_i f_i) \leq \mathrm{mdeg}(f)$, and either $r = 0$ or r is a linear combination, with coefficients in $\mathbf{K}$, of monomials, none of which is divisible by any of $\mathrm{LT}(f_1),\ldots,\mathrm{LT}(f_s)$. The polynomial r is called *a remainder of* f on division by $F := [f_1,\ldots,f_s]$, and is denoted by $\bar{f}^F$.

Initialization: $q_1 := 0; \cdots; q_s := 0; r := 0; p := f$
WHILE $p \neq 0$ DO
   $i := 1;$
   div := false
   WHILE $i \leq s$ AND div $=$ false DO
      IF $\mathrm{LT}(f_i)$ divides $\mathrm{LT}(p)$ THEN
         $q_i := q_i + \frac{\mathrm{LT}(p)}{\mathrm{LT}(f_i)}$
         $p := p - \frac{\mathrm{LT}(p)}{\mathrm{LT}(f_i)} f_i$
         div := true
      ELSE
         $i := i + 1$
   IF div $=$ false THEN
      $r := r + \mathrm{LT}(p)$
      $p := p - \mathrm{LT}(p)$

Proposition 15. *Algorithm 14 terminates and is correct.*

Proof. The fact that Algorithm 14 terminates is constructively proven by Theorem 12 since $\mathrm{LM}(p)$ decreases until reaching $p = 0$. The correctness of Algorithm 14 is obvious.

Example 16. Let $f = X^2Y^2$, $f_1 = 2 + 12XY$, $f_2 = 8Y^2 \in \mathbb{Q}[X,Y]$, and fix any monomial order $>$ on $\mathbb{Q}[X,Y]$. Then $\bar{f}^{\,[f_1,f_2]} = \frac{1}{36}$ while $\bar{f}^{\,[f_2,f_1]} = 0$.

We conclude that, contrary to the univariate case, the remainder is not unique. Moreover, from the second division, we infer that $f \in \langle f_1, f_2 \rangle$ despite that the remainder of the first division is not null (is a unit). It can be seen that the division algorithm (Algorithm 14) is not a satisfactory generalization of its univariate counterpart. The remedy for this is the notion of "Gröbner basis".

Definition 17. Let **R** be a ring.

(1) For a nonzero ideal $I = \langle f_1, \ldots, f_s \rangle$ of $\mathbf{R}[X_1, \ldots, X_n]$, choosing a monomial order on $\mathbf{R}[X_1, \ldots, X_n]$, we set $\mathrm{LT}(I) := \langle \mathrm{LT}(g), g \in I \setminus \{0\} \rangle$. It is an ideal of $\mathbf{R}[X_1, \ldots, X_n]$.

(2) Let I be a nonzero ideal of $\mathbf{R}[X_1, \ldots, X_n]$, and $f_1, \ldots, f_s \in I$. We say that $G = \{f_1, \ldots, f_s\}$ or $G = [f_1, \ldots, f_s]$ or $G = (f_1, \ldots, f_s)$ is a *Gröbner basis* for I accordingly to a monomial order if

$$\mathrm{LT}(I) = \langle \mathrm{LT}(f_1), \ldots, \mathrm{LT}(f_s) \rangle.$$

We convene that $\emptyset$ is a Gröbner basis for $\{0\}$.

(3) If for any monomial order, every finitely-generated ideal of $\mathbf{R}[X_1, \ldots, X_n]$ has a Gröbner basis, we say that **R** is a *Gröbner*.

By the following, we see that a Gröbner basis is necessarily a generating set.

Proposition 18. *Fix a monomial order on $\mathbf{K}[X_1, \ldots, X_n]$ where $\mathbf{K}$ is a field, and consider a nonzero ideal I of $\mathbf{K}[X_1, \ldots, X_n]$. If $G = \{g_1, \ldots, g_s\}$ is a Gröbner basis of I, then $I = \langle G \rangle = \langle g_1, \ldots, g_s \rangle$.*

Proof. We have $\langle g_1, \ldots, g_s \rangle \subseteq I$ since $g_i \in I$. Conversely, let $f \in I$. Dividing by f G we obtain $f = q_1 g_1 + \cdots + q_s g_s + r$, where either $r = 0$ or none of the terms of r is divisible by any of the $\mathrm{LT}(g_i)$.
Suppose that $r \neq 0$. Since $r \in I$, we have

$$\mathrm{LT}(r) = h_1 \mathrm{LT}(g_1) + \cdots + h_s \mathrm{LT}(g_s),$$

with $h_i \in \mathbf{K}[X_1, \ldots, X_n]$. Expanding each polynomial h_i as a sum of terms, we see that the right side of the above equation is a sum of terms each of which is divisible by a certain $\mathrm{LT}(g_i)$. This, however, contradicts our assumption on r. It follows that $r = 0$, and hence $f \in \langle g_1, \ldots, g_s \rangle$.

From Dickson's lemma (Theorem 12) ensues that every monomial ideal of $K[X_1,\ldots,X_n]$ (**K** a field) is finitely-generated and this directly yields to the following theorem which, in particular, implies that fields are Gröbner rings.

Theorem 19. *Fix a monomial order on* $K[X_1,\ldots,X_n]$ *where* **K** *is a field. Then every ideal of* $K[X_1,\ldots,X_n]$ *has a Gröbner basis (and, thus, finitely-generated).*

Corollary 20. *For any field* **K**, *the ring* $K[X_1,\ldots,X_n]$ *is Noetherian, that is, every ideal of* $K[X_1,\ldots,X_n]$ *is finitely-generated.*

The following gives an answer to the IMP via the uniqueness of the remainder on division by a Gröbner basis.

Proposition 21. *Let* $G = \{g_1,\ldots,g_s\}$ *be a Gröbner basis of a nonzero ideal* I *of* $K[X_1,\ldots,X_n]$ *and let* $f \in K[X_1,\ldots,X_n]$ (**K** *a field). There exists a unique* $r \in K[X_1,\ldots,X_n]$ *such that:*
(i) $r = 0$ *or none of the terms of* r *is divisible by any of the* $LT(g_i)$.
(ii) $\exists\ g \in I$ *such that* $f = g + r$. *In particular,* r *is the remainder of* f *on division by* $\{g_1,\ldots,g_s\}$ *(regardless of how the* g_i's *are ordered).*

Proof. Existence: By Division Algorithm 14.
Uniqueness: Suppose that $f = g_1 + r_1 = g_2 + r_2$. This implies $r_2 - r_1 \in I$.
If $r_2 \neq r_1$, then $LT(r_2 - r_1) \in \langle LT(g_1),\ldots,LT(g_s)\rangle$. This implies that there is i such that $LT(g_i)/LT(r_2 - r_1)$. However, this is impossible due to the fact that the terms of both r_1 and r_2 are not divisible by any of the $LT(g_i)$. It follows that $r_1 = r_2$ and, thus, $g_1 = g_2$.

Corollary 22. (Ideal Membership Problem)

Let $G = \{g_1,\ldots,g_s\}$ be a Gröbner basis of a nonzero ideal $I \subseteq K[X_1,\ldots,X_n]$ (**K** a field) and let $f \in K[X_1,\ldots,X_n]$. Then

$$f \in I \Longleftrightarrow \bar{f}^G = 0.$$

Now, for a complete solution to the IMP, there is only one issue remaining to be addressed: how to compute effectively a Gröbner basis?
Buchberger brought an answer to this problem by inventing the notion of *S*-polynomial.

Definitions 23. Let $f, g \in K[X_1,\ldots,X_n]\setminus\{0\}$, where **K** is a field.

1. If $mdeg(f) = \alpha$ and $mdeg(g) = \beta$, then let $\gamma = (\gamma_1,\ldots,\gamma_n)$ where $\gamma_i = \max(\alpha_i,\beta_i)$ for each i. We call $X^\gamma = LCM(LM(f),LM(g))$ the least common multiple of X^α and X^β.

2. The *S*-polynomial $S(f,g)$ of f and g is the combination

$$S(f,g) = \frac{X^\gamma}{LT(f)}f - \frac{X^\gamma}{LT(g)}g.$$

Notice that $\mathrm{mdeg}(S(f,g)) < \gamma$. This is a special feature of S-polynomials.

Example 24. Let $f = 2Y - X^2$, $g = Z - X^4 \in \mathbb{Q}[X,Y,Z]$ equipped with the lexicographic monomial order with $Y > Z > X$. Then $\mathrm{LT}(f) = 2Y$, $\mathrm{LT}(g) = Z$, and

$$S(f,g) = \frac{1}{2}Zf - Yg = -\frac{1}{2}ZX^2 + YX^4.$$

Theorem 25. (Buchberger's Criterion)

Let $I = \langle g_1,\ldots,g_s \rangle$ be a nonzero ideal of $\mathbf{K}[X_1,\ldots,X_n]$ ($\mathbf{K}$ a field), and fix a monomial order. Then $G = \{g_1,\ldots,g_s\}$ is a Gröbner basis of I if and only if for all pairs $i \neq j$, the remainder on division of $S(g_i,g_j)$ by G is zero.

Proof. See the proof of the more general Theorem 37.

From Theorem 25 ensues the following algorithm for constructing Gröbner bases over fields. Note that its correctness is guaranteed by Theorem 25 while its termination is ensured by Dickson's lemma (Theorem 12). Note also that the version below is the strain from which stemmed many refined versions.

Algorithm 26. (Buchberger's Algorithm for fields)

Input: $g_1,\ldots,g_s \in \mathbf{K}[X_1,\ldots,X_n]$, $\mathbf{K}$ a field, and $>$ a monomial order.

Output: a Gröbner basis G for $\langle g_1,\ldots,g_s \rangle$ with $\{g_1,\ldots,g_s\} \subseteq G$

$G := \{g_1,\ldots,g_s\}$
REPEAT
$\quad G' := G$
$\quad$For each pair $f \neq g$ in G' DO
$\quad\quad S := \overline{S(f,g)}^{G'}$
$\quad\quad$If $S \neq 0$ THEN $G := G' \cup \{S\}$
UNTIL $G = G'$

Example 27. Let $I = \langle f_1 = 3XY + 4, f_2 = 2X^2 + 3 \rangle \subseteq \mathbb{Q}[X,Y]$, and consider the lexicographic monomial order with $X > Y$. We want to compute a Gröbner basis for I. We have:

$$S(f_1,f_2) = \frac{1}{3}Xf_1 - \frac{1}{2}Yf_2 = \frac{4}{3}X - \frac{3}{2}Y =: f_3,$$
$$S(f_1,f_3) = \frac{1}{3}f_1 - \frac{3}{4}Yf_3 = \frac{9}{8}Y^2 + \frac{4}{3} =: f_4,$$
$$S(f_1,f_4) = \frac{1}{3}Yf_1 - \frac{8}{9}Xf_4 = \frac{-8}{9}f_3 \xrightarrow{f_3} 0,$$
$$S(f_2,f_3) = \frac{1}{2}f_2 - \frac{3}{4}Xf_3 = \frac{3}{8}f_1 \xrightarrow{f_1} 0,$$
$$S(f_2,f_4) = \frac{1}{2}Y^2f_2 - \frac{8}{9}X^2f_4 = -\frac{32}{27}X^2 + \frac{3}{2}Y^2 \xrightarrow{f_2,f_4} 0,$$
$$S(f_3,f_4) = \frac{3}{4}Y^2f_3 - \frac{8}{9}Xf_4 = \frac{-8}{9}f_3 - 4f_4 \xrightarrow{f_3,f_4} 0.$$

Thus, $G = \{f_1,f_2,f_3,f_4\}$ is a Gröbner basis for I.

Question: $f = -6XY + 8X - 9Y - 8 \in I$?

Answer: Dividing f by G, we obtain $f = -2f_1 + 6f_3 = -2f_1 + 6(\frac{1}{3}Xf_1 - \frac{1}{2}Yf_2) = (2X - 2)f_1 - 3Yf_2 \in I$.

Of course, the IMP is one important application of Gröbner bases. For more applications, see Section 2.2.

1.2   Gröbner bases over coherent valuation rings

As already said, it was Buchberger [10] that first constructed Gröbner bases over fields (i.e., in $\mathbf{K}[X_1, \dots, X_n]$, where $\mathbf{K}$ is a field). In this section we will explain how to construct Gröbner bases in the more general setting of valuation rings possibly with zero-divisors. We start with recalling the following definitions.

Definition 28.

- Let U be an $\mathbf{R}$-module, where $\mathbf{R}$ is a ring. The *syzygy* module of $(v_1, \dots, v_n) \in U^n$ is $\mathrm{Syz}(v_1, \dots, v_n) := \{(b_1, \dots, b_n) \in \mathbf{R}^n \mid b_1 v_1 + \cdots + b_n v_n = 0\}$. It is a submodule of $\mathbf{R}^n$. The syzygy module of a single element v is the *annihilator* $\mathrm{Ann}(v)$ of v.

- An $\mathbf{R}$-module U is *coherent* if all syzygy modules $\mathrm{Syz}(v_1, \dots, v_n)$ with $v_1, \dots, v_n \in U$ are finitely-generated. The ring $\mathbf{R}$ is *coherent* if it is coherent as an $\mathbf{R}$-module.

  It is well known that a module is coherent iff, on the one hand, any intersection of two finitely-generated submodules is finitely-generated, and, on the other hand, the annihilator of every element is a finitely-generated ideal.

- A ring $\mathbf{R}$ is *local* if for every element $x \in \mathbf{R}$, either x or $1 + x$ is invertible. The unit group of $\mathbf{R}$ is denoted by $\mathbf{R}^\times$. A local ring has a unique maximal ideal $\mathfrak{m}$ with $\mathbf{R} \setminus \mathfrak{m} = \mathbf{R}^\times$.

- The (Jacobson) *radical* $\mathrm{Rad}(\mathbf{R})$ of a ring $\mathbf{R}$ is the ideal $\{x \in \mathbf{R} \mid 1 + x\mathbf{R} \subseteq \mathbf{R}^\times\}$. When $\mathbf{R}$ is local, $\mathrm{Rad}(\mathbf{R})$ is its unique maximal ideal. The *residue field* of a local ring $\mathbf{R}$ is the quotient $\mathbf{R}/\mathrm{Rad}(\mathbf{R})$.

- A ring $\mathbf{R}$ is a *valuation ring* if every two elements are comparable w.r.t. division, i.e. if, given $a, b \in \mathbf{R}$, either a divides b or b divides a ($\mathbf{R}$ may have nonzero zero-divisors). A valuation ring is a local ring; it is coherent iff the annihilator of any element is principal. A valuation domain is coherent. Of course, a field is a valuation domain. A typical example of a valuation domain which is not a field is $\mathbb{Z}_{p\mathbb{Z}} := \{\frac{a}{b} \in \mathbb{Q} \mid a \in \mathbb{Z} \text{ and } b \in \mathbb{Z} \setminus p\mathbb{Z}\}$, where p is a prime number. A typical example of a valuation ring with nonzero zero-divisors is $\mathbb{Z}/p^\alpha\mathbb{Z}$, where p is a prime number and $\alpha \geq 2$.

- A valuation ring $\mathbf{V}$ is *archimedean* if

$$\forall a, b \in \mathrm{Rad}(\mathbf{V}) \setminus \{0\} \ \exists n \in \mathbb{N} \mid a \text{ divides } b^n.$$

It is folklore that A valuation domain $\mathbf{V}$ is archimedean iff its valuation group is archimedean iff its Krull dimension is ≤ 1. Moreover, we know that a valuation ring $\mathbf{V}$ that contains a nonzero zero-divisor is archimedean if and only if $\dim \mathbf{V} \leq 0$ (see [49]).

Definition 29. Let $\mathbf{R}$ be a valuation ring, $I = \langle f_1, \ldots, f_s \rangle$ a nonzero finitely-generated ideal of $\mathbf{R}[X_1, \ldots, X_n]$, and $>$ a monomial order on $\mathbf{R}[X_1, \ldots, X_n]$.

As in the classical division algorithm in $\mathbf{F}[X_1, \ldots, X_n]$ ($\mathbf{F}$ a field, see Algorithm 14), for each polynomials $h, h_1, \ldots, h_m \in \mathbf{R}[X_1, \ldots, X_n]$, there exist $q_1, \ldots, q_m, r \in \mathbf{R}[X_1, \ldots, X_n]$ such that

$$h = q_1 h_1 + \cdots + q_m h_m + r,$$

with $\mathrm{mdeg}(h) \geq \mathrm{mdeg}(q_i h_i)$, and either $r = 0$ or r is a sum of terms none of which is divisible by any of $\mathrm{LT}(h_1), \ldots, \mathrm{LT}(h_m)$. The polynomial r is called a remainder of h on division by $H = \{h_1, \ldots, h_m\}$ and denoted $r = \overline{h}^H$.

The following lemma gives a sufficient and necessary condition for a term to belong to a module generated by terms over a valuation ring.

Lemma 30. *Let $\mathbf{R}$ be a valuation ring, H a free $\mathbf{R}[X_1, \ldots, X_n]$-module with basis $e_1, \ldots, e_m$, and $>$ a monomial order on H. Let $U = \langle a_\alpha X^\alpha e_{i_\alpha}, \alpha \in A \rangle$ be a submodule of H generated by a collection of terms. Then a term $b X^\beta e_j$ lies in U if and only if $X^\beta e_j$ is divisible by $X^\alpha e_{i_\alpha}$ (i.e., $e_j = e_{i_\alpha}$ and $X^\alpha \mid X^\beta$) and b is divisible by a_α for some $\alpha \in A$.*

Proof. It is obvious that the condition is sufficient. For proving the necessity, write $b X^\beta e_j = \sum_{i=1}^s c_i a_{\alpha_i} X^{\gamma_i} X^{\alpha_i} e_j$ for some $\alpha_1, \ldots, \alpha_s \in A$, $c_i, a_{\alpha_i} \in \mathbf{R} \setminus \{0\}$, and $X^{\gamma_i} \in \mathbb{M}_n$. Ignoring the superfluous terms, we have $\gamma_i + \alpha_i = \beta$ for $1 \leq i \leq s$, and $b = \sum_{i=1}^s c_i a_{\alpha_i}$. It is clear that for each $1 \leq i \leq s$, X^β is divisible by X^{α_i}. Since all the coefficients are comparable under division, we can suppose that a_{α_1} divides all the a_{α_i} and, thus, divides b.

The following proposition shows that the rather "disappointing" behavior of the division algorithm (Algorithm 14) detected in Example 16 (nonuniqueness of the remainder) does not occur when one divides by a Gröbner basis.

Proposition 31. *Let $\mathbf{R}$ be a valuation ring, $>$ a monomial order on $\mathbf{R}[X_1, \ldots, X_n]$, I a nonzero ideal of $\mathbf{R}[X_1, \ldots, X_n]$, $f_1, \ldots, f_s \in I$ such that $G = \{f_1, \ldots, f_s\}$ is a Gröbner basis for I, and $f \in \mathbf{R}[X_1, \ldots, X_n]$. Then:*

(1) *There is a unique $r \in \mathbf{R}[X_1, \ldots, X_n]$ with the following two properties:*

(i) *No term of r is divisible by any of the* LT(f_i)*'s.*

(ii) *There is $g \in I$ such that $f = g + r$.*

In particular, r is the remainder on division of f by G regardless how the elements of G are listed when using the division algorithm (Algorithm 14).

(2) $f \in I \Leftrightarrow \bar{f}^G = 0$ (the ideal membership test).

(3) $I = \langle f_1, \ldots, f_s \rangle$.

Proof. (1) The division algorithm (see Definition 29) gives $r = \bar{f}^G \in \mathbf{R}[X_1, \ldots, X_n]$ satisfying the property (i) and $q_1, \ldots, q_s \in \mathbf{R}[X_1, \ldots, X_n]$ such that $f = q_1 f_1 + \cdots + q_s f_s + r$. Then take $g = q_1 f_1 + \cdots + q_s f_s \in I$.

For the uniqueness, let $f = g_1 + r_1 = g_2 + r_2$ satisfying (i) and (ii). As $r_1 - r_2 = g_2 - g_1 \in I$ then either it is null or LT($r_1 - r_2$) is divisible by one of the LT(f_i)'s (because G is a Gröbner basis for I and taking into account Lemma 30). The latter case is impossible since no term of r_1, r_2 is divisible any of the LT(f_i)'s. We conclude that $r_1 = r_2$.

(2) If $\bar{f}^G = 0$ then clearly $f \in \langle f_1, \ldots, f_s \rangle \subseteq I$. Conversely, if $f \in I$, then $f = f + 0$ satisfies the two conditions of (1), and thus $\bar{f}^G = 0$.

(3) This is an immediate consequence of (2).

We now consider the problem of the construction of a Gröbner basis. A key tool introduced by Buchberger [10] in the case where the base ring is a field is the notion of S-polynomial of two polynomials. Now we extend this notion to valuation rings.

Definition 32. Let $\mathbf{R}$ be a coherent valuation ring, $f \neq g \in \mathbf{R}[X_1, \ldots, X_n] \setminus \{0\}$, and $>$ a monomial order on $\mathbf{R}[X_1, \ldots, X_n]$.

(1) If $\mathrm{mdeg}(f) = \alpha$ and $\mathrm{mdeg}(g) = \beta$ then let $\gamma = (\gamma_1, \ldots, \gamma_n)$, where $\gamma_i = \max(\alpha_i, \beta_i)$ for each i. Perform the test $\mathrm{LC}(f) \mid \mathrm{LC}(g)$ or $\mathrm{LC}(g) \mid \mathrm{LC}(f)$.

$$S(f, g) = \frac{X^\gamma}{\mathrm{LM}(f)} f - \frac{\mathrm{LC}(f)}{\mathrm{LC}(g)} \frac{X^\gamma}{\mathrm{LM}(g)} g \quad \text{if } \mathrm{LC}(g) \text{ divides } \mathrm{LC}(f).$$

$$S(f, g) = \frac{\mathrm{LC}(g)}{\mathrm{LC}(f)} \frac{X^\gamma}{\mathrm{LM}(f)} f - \frac{X^\gamma}{\mathrm{LM}(g)} g \quad \text{if } \mathrm{LC}(f) \text{ divides } \mathrm{LC}(g) \text{ and } \mathrm{LC}(g)$$
does not divide $\mathrm{LC}(f)$.

$S(f, g)$ is called the *S-polynomial* of f and g. It is "designed" to produce cancellation of leading terms. Here, it is worth pointing out that $S(f, g)$ is not uniquely determined when $\mathbf{R}$ has zero-divisors. This minor technical issue will be repaired through the consideration in (2) of $S(f, f)$ and $S(g, g)$.

(2) Let d be a generator of the annihilator of $LC(f)$ (note that this annihilator is principal because **R** is a coherent valuation ring). We set

$$S(f,f) := df$$

(it is defined up to a unit). Note that $S(f,f)$ behaves exactly like usual S-polynomials in the sense that $\text{mdeg}(S(f,f)) < \text{mdeg}(f)$ and $S(X^\delta f, X^\delta f) = X^\delta S(f,f) \; \forall \delta \in \mathbb{N}^n$. In addition, if the leading coefficient of f is not a zero-divisor then automatically $S(f,f) = 0$ (as in the case where **R** is a field).

$S(f,f)$ is called the *auto-S-polynomial* of f. It is "designed" to cover the cancellation of the leading term of f produced by a mutiplication of f by an element of the annihilator of $LC(f)$.

Example 33. (Example 16 continued)

Let $f_1 = 2 + 12XY$, $f_2 = 8Y^2 \in \mathbb{Q}[X,Y]$, and fix any monomial order $>$ on $\mathbb{Q}[X,Y]$. Then

$$S(f_1,f_2) = Yf_1 - \frac{3}{2}Xf_2 = 2Y.$$

Of course, $S(f_1,f_1) = S(f_2,f_2) = 0$ as we are on an integral ground.

Example 34. (S-polynomials over $\mathbb{F}_2[Y]/\langle Y^2 \rangle$, a useful ring in coding theory)

The ring $\mathbf{V} := \mathbb{F}_2[Y]/\langle Y^2 \rangle = \mathbb{F}_2[y]$ (where $y = \bar{Y}$) is a zero-dimensional coherent valuation ring with zero-divisors (as $y^2 = 0$).

Let $f \neq g \in \mathbf{V}[X_1, \ldots, X_n] \setminus \{0\}$, and $>$ a monomial order. Denoting by $\text{mdeg}(f) = \alpha = (\alpha_1, \ldots, \alpha_n)$, $\text{mdeg}(g) = \beta = (\beta_1, \ldots, \beta_n)$, $\gamma = (\gamma_1, \ldots, \gamma_n)$, where $\gamma_k = \max(\alpha_k, \beta_k)$ for each k, the only case where $S(f,g)$ is not equal to $LC(g)\frac{X^\gamma}{X^\alpha}f - LC(f)\frac{X^\gamma}{X^\beta}g$ (up to a unit) is when $LC(f) = LC(g) = y$. In that case $S(f,g)$ is simply equal to $\frac{X^\gamma}{X^\alpha}f - \frac{X^\gamma}{X^\beta}g$. On the other hand, for the computation of $S(f,f)$, two cases may arise:

If $LC(f) = 1$ or $1 + y$ then $S(f,f) = 0$.
If $LC(f) = y$ then $S(f,f) = yf$.

For example, fixing the lexicographic order with $X_1 > X_2$ as monomial order, and considering the polynomials $f_1 := yX_1 + X_2$ and $f_2 = y + yX_2$, we have:

$$S(f_1,f_2) = X_2f_1 - X_1f_2 = X_2^2 + yX_1,$$

$$S(f_1,f_1) = yf_1 = yX_2, S(f_2,f_2) = yf_2 = 0.$$

Example 35. (The ring $\mathbb{Z}_{p\mathbb{Z}}$, where p is a prime number)

(1) Recall that the ring $\mathbb{Z}_{p\mathbb{Z}}$ is the following localization of $\mathbb{Z}$:

$$\mathbb{Z}_{p\mathbb{Z}} := \left\{ \frac{a}{b} \in \mathbb{Q} \mid a \in \mathbb{Z} \text{ and } b \in \mathbb{Z} \setminus p\mathbb{Z} \right\}.$$

For $a \in \mathbb{Z} \setminus \{0\}$, we denote by $v_p(a) = \max\{k \in \mathbb{N} \mid p^k \text{ divides } a\}$ (the valuation of a at p), so that, $a = p^{v_p(a)}a'$ with $a' \wedge p = 1$, that is, $a' \in \mathbb{Z} \setminus p\mathbb{Z}$.

For $x = \frac{a}{b} \in \mathbb{Q} \setminus \{0\}$, we denote by $v_p(x) = v_p(a) - v_p(b) \in \mathbb{Z}$ (*the valuation of x at p*). We convene that $v_p(0) = +\infty$. We have $\mathbb{Z}_{p\mathbb{Z}} = \{x \in \mathbb{Q} \mid v_p(x) \geq 0\}$.

Any element $x \in \mathbb{Z}_{p\mathbb{Z}}$ can be written in the form $x = p^k \frac{a'}{b'}$ with $k = v_p(x) \in \mathbb{N}$, and $a', b' \in \mathbb{Z} \setminus p\mathbb{Z}$. It follows that for $x, y \in \mathbb{Z}_{p\mathbb{Z}} \setminus \{0\}$,

$$x \mid y \iff v_p(x) \leq v_p(y),$$

and, thus, $\mathbb{Z}_{p\mathbb{Z}}$ is a valuation domain. It follows also, that any nondecreasing sequence $(\langle x_n \rangle)_{n \in \mathbb{N}}$ of principal ideals of $\mathbb{Z}_{p\mathbb{Z}}$ (finitely-generated ideals of $\mathbb{Z}_{p\mathbb{Z}}$ are principal as $\mathbb{Z}_{p\mathbb{Z}}$ is valuation domain) pauses after at most $(v_p(x_0) + 1)$ iterations, and, thus, $\mathbb{Z}_{p\mathbb{Z}}$ is Noetherian. In classical literature, $\mathbb{Z}_{p\mathbb{Z}}$ is called a "discrete valuation domain" (discrete because its valuation group is $\mathbb{Z}$) but we prefer to call it a Noetherian valuation domain.

(2) In $\mathbb{Z}_{p\mathbb{Z}}[X_1, \ldots, X_n]$, for $x, y \in \mathbb{Z}_{p\mathbb{Z}} \setminus \{0\}$, a term xX^α divides a term yX^β if and only if $v_p(x) \leq v_p(y)$ and $X^\alpha \mid X^\beta$.

(3) We will specify the definition of S-polynomials given in Definition 29 to the case where the base ring is $\mathbb{Z}_{p\mathbb{Z}}$.

Let $f_i, f_j \in \mathbb{Z}_{p\mathbb{Z}}[X_1, \ldots, X_n] \setminus \{0\}$ ($i \neq j$), and fix a monomial order $>$ on $\mathbb{Z}_{p\mathbb{Z}}[X_1, \ldots, X_n]$. Denote by $\mathrm{mdeg}(f_i) = \beta = (\beta_1, \ldots, \beta_n)$, $\mathrm{mdeg}(f_j) = \beta' = (\beta_1', \ldots, \beta_n')$, $\gamma = (\gamma_1, \ldots, \gamma_n)$, where $\gamma_k = \max(\beta_k, \beta_k')$ for each k.

Moreover, denoting by $\mathrm{LC}(f_i) = c_i = p^{v_p(c_i)} \frac{a_i}{b_i}$, $\mathrm{LC}(f_j) = c_j = p^{v_p(c_j)} \frac{a_j}{b_j}$, with $a_i, b_i, a_j, b_j \in \mathbb{Z} \setminus p\mathbb{Z}$, we have:

(i) $S(f_i, f_j) := \frac{X^\gamma}{\mathrm{LM}(f_i)} f_i - \frac{a_i b_j}{b_i a_j} p^{v_p(c_i) - v_p(c_j)} \frac{X^\gamma}{\mathrm{LM}(f_j)} f_j$ if $v_p(c_j) \leq v_p(c_i)$.

$S(f_i, f_j) := \frac{a_j b_i}{b_j a_i} p^{v_p(c_j) - v_p(c_i)} \frac{X^\gamma}{\mathrm{LM}(f_i)} f_i - \frac{X^\gamma}{\mathrm{LM}(f_j)} f_j$ if $v_p(c_j) > v_p(c_i)$.

(ii) $S(f_i, f_i) = S(f_j, f_j) = 0$ (as we are on an integral ground).

(4) Let $f_1 = 2 + 12XY$, $f_2 = 8Y^2 \in \mathbb{Z}_{2\mathbb{Z}}[X, Y]$. Fixing any monomial order $>$ on $\mathbb{Z}_{2\mathbb{Z}}[X, Y]$, we have:

$$S(f_1, f_2) = \frac{2}{3} Y f_1 - X f_2 = \frac{4}{3} Y.$$

The following lemma will be of great use since it is a key result for the characterization of Gröbner bases by means of S-polynomials.

Lemma 36. *Let $\mathbf{R}$ be a coherent valuation ring, $>$ a monomial order, and $f_1, \ldots, f_s \in \mathbf{R}[X_1, \ldots, X_n]$ such that $\mathrm{mdeg}(f_i) = \gamma$ for each $1 \leq i \leq s$. If $\mathrm{mdeg}(\sum_{i=1}^s a_i f_i) < \gamma$ for some $a_1, \ldots, a_s \in \mathbf{R}$, then $\sum_{i=1}^s a_i f_i$ is a linear combination with coefficients in $\mathbf{R}$ of the S-polynomials $S(f_i, f_j)$ for $1 \leq i, j \leq s$. Furthermore, each $S(f_i, f_j)$ has multidegree $< \gamma$.*

Proof. Since **R** is a valuation ring, we can suppose that $LC(f_s)/LC(f_{s-1})/\cdots$
$/LC(f_1)$. Thus, for $i < j$, $S(f_i, f_j) = f_i - \frac{LC(f_i)}{LC(f_j)}f_j$. We have:

$$\sum_{i=1}^{s} a_i f_i = a_1 \left(f_1 - \frac{LC(f_1)}{LC(f_2)}f_2 \right) + \left(a_2 + \frac{LC(f_1)}{LC(f_2)}a_1 \right) \left(f_2 - \frac{LC(f_2)}{LC(f_3)}f_3 \right)$$

$$+ \cdots + \left(a_{s-1} + \frac{LC(f_{s-2})}{LC(f_{s-1})}a_{s-2} + \cdots + \frac{LC(f_1)}{LC(f_{s-1})}a_1 \right) \left(f_{s-1} - \frac{LC(f_{s-1})}{LC(f_s)}f_s \right)$$

$$+ \left(a_s + \frac{LC(f_{s-1})}{LC(f_s)}a_{s-1} + \cdots + \frac{LC(f_1)}{LC(f_s)}a_1 \right) f_s.$$

But $(a_s + \frac{LC(f_{s-1})}{LC(f_s)}a_{s-1} + \cdots + \frac{LC(f_1)}{LC(f_s)}a_1)LC(f_s) = 0$ since $mdeg(\sum_{i=1}^{s} a_i f_i) < \gamma$,
and, thus,

$$\left(a_s + \frac{LC(f_{s-1})}{LC(f_s)}a_{s-1} + \cdots + \frac{LC(f_1)}{LC(f_s)}a_1 \right) f_s \in \mathbf{R} S(f_s, f_s).$$

Using Lemma 30 and Lemma 36, we generalize some classical results about the existence and characterization of Gröbner basis over coherent valuation rings (the proof for modules is analogous to that for ideals).

Theorem 37. *Let* **R** *be a coherent valuation ring,* $I = \langle g_1, \ldots, g_s \rangle$ *a submodule of a free* $\mathbf{R}[X_1, \ldots, X_n]$*-module H, and fix a monomial order* $>$ *on H. Then,* $G = \{g_1, \ldots, g_s\}$ *is a Gröbner basis for I if and only if for all pairs* $i \leq j$*, the remainder on division of* $S(g_i, g_j)$ *by G is zero.*

Proof. For sake of simplicity, we will treat the ideal case.
"$\Rightarrow$" As $S(g_i, g_j) \in \langle g_i, g_j \rangle \subseteq I$, then, by virtue of Proposition 31.(2), $\overline{S(g_i, g_j)}^G = 0$.
"$\Leftarrow$" Instead of going through the details of the proof, we prefer to give the idea behind it. This is nicely explained in [14] (page 83) in case the base ring is a field. The same proof holds in our situation as we have all the necessary ingredients. Letting $f \in I = \langle g_1, \ldots, g_s \rangle$, there are polynomials $h_i \in \mathbf{R}[X_1, \ldots, X_n]$ such that

$$f = \sum_{i=1}^{s} h_i g_i, \tag{1.1}$$

with $mdeg(f) \leq \max_{1 \leq i \leq s}(mdeg(h_i g_i))$.

Case 1: $mdeg(f) = \max_{1 \leq i \leq s}(mdeg(h_i g_i))$, say $mdeg(f) = mdeg(h_{i_0} g_{i_0})$ for some $i_0 \in \{1, \ldots, s\}$. As the leading coefficients of the $h_i g_i$'s such that $mdeg(f) = mdeg(h_i g_i)$ are comparable under division, we can suppose that all of them are divisible by the leading coefficient of $h_{i_0} g_{i_0}$. It follows that $LT(f) \in \langle LT(g_{i_0}) \rangle \subseteq \langle LT(g_1), \ldots, LT(g_s) \rangle$.

Case 2: $\mathrm{mdeg}(f) < \max_{1 \leq i \leq s}(\mathrm{mdeg}(h_i g_i))$. Then, roughly speaking, some cancellation must occur among the leading terms of (1.2). Using Lemma 36, we can rewrite this in terms of S-polynomials. Then, the assumption that S-polynomials have zero remainders modulo G allows to replace the S-polynomials by expressions involving less cancellation. Thus, we obtain an expression for f that has less cancellation of leading terms. An so on, as the set of monomials is well-ordered (by virtue of Corollary 13), we end up with a situation like that of Case 1.

From Theorem 37 ensues the following algorithm for constructing Gröbner bases over coherent valuation rings.

Algorithm 38. (Buchberger's Algorithm for coherent valuation rings)

Input: $g_1, \ldots, g_s \in H$ where H is a finite-rank free $\mathbf{V}[X_1, \ldots, X_n]$-module, $\mathbf{V}$ a coherent valuation ring, and $>$ a monomial order on H.

Output: a Gröbner basis G for $\langle g_1, \ldots, g_s \rangle$ with $\{g_1, \ldots, g_s\} \subseteq G$

$G := \{g_1, \ldots, g_s\}$
REPEAT
$G' := G$
For each pair f, g in G' DO
$\quad S := \overline{S(f,g)}^{G'}$
$\quad$ If $S \neq 0$ THEN $G := G' \cup \{S\}$
UNTIL $G = G'$

The algorithm above is exactly the same algorithm as in the case where the base ring is a field. The only modifications are in the definition of S-polynomials, in the consideration of the auto-S-polynomials, and in the divisions of terms. Note that the precise reason why Buchberger's Algorithm 38 terminates (at least for the lexicographic monomial order) is that the valuation ring is archimedean, or equivalently, either it is a valuation domain of Krull dimension ≤ 1 or a zero-dimensional valuation ring containing nonzero zero-divisors (the module case being analogous to the ideal case) (for more details, see [66]).

Theorem 39. *Let $\mathbf{V}$ be a coherent valuation ring, $I = \langle f_1, \ldots, f_s \rangle$ with $f_1, \ldots, f_s \in H \setminus \{0\}$ and H is a finite-rank free $\mathbf{V}[X_1, \ldots, X_n]$-module, and let us fix a monomial order $>$ on H. If $\mathrm{LT}(I)$ is finitely-generated then a finite Gröbner basis for I can be computed by Algorithm 38.*

Proof. For sake of simplicity, we will treat the ideal case.
Denote by $\mathrm{LT}(I) = \langle \mathrm{LT}(g_1), \ldots, \mathrm{LT}(g_r) \rangle$ with $g_i \in I$. Let $1 \leq k \leq r$. As $g_k \in I$, there exist $h_1, \ldots, h_s \in \mathbf{V}[X_1, \ldots, X_n]$ such that

$$g_k = \sum_{i=1}^{s} h_i f_i, \tag{1.2}$$

with $\mathrm{mdeg}(g_k) \leq \max_{1\leq i\leq s}(\mathrm{mdeg}(M_iN_i)) =: \gamma_1$ (we call it the multidegree of the expression (1.2) of g_k w.r.t. the generating set $\{f_1,\ldots,f_s\}$ of I), where $M_i = \mathrm{LM}(h_i)$ and $N_i = \mathrm{LM}(f_i)$.

Case 1: $\mathrm{mdeg}(g_k) = \gamma_1$, say $\mathrm{mdeg}(g_k) = \mathrm{mdeg}(M_{i_0}N_{i_0}) = \gamma_1$ for some $i_0 \in \{1,\ldots,s\}$. As the leading coefficients of the h_if_i's with $\mathrm{mdeg}(g_k) = \mathrm{mdeg}(M_iN_i)$ are comparable w.r.t. division, we can suppose that all of them are divisible by the leading coefficient of $h_{i_0}f_{i_0}$. It follows that $\mathrm{LT}(g_k) \in \langle \mathrm{LT}(f_{i_0})\rangle \subseteq \langle \mathrm{LT}(f_1),\ldots,\mathrm{LT}(f_s)\rangle$.

Case 2: $\mathrm{mdeg}(g_k) < \gamma_1$. We have

$$g_k = \sum_{\mathrm{mdeg}(M_iN_i)<\gamma_1} h_if_i + \sum_{\mathrm{mdeg}(M_iN_i)=\gamma_1} h_if_i$$

$$= \sum_{\mathrm{mdeg}(M_iN_i)<\gamma_1} h_if_i + \sum_{\mathrm{mdeg}(M_iN_i)=\gamma_1} (h_i - \mathrm{LT}(h_i))f_i + \sum_{\mathrm{mdeg}(M_iN_i)=\gamma_1} \mathrm{LT}(h_i)f_i.$$

Letting $E = \{i \mid \mathrm{mdeg}(M_iN_i) = \gamma_1\}$ and $c_i = \mathrm{LC}(h_i)$, we get

$$\mathrm{mdeg}\left(\sum_{i\in E} c_iM_if_i\right) < \gamma_1.$$

By virtue of Lemma 36, there exists a finite family $(a_{i,j})$ of elements of $\mathbf{V}$ such that

$$\sum_{i\in E} c_iM_if_i = \sum_{i\leq j\in E} a_{i,j}S(M_if_i,M_jf_j).$$

But, for $i \leq j \in E$, letting $N_{i,j} = \mathrm{LCM}(N_i,N_j)$ and writing $S(f_i,f_j) = a\frac{N_{i,j}}{N_i}f_i + b\frac{N_{i,j}}{N_j}f_j$ for some $a,b \in \mathbf{V}$, we have $S(M_if_i,M_jf_j) = a\frac{X^{\gamma_1}}{M_iN_i}M_if_i + b\frac{X^{\gamma_1}}{M_jN_j}M_jf_j = \frac{X^{\gamma_1}}{N_{i,j}}S(f_i,f_j)$. It follows that

$$\sum_{i\in E} c_iM_if_i = \sum_{i\leq j\in E} a_{i,j}m_{i,j}S(f_i,f_j),$$

where the $m_{i,j}$'s are monomials. Thus, we obtain another expression for g_k,

$$g_k = \sum_{\mathrm{mdeg}(M_iN_i)<\gamma_1} h_if_i + \sum_{i\in E}(h_i - \mathrm{LT}(h_i))f_i + \sum_{i\leq j\in E} a_{i,j}m_{i,j}S(f_i,f_j),$$

and the multidegree of this expression, now w.r.t. the generating set of I obtained by adding the elements $S(f_i,f_j)$, $i \leq j \in E$, to the $f_1,\ldots,f_s$, is $< \gamma_1$. Reiterating this, we end up with a situation like that of Case 1 for all the g_k's because the set of monomials is well-ordered (by virtue of Corollary 13). So we reach the termination condition in Algorithm 38 after a finite number of steps.

Corollary 40. *When $\mathbf{V}$ is a coherent archimedean valuation ring (i.e., a valuation domain of Krull dimension ≤ 1 or a coherent valuation ring containing a nonzero zero-divisor and with Krull dimension ≤ 0), Buchberger's Algorithm 38 terminates for $\mathbf{V}[X_1,\ldots,X_n]$ endowed with the lexicographic monomial order.*

Proof. If the algorithm ends, Theorem 49 guarantees its correctness. We know that if **V** is a coherent archimedean valuation ring, choosing the lexicographic monomial order, $LT(I)$ is also finitely-generated (see [66, Theorem 272]). The desired result follows from Theorem 39.

Remark 41. We don't know whether Corollary 40 remains true for any monomial order. Of course, when the valuation ring is Noetherian (and, thus, Gröbner), Buchberger's Algorithm 38 always terminates.

Example 42. Examples of archimedean coherent valuation rings include:

- Fields,

- $\mathbb{Z}_{p\mathbb{Z}}$, where p is a prime number,

- $\mathbf{D}/\langle a^k \rangle$ with **D** a PID Principal Ideal Domain (such as $\mathbb{Z}$), and a an irreducible element. When $k \geq 2$, it has zero-divisors ($\mathbb{F}_2[Y]/\langle Y^2 \rangle$ for example),

- Galois rings $GR(p^k, n) = (\mathbb{Z}/p^k\mathbb{Z})[t]/\langle f \rangle$, where f is a monic irreducible polynomial in $(\mathbb{Z}/p^k\mathbb{Z})[t]$ (p a prime number) of degree n whose image modulo p is irreducible. When $k \geq 2$, they have zero-divisors.

Keeping the above notation, it is obvious that if G is a Gröbner basis for I, then for any $p \in G$ such that $LT(p) \in \langle LT(G \setminus \{p\}) \rangle$, $G \setminus \{p\}$ is also a Gröbner basis for I. So, using Algorithm 38 and removing any p with $LT(p) \in \langle LT(G \setminus \{p\}) \rangle$, we can construct for I a Gröbner basis G such that

$$\forall p \in G, \ LT(p) \notin \langle LT(G \setminus \{p\}) \rangle.$$

Such a Gröbner basis will be called a *pseudo-minimal* Gröbner basis. But even more, one can ask that the Gröbner basis G satisfies the following property:

$$\forall p \in G, \ \text{no term of } p \text{ lies in } \langle LT(G \setminus \{p\}) \rangle. \tag{1.3}$$

Such a Gröbner basis will be called a *pseudo-reduced* Gröbner basis, and can be computed by exhausting all the possible reductions.

Remark 43. In search of uniqueness of a pseudo-reduced Gröbner basis, we need a "normalization" of the elements of a pseudo-reduced Gröbner basis. In case the base ring is a field, this is easily done by requiring the leading coefficients of the elements of a pseudo-reduced Gröbner basis to be 1. This gives birth to the notion of *reduced* Gröbner basis. For rings which are not fields, this has to be done on a case-by-case basis. Hereafter a few examples:

(1) Consider the case where the base ring is $\mathbb{Z}/p^\alpha\mathbb{Z}$, with p is a prime number and $\alpha \geq 2$. For $f \in (\mathbb{Z}/p^\alpha\mathbb{Z})[X_1, \ldots, X_n] \setminus \{0\}$, fixing a monomial order on $(\mathbb{Z}/p^\alpha\mathbb{Z})[X_1, \ldots, X_n]$, and denoting by $LC(f) = a$, where $a = p^m c, 0 \leq m \leq$

$\alpha - 1$ and $c \wedge p = 1$, the normalization of f, denoted by $\tilde{f}$, can be defined as

$$\tilde{f} := c^{-1}f, \text{ with } LC(\tilde{f}) = p^m, \ 0 \leq m \leq \alpha - 1.$$

(2) Consider the case where the base ring is $\mathbb{Z}_{p\mathbb{Z}} = \{\frac{a}{b} \mid a \in \mathbb{Z} \ \& \ b \in Z \setminus p\mathbb{Z}\}$ (p a prime number). Let $f \in \mathbb{Z}_{p\mathbb{Z}}[X_1, \ldots, X_n] \setminus \{0\}$, and fix a monomial order on $\mathbb{Z}_{p\mathbb{Z}}[X_1, \ldots, X_n]$. Denote by $LC(f) = \frac{a}{b}$, where $a = p^m c$, $b = p^r d$, $0 \leq r \leq m$, and $c, d \in \mathbb{Z}$ with $c \wedge p = d \wedge p = 1$. The normalization of f, denoted by $\tilde{f}$, can be defined as

$$\tilde{f} := \frac{d}{c}f, \text{ with } LC(\tilde{f}) \in p^{\mathbb{N}}.$$

(3) Consider the ring $\mathbf{V} := \mathbb{F}_2[Y]/\langle Y^2 \rangle = \mathbb{F}_2[y] = \{0, 1, 1+y, y\}$, with $\mathbf{V}^{\times} = \{1, 1+y\}$. For $f \in \mathbf{V}[X_1, \ldots, X_n] \setminus \{0\}$, fixing a monomial order, the normalization of $\tilde{f}$ of f can be defined as:

$$\tilde{f} = \begin{cases} f & \text{if } LC(f) = y \text{ or } 1 \\ (1+y)f & \text{if } LC(f) = 1+y, \end{cases}$$

with $LC(\tilde{f}) = 1$ or y.

Let us specify the previous definitions to the case where the base ring is a field.

Definition 44. Let $\mathbf{K}$ be a field and fix a monomial order on $\mathbf{K}[X_1, \ldots, X_n]$.

1) A minimal Gröbner basis for an ideal I of $\mathbf{K}[X_1, \ldots, X_n]$ is a Gröbner basis for I such that
   (i) $LC(p) = 1$ for all $p \in G$.
   (ii) $LT(p) \notin \langle LT(G \setminus \{p\}) \rangle$ for all $p \in G$.

2) A reduced Gröbner basis for an ideal I of $\mathbf{K}[X_1, \ldots, X_n]$ is a Gröbner basis for I such that:
   (i) $LC(p) = 1$ for all $p \in G$.
   (ii) For all $p \in G$, no monomial of p lies in $\langle LT(G \setminus \{p\}) \rangle$.

As explained above, such bases always exist.

Example 45. (Example 33 continued)

Let $f_1 = 2 + 12XY$, $f_2 = 8Y^2 \in \mathbb{Q}[X, Y]$, and fix any monomial order $>$ on $\mathbb{Q}[X, Y]$. Then

$$S(f_1, f_2) = Yf_1 - \frac{3}{2}Xf_2 = 2Y \xrightarrow{\text{normalization}} Y =: f_3; \ f_1 \xrightarrow{f_3} 1.$$

Thus, $\{1\}$ is the reduced Gröbner basis of $\langle f_1, f_2 \rangle$ in $\mathbb{Q}[X, Y]$.

Example 46. (Example 35.(4) continued)

Let $f_1 = 2 + 12XY$, $f_2 = 8Y^2 \in \mathbb{Z}_{2\mathbb{Z}}[X,Y]$, and fix any monomial order $>$ on $\mathbb{Z}_{2\mathbb{Z}}[X,Y]$. Then

$$S(f_1,f_2) = \frac{2}{3}Yf_1 - Xf_2 = \frac{4}{3}Y \xrightarrow{\text{normalization}} 4Y =: f_3;\ f_1 \xrightarrow{f_3} 2,\ f_2 \xrightarrow{2} 0,\ f_3 \xrightarrow{2} 0.$$

Thus, $\{2\}$ is the reduced Gröbner basis of $\langle f_1,f_2 \rangle$ in $\mathbb{Z}_{2\mathbb{Z}}[X,Y]$.

Example 47. (Gröbner bases over $\mathbb{F}_2[Y]/\langle Y^2 \rangle$, Example 34 continued)

Consider the ring $\mathbf{V} := \mathbb{F}_2[Y]/\langle Y^2 \rangle = \mathbb{F}_2[y]$, and the ideal

$$I = \langle f_1 = yX_1 + X_2, f_2 = y + yX_2 \rangle \subseteq \mathbf{V}[X_1, X_2].$$

Let us compute a Gröbner basis for I accordingly to the lexicographic order with $X_1 > X_2$. We have:

$$S(f_1,f_1) = yX_2 \xrightarrow{f_2} y =: f_3;\ f_2 \xrightarrow{f_3} 0;\ f_1 \xrightarrow{f_3} X_2.$$

Thus, $\{y, X_2\}$ is the reduced Gröbner basis of I.

Gröbner bases are a powerful tool to eliminate variables. To see this, let us first recall the definition of the elimination ideals of an ideal of $\mathbf{R}[X_1, \ldots, X_n]$.

Definition 48. Let $\mathbf{R}$ be a ring, $I = \langle f_1, \ldots, f_s \rangle$ an ideal of $\mathbf{R}[X_1, \ldots, X_n]$, and $1 \leq k \leq n$. The *kth elimination ideal* of I is

$$I_k := I \cap \mathbf{R}[X_{k+1}, \ldots, X_n].$$

It is an ideal of $\mathbf{R}[X_{k+1}, \ldots, X_n]$ consisting in all combinations of $f_1, \ldots, f_s$ eliminating the variables $X_1, \ldots, X_k$. Note that I_n is nothing but $I \cap \mathbf{R}$.

The following theorem shows that Gröbner bases with respect to the lexicographic monomial order allow the computation of the elimination ideals.

Theorem 49. *Let $\mathbf{R}$ be a valuation ring, $G = \{g_1, \ldots, g_s\}$ a Gröbner basis for an ideal $I = \langle g_1, \ldots, g_s \rangle$ of $\mathbf{R}[X_1, \ldots, X_n]$ with respect to the lexicographic monomial order with $X_1 > X_2 > \cdots > X_n$. Then, for all $1 \leq k \leq n$,*

$$G_k := G \cap \mathbf{R}[X_{k+1}, \ldots, X_n]$$

is a Gröbner basis for the kth elimination ideal I_k of I.

Proof. As $G_k \subseteq I_k$, it suffices to show that $\mathrm{LT}(I_k) \subseteq \langle \mathrm{LT}(G_k) \rangle$. Let $f \in I_k \subseteq I$. Since G is a Gröbner basis for I, there exists $1 \leq i \leq s$ such that $\mathrm{LT}(g_i)$ divides $\mathrm{LT}(f)$. It follows that $\mathrm{LT}(g_i)$ involves only the variables $X_{k+1}, \ldots, X_n$. As any monomial involving one of the variables $X_1, \ldots, X_k$ is greater that $\mathrm{LT}(g_i)$, we infer that $g_i \in \mathbf{R}[X_{k+1}, \ldots, X_n]$, and, thus, $g_i \in G_k$.

Example 50. (Example 46 continued)

Let $f_1 = 2 + 12XY$, $f_2 = 8Y^2 \in \mathbb{Z}_{2\mathbb{Z}}[X,Y]$, and fix any monomial order $>$ on $\mathbb{Z}_{2\mathbb{Z}}[X,Y]$. Then $\{2\}$ is the reduced Gröbner basis of $I = \langle f_1, f_2 \rangle$ in $\mathbb{Z}_{2\mathbb{Z}}[X,Y]$. It follows that $I \cap \mathbb{Z}_{2\mathbb{Z}}[X] = \langle 2 \rangle = 2\mathbb{Z}_{2\mathbb{Z}}[X]$, $I \cap \mathbb{Z}_{2\mathbb{Z}}[Y] = \langle 2 \rangle = 2\mathbb{Z}_{2\mathbb{Z}}[Y]$, and $I \cap \mathbb{Z}_{2\mathbb{Z}} = \langle 2 \rangle = 2\mathbb{Z}_{2\mathbb{Z}}$.

Example 51. (Example 47 continued)

Consider the ring $\mathbf{V} := \mathbb{F}_2[Y]/\langle Y^2 \rangle = \mathbb{F}_2[y]$, and the ideal

$$I = \langle f_1 = yX_1 + X_2, f_2 = y + yX_2 \rangle \subseteq \mathbf{V}[X_1, X_2].$$

We know that $\{y, X_2\}$ is a Gröbner basis for I with respect to the lexicographic order with $X_1 > X_2$. Thus, $I \cap \mathbf{V}[X_2] = \langle y, X_2 \rangle$ and $I \cap \mathbf{V} = \langle y \rangle = \{0, y\}$.

If one wants to compute $I \cap \mathbf{V}[X_1]$, then he has to consider the lexicographic order with $X_2 > X_1$. For this order, $\{y, X_2\}$ is again a Gröbner basis for I, and thus,

$$I \cap \mathbf{V}[X_1] = \langle y \rangle.$$

At the end of this section, let us point out that the versions of Buchberger's algorithm presented in this course did not take into account the considerable mass of optimizations made these last years for the purpose of speeding up Buchberger's algorithm in case the base ring is a field. The faster versions are Faugère's F4 and F5 algorithms [21] allowing to compute routinely Gröbner bases consisting of several hundreds of polynomials, having each several hundreds of terms and coefficients of several hundreds of digits.

Almost all the improvements that have been made in case where the base ring is a field will prove to be easily adaptable to arithmetical rings.

1.2.1 Gröbner bases over $\mathbb{Z}/p^\alpha\mathbb{Z}$

Recently Gröbner basis techniques in polynomial rings over $\mathbb{Z}/m\mathbb{Z}$ and $(\mathbb{Z}/p^\alpha\mathbb{Z}) \times (\mathbb{Z}/p^\alpha\mathbb{Z})$ (in particular $\mathbb{Z}/2^\alpha\mathbb{Z}$ and $(\mathbb{Z}/2^\alpha\mathbb{Z}) \times (\mathbb{Z}/2^\alpha\mathbb{Z})$) have attracted some attention due to their potential applications in formal verification of data paths, and coding theory. Also, many authors have been interested in computing Gröbner bases over $\mathbb{Z}/p^\alpha\mathbb{Z}$ (where p is a "lucky" prime number), because modular methods give a satisfactory way to avoid intermediate coefficients swell with Buchberger's algorithm for computing Gröbner bases over the rational numbers.

We will specify the definition of S-polynomials given in Definition 29 to the important case where the base ring is $\mathbb{Z}/p^\alpha\mathbb{Z}$, where p is a prime number and $\alpha \geq 2$. To lighten the notation, the class of $a \in \mathbb{Z}$ modulo $p^\alpha\mathbb{Z}$ will also be denoted by a.

Definition 52. Recall that the valuation of $a \in \mathbb{Z} \setminus \{0\}$ at p is $v_p(a) := \max\{k \in \mathbb{N} \mid p^k \text{ divides } a\}$, so that $a = p^{v_p(a)}a'$ with $a' \wedge p = 1$.

Note that, in $\mathbb{Z}/p^{\alpha}\mathbb{Z}$, a' is a unit and, writing $ua' + vp^{\alpha} = 1$ (Bézout identity) for some $u,v \in \mathbb{Z}$, we have $a'^{-1} = u$ in $\mathbb{Z}/p^{\alpha}\mathbb{Z}$.

Definition 53. (S-polynomials over $\mathbb{Z}/p^{\alpha}\mathbb{Z}$)

Let p be a prime number, $f_i, f_j \in (\mathbb{Z}/p^{\alpha}\mathbb{Z})[X_1,\ldots,X_n] \setminus \{0\}$ $(i \neq j)$, and fix a monomial order $>$ on $(\mathbb{Z}/p^{\alpha}\mathbb{Z})[X_1,\ldots,X_n]$. Denote by $\mathrm{mdeg}(f_i) = \beta = (\beta_1,\ldots,\beta_n)$, $\mathrm{mdeg}(f_j) = \beta' = (\beta'_1,\ldots,\beta'_n)$, $\gamma = (\gamma_1,\ldots,\gamma_n)$, where $\gamma_k = \max(\beta_k, \beta'_k)$ for each k.

Moreover, denote by $\mathrm{LC}(f_i) = a_i$, $a_i = p^{m_i} c_i$, $\mathrm{LC}(f_j) = a_j$, $a_j = p^{m_j} c_j$ with $0 \leq m_i, m_j \leq \alpha - 1$ and $c_i \wedge p = c_j \wedge p = 1$.

(i) $S(f_i, f_j) := \frac{X^{\gamma}}{\mathrm{LM}(f_i)} f_i - p^{m_i - m_j} c_i c_j^{-1} \frac{X^{\gamma}}{\mathrm{LM}(f_j)} f_j$ if $m_j \leq m_i$.

$S(f_i, f_j) := p^{m_j - m_i} c_j c_i^{-1} \frac{X^{\gamma}}{\mathrm{LM}(f_i)} f_i - \frac{X^{\gamma}}{\mathrm{LM}(f_j)} f_j$ if $m_j > m_i$.

(ii) $S(f_i, f_i) := p^{\alpha - m_i} f_i$.

Example 54. Let $\mathbf{V}[X] = (\mathbb{Z}/16\mathbb{Z})[X]$ and consider the ideal $I = \langle f_1 \rangle$, where $f_1 = 2 + 4X + 8X^2$.

$S(f_1, f_1) = 2f_1 = 4 + 8X =: f_2$,
$S(f_1, f_2) = 2 =: f_3$,
$S(f_2, f_2) = 2f_2 = 8 \xrightarrow{f_3} 0$, $S(f_3, f_3) = 0$,
$f_2 \xrightarrow{f_3} 0$.

Thus, $\mathscr{G} = \{2\}$ is the reduced Gröbner basis of I.

Example 55. Let $\mathbf{V}[X,Y] = (\mathbb{Z}/27\mathbb{Z})[X,Y]$ and consider $\mathscr{G} = \{g_i\}_{i=1}^4$, where $g_1 = 9, g_2 = X + 1, g_3 = 3Y^2, g_4 = Y^3 + 13Y^2 - 12$. Let us fix the lexicographic order as monomial order with $X > Y$.

$S(g_1, g_2) = Xg_1 - 9g_2 = -9 \xrightarrow{g_1} 0$,
$S(g_1, g_3) = Y^2 g_1 - 3g_3 = 0$,
$S(g_1, g_4) = -9Y^2 \xrightarrow{g_1} 0$,
$S(g_2, g_3) = 3Y^2 g_2 - Xg_3 = 3Y^2 \xrightarrow{g_3} 0$,
$S(g_2, g_4) = Y^3 g_2 - Xg_4 = -13XY^2 + 12X + Y^3 \xrightarrow{g_2} 12X + Y^3 + 13Y^2$
$\xrightarrow{g_2} Y^3 + 13Y^2 - 12 \xrightarrow{g_3} 0$,
$S(g_3, g_4) = Yg_3 - 3g_4 = -12Y^3 + 9 \xrightarrow{g_3} 9 \xrightarrow{g_1} 0$.

Thus, $\mathscr{G}$ is a Gröbner basis for $\langle g_1, g_2, g_3, g_4 \rangle$ in $\mathbf{V}[X,Y]$.

Example 56. Let $\mathbf{V}[X,Y] = (\mathbb{Z}/4\mathbb{Z})[X,Y]$ and consider the ideal $I = \langle f_1, f_2, f_3 \rangle$, where $f_1 = X^4 - X, f_2 = Y^3 - 1, f_3 = 2XY$. Let us fix the lexicographic order as monomial order with $X > Y$.

$S(f_1, f_2) = Y^3 f_1 - X^4 f_2 = X^4 - XY^3 \xrightarrow{f_1} X - XY^3 \xrightarrow{f_2} 0$,

$$S(f_1, f_3) = 2Yf_1 - X^3 f_3 = -2XY \xrightarrow{f_3} 0,$$
$$S(f_2, f_3) = 2Xf_2 - Y^2 f_3 = -2X =: f_4,$$
$$S(f_2, f_4) = 2Xf_2 + Y^3 f_4 = -2X \xrightarrow{f_4} 0,$$
$$S(f_1, f_4) = 2f_1 + X^3 f_4 = -2X \xrightarrow{f_4} 0,$$
$$f_3 \xrightarrow{f_4} 0.$$

Thus, $\mathscr{G} = \{f_1, f_2, f_4\}$ is a Gröbner basis for I in $\mathbf{V}[X,Y]$.

Example 57. Take $p = 2$ and $\alpha = 4$. Let $I = \langle f_1 \rangle \subseteq (\mathbb{Z}/16\mathbb{Z})[X,Y]$, where $f_1 = 8X + 2Y + 1$. Let us fix the lexicographic order as monomial order with $X > Y$.

$$S(f_1, f_1) = 2f_1 = 4Y + 2 =: f_2, \; S(f_2, f_2) = 4f_2 = 8 =: f_3, \; S(f_2, f_3) = 4 =: f_4,$$
$$S(f_2, f_4) = 2 =: f_5, \; S(f_2, f_5) = 1. \text{ We conclude that } \{1\} \text{ is the reduced Gröbner}$$
of I.

Here, it is worth pointing out that, contrary to the integral case, in the presence of zero-divisors, $\{f\}$ need not be a Gröbner basis of $\langle f \rangle$.

1.3 Gröbner bases over coherent arithmetical rings

The ideal membership problem for multivariate polynomials over rings has received considerable attention from the constructive algebra community resulting in algorithms that generalize the work of Buchberger. A dynamical approach to Gröbner bases over PID was first introduced in [64]. Our goal in this section is to introduce the notion of dynamical Gröbner basis for coherent arithmetical rings (rings which are locally coherent valuation rings). First note that for $\mathbb{Z}$, a necessary condition so that $f \in \langle f_1, \ldots, f_s \rangle$ in $\mathbb{Z}[X_1, \ldots, X_n]$ is: $f \in \langle f_1, \ldots, f_s \rangle$ in $\mathbb{Q}[X_1, \ldots, X_n]$.

Suppose that this condition is fulfilled, that is, there exists $d \in \mathbb{N} \setminus \{0\}$ such that

$$df \in \langle f_1, \ldots, f_s \rangle \text{ in } \mathbb{Z}[X_1, \ldots, X_n] \quad (0).$$

If $d = 1$ then $f \in \langle f_1, \ldots, f_s \rangle$ in $\mathbb{Z}[X_1, \ldots, X_n]$ and we are done. Otherwise, we can write $d = \prod_{i=1}^{\ell} p_i^{n_i}$, where the p_i's are distinct prime numbers, and $n_i \in \mathbb{N} \setminus \{0\}$. Other necessary conditions so that $f \in \langle f_1, \ldots, f_s \rangle$ in $\mathbb{Z}[X_1, \ldots, X_n]$ is: $f \in \langle f_1, \ldots, f_s \rangle$ in $\mathbb{Z}_{p_i \mathbb{Z}}[X_1, \ldots, X_n]$ for each $1 \leq i \leq \ell$. Here the polynomial ring is over the Noetherian valuation domain $\mathbb{Z}_{p_i \mathbb{Z}}$. Write:

$$d_i f \in \langle f_1, \ldots, f_s \rangle \text{ in } \mathbb{Z}[X_1, \ldots, X_n] \text{ for some } d_i \in \mathbb{Z} \setminus p_i \mathbb{Z} \quad (i).$$

Since no prime number divides all of $d, d_1, \ldots, d_\ell$, we have $\gcd(d, d_1, \ldots, d_\ell) = 1$, and, thus, we can find a Bézout identity $\alpha d + \alpha_1 d_1 + \cdots + \alpha_\ell d_\ell = 1$, $\alpha, \alpha_i \in \mathbb{Z}$. Using this identity, we can find an equality asserting that $f \in \langle f_1, \ldots, f_s \rangle$ in $\mathbb{Z}[X_1, \ldots, X_n]$. Thus, the necessary conditions are sufficient and it suffices to treat the problem in case the base ring is a Noetherian valuation domain.

This method raises the following question:

How to avoid the obstacle of complete prime factorization when it is expensive or infeasible?

The fact that the method explained above is based on gluing "local realizability" appeals to the use of dynamical methods, namely, the use of the notion of "dynamical Gröbner basis". Our goal is to mimic dynamically as much as we can the method explained above.

We first need to recall the following definitions.

Definition 58.

- A ring **R** is *arithmetical* if it is locally a valuation ring. This is equivalent to the fact that it satisfies the following property:

$$\forall x, y \in \mathbf{R} \quad \exists t, a, b \in \mathbf{R} \left\{ \begin{array}{rcl} (1-t)x & = & ay \\ bx & = & ty \end{array} \right. \tag{1.4}$$

Thus, x divides y in the ring $\mathbf{R}_t := \mathbf{R}[\frac{1}{t}]$ and y divides x in the ring $\mathbf{R}_{1-t} := \mathbf{R}[\frac{1}{1-t}]$ (we see here that, locally, **R** behaves like a valuation ring).

- An integral domain is called a *Prüfer domain* if it is arithmetical.

Definition 59. (Comaximal elements and monoids)

Let **R** be a ring.

(1) Let $s_1, \ldots, s_k \in \mathbf{R}$. We say that the elements $s_1, \ldots, s_k$ are *comaximal* if $\langle s_1, \ldots, s_k \rangle = \mathbf{R}$.

(2) Let $S_1, \ldots, S_n$ be monoids of **R** (that is, subsets of **R**, stable under multiplication and containing 1). We say that the monoids $S_1, \ldots, S_n$ are *comaximal* if any ideal of **R** meeting all the S_i must contain 1. In other words, if we have:

$$\forall s_1 \in S_1 \cdots \forall s_n \in S_n \ \exists a_1, \ldots, a_n \in \mathbf{R} \quad | \quad \sum_{i=1}^{n} a_i s_i = 1,$$

that is, $s_1, \ldots, s_k$ are comaximal elements in **R**.

If $u_1, \ldots, u_m$ are comaximal elements in **R**, then the monoids $u_1^{\mathbb{N}}, \ldots, u_m^{\mathbb{N}}$ are comaximal. For example, for $u \in \mathbf{R}$, the monoids $u^{\mathbb{N}}$ and $(1-u)^{\mathbb{N}}$ are comaximal. Remark that comaximal monoids remain comaximal when you replace the ring by a bigger one or the multiplicative subsets by smaller ones.

Dynamical version of Buchberger's Algorithm for coherent arithmetical rings

Let **R** be a coherent arithmetical ring which is locally Gröbner, and fix a monomial order $>$. For example, take **R** a Prüfer domain of Krull dimension ≤ 1 and consider the lexicographic monomial order. The purpose is to construct a dynamical Gröbner basis G for a nonzero finitely-generated ideal $I = \langle f_1, \ldots, f_s \rangle$ of $\mathbf{R}[X_1, \ldots, X_n]$. The algorithm for constructing such a Gröbner basis works like Buchberger's Algorithm 38 for coherent valuation rings. The only difference is when it has to handle two incomparable (under division) elements a, b in **R**. In this situation, one should first compute $u, v, w \in \mathbf{R}$ such that

$$\begin{cases} ub = va \\ wb = (1-u)a. \end{cases}$$

Now, one opens two branches: the computations are pursued in $\mathbf{R}_u = \mathbf{R}[\frac{1}{u}]$ and $\mathbf{R}_{1-u} = \mathbf{R}[\frac{1}{1-u}]$.

- First possibility: the two incomparable elements a and b are encountered when performing the division algorithm (analogous to the division algorithm in the case of a valuation ring). Suppose that one has to divide a term $aX^\alpha = \mathrm{LT}(f)$ by another term $bX^\beta = \mathrm{LT}(g)$ with X^β divides X^α.

  In the ring $\mathbf{R}_{1-u}$: $f = \frac{w}{1-u} \frac{X^\alpha}{X^\beta} g + r$ (mdeg$(r) <$ mdeg(f)) and the division is pursued with f replaced by r.

  In the ring $\mathbf{R}_u$: $\mathrm{LT}(f)$ is not divisible by $\mathrm{LT}(g)$ and thus $f = \overline{f}^{\{g\}}$.

- Second possibility: the two incomparable elements a and b are encountered when computing $S(f, g)$ with $\mathrm{LT}(f) = aX^\alpha$ and $\mathrm{LT}(g) = bX^\beta$. Denote $\gamma = (\gamma_1, \ldots, \gamma_n)$, with $\gamma_i = \max(\alpha_i, \beta_i)$ for each i.

  In the ring $\mathbf{R}_{1-u}$: $S(f, g) = \frac{X^\gamma}{X^\alpha} f - \frac{w}{1-u} \frac{X^\gamma}{X^\beta} g$.

  In the ring $\mathbf{R}_u$: $S(f, g) = \frac{v}{u} \frac{X^\gamma}{X^\alpha} f - \frac{X^\gamma}{X^\beta} g$.

At each new branch, if $S = \overline{S(f,g)}^{G'} \neq 0$ where G' is the current Gröbner basis, then S must be added to G'. At the end of the computation, we obtain a binary tree whose leaves (terminal nodes) correspond to comaximal localizations of the initial ring and at each leaf we have computed a Gröbner basis. The collection of these Gröbner bases computed at each leaf is the so-called *dynamical Gröbner basis*. It is worth mentioning that in some cases the leaves are known in advance (see Examples 61, 62, 65, 66), and in case the base ring is Bézout we can avoid branching (see Section 1.4). Our goal now is to give a series of examples of dynamical Gröbner bases to illustrate this approach.

Example 60. Let $I = \langle f_1 = 3XY + 1, \ f_2 = (4 + 2\theta)Y + 9 \rangle$ in $\mathbb{Z}[\theta][X,Y]$ where $\theta = \sqrt{-5}$. The ring $\mathbb{Z}[\theta]$ is a Dedekind domain which is not Bézout. Let us fix the lexicographic order with $X > Y$ as monomial order.

a) Computing a dynamical Gröbner basis: we will first compute a dynamical Gröbner basis for I in $\mathbb{Z}[\theta][X,Y]$. We will give all the details of the computations only for one leaf. Since $x_1 = 3$ and $x_2 = 4 + 2\theta$ are not comparable, we have to find $u,v,w \in \mathbb{Z}[\theta]$ such that:

$$\begin{cases} ux_2 = vx_1 \\ wx_2 = (1-u)x_1. \end{cases}$$

Note that as the ring $\mathbb{Z}[\theta]$ has a $\mathbb{Z}$-basis (it is a rank 2 free $\mathbb{Z}$-module), u,v,w can be computed by solving an underdetermined linear system over the integers. A solution is given by: $u = 5 + 2\theta$, $v = 6\theta$, $w = -3$. Then we can open two branches:

$$\mathbb{Z}[\theta]$$
$$\swarrow \qquad \searrow$$
$$\mathbb{Z}[\theta]_{4+2\theta} \qquad \mathbb{Z}[\theta]_{5+2\theta}$$

For $\alpha \in \mathbb{Z}[\theta]$, we will denote by $\mathcal{M}(\alpha)$ the monoid $\alpha^{\mathbb{N}}$ generated by α in such a way $\mathbb{Z}[\theta]_\alpha := \mathbb{Z}[\theta][\frac{1}{\alpha}] = \mathcal{M}(\alpha)^{-1}\mathbb{Z}[\theta]$.

<u>In $\mathbb{Z}[\theta]_{5+2\theta}$:</u>

$$S(f_1,f_2) = \tfrac{6\theta}{5+2\theta}f_1 - Xf_2 = -9X + \tfrac{6\theta}{5+2\theta} =: f_3,$$
$$S(f_1,f_3) = -3f_1 - Yf_3 = -\tfrac{6\theta}{5+2\theta}Y - 3 =: f_4,$$
$$S(f_1,f_4) = -\tfrac{2\theta}{5+2\theta}f_1 - Xf_4 = 3X - \tfrac{2\theta}{5+2\theta} =: f_5,$$
$$f_2 \xrightarrow{f_4} 0, \ f_3 \xrightarrow{f_5} 0,$$
$$S(f_1,f_5) = f_1 - Yf_5 = \tfrac{2\theta}{5+2\theta}Y + 1 =: f_6,$$
$$f_4 \xrightarrow{f_6} 0, \ S(f_2,f_5) = Xf_2 - \tfrac{6\theta}{5+2\theta}Yf_5 \xrightarrow{f_5,f_6} 0.$$

As 2 and 3 are not comparable under division in $\mathbb{Z}[\theta]_{5+2\theta}$, we open two news branches:

$$\mathbb{Z}[\theta]_{5+2\theta}$$
$$\swarrow \qquad \searrow$$
$$\mathbb{Z}[\theta]_{(5+2\theta).3} \qquad \mathbb{Z}[\theta]_{(5+2\theta).2}$$

Recall that $\mathbb{Z}[\theta]_{\alpha.\beta} := \mathbb{Z}[\theta][\frac{1}{\alpha}][\frac{1}{\beta}] = \mathbb{Z}[\theta][\frac{1}{\alpha\beta}]$ for $\alpha, \beta \in \mathbb{Z}[\theta]$.

<u>In $\mathbb{Z}[\theta]_{(5+2\theta).3}$:</u>

$$S(f_1,f_6) = \tfrac{2\theta}{3(5+2\theta)}f_1 - Xf_6 = -\tfrac{1}{3}f_5 \xrightarrow{f_5} 0,$$

$$S(f_5, f_6) = \frac{2\theta}{3(5+2\theta)}Yf_5 - Xf_6 = \frac{20}{3(5+2\theta)^2}Y - X \xrightarrow{f_5} \frac{20}{3(5+2\theta)^2}Y - \frac{2\theta}{3(5+2\theta)} \xrightarrow{f_6}$$

0.

Thus, $G_1 = \{3XY + 1, 3X - \frac{2\theta}{5+2\theta}, \frac{2\theta}{5+2\theta}Y + 1\}$ is a Gröbner basis for $\langle 3XY + 1, (4+2\theta)Y + 9 \rangle$ in $\mathscr{M}(5+2\theta, 3)^{-1}\mathbb{Z}[\theta] = \mathbb{Z}[\theta]_{(5+2\theta).3}$.

In $\mathbb{Z}[\theta]_{(5+2\theta).2}$:

$G_2 = \{3XY + 1, 3X - \frac{2\theta}{5+2\theta}, \frac{2\theta}{5+2\theta}Y + 1\}$ is a Gröbner basis for $\langle 3XY + 1, (4+2\theta)Y + 9 \rangle$.

In $\mathbb{Z}[\theta]_{(4+2\theta)}$:

$G_3 = \{3XY + 1, (4+2\theta)Y + 9, \frac{-27}{4+2\theta}X + 1\}$ is a Gröbner basis for $\langle 3XY + 1, (4+2\theta)Y + 9 \rangle$.

As a conclusion, the dynamical evaluation of the problem of constructing a Gröbner basis for I produces the following evaluation tree:

$$\mathbb{Z}[\theta]$$

$$\mathbb{Z}[\theta]_{4+2\theta} \qquad \mathbb{Z}[\theta]_{5+2\theta}$$

$$\mathbb{Z}[\theta]_{(5+2\theta).3} \quad \mathbb{Z}[\theta]_{(5+2\theta).2}$$

The obtained dynamical Gröbner basis of I is $G = \{(\mathscr{M}(5+2\theta), G_1), (\mathscr{M}(4+2\theta), G_2)\}$.

b) The ideal membership problem: suppose that we have to deal with the ideal membership problem: $f = (4\theta - 1)X^2Y + 6\theta XY^2 + 9\theta X^2 + 3X - 4Y - 9 \in ? I$

Let us first execute the dynamical division algorithm of f by $G_1 = \{f_1 = 3XY + 1, f_5 = -3X + \frac{2\theta}{5+2\theta}, f_6 = \frac{2\theta}{5+2\theta}Y + 1\}$ in the ring $\mathbb{Z}[\theta]_{(5+2\theta).3}[X,Y]$. With the same notation as in the Division Algorithm 14, one obtains:

q_1	q_5	q_6	p
$\frac{4\theta-1}{3}X$	0	0	$6\theta XY^2 + 9\theta X^2 + \frac{10-4\theta}{3}X - 4Y - 9$
$\frac{4\theta-1}{3}X + 2\theta Y$	0	0	$9\theta X^2 + \frac{10-4\theta}{3}X - (4+2\theta)Y - 9$
$\frac{4\theta-1}{3}X + 2\theta Y$	$-3\theta X$	0	$-(4+2\theta)Y - 9$
$\frac{4\theta-1}{3}X + 2\theta Y$	$-3\theta X$	-9	0

Thus, the answer to this ideal membership problem in the ring $\mathbb{Z}[\theta]_{(5+2\theta).3}[X,Y]$ is positive and one obtains $f = (\frac{4\theta-1}{3}X + 2\theta Y)f_1 - 3\theta Xf_5 - 9f_6$. But since

$$f_5 = \left(\frac{-6\theta}{5+2\theta}XY - 3X + \frac{2\theta}{5+2\theta}\right)f_1 - X^2Yf_2, \text{ and}$$

$$f_6 = \left(\frac{-6\theta}{5+2\theta}XY^2 - 3XY + \frac{2\theta}{5+2\theta}Y + 1\right)f_1 - X^2Y^2f_2, \text{ one infers that}$$

$$f = \left[\frac{-90}{5+2\theta}X^2Y + 9\theta X^2 + \frac{54\theta}{5+2\theta}XY^2 + 27XY + \frac{6\theta+15}{5+2\theta}X - 4Y - 9\right]f_1$$
$$+ [3\theta X^3Y + 9X^2Y^2]f_2.$$

Seeing that 3 does not appear in the denominators of the relation above, we can say that we have a positive answer to our ideal membership problem in the ring $\mathbb{Z}[\theta]_{5+2\theta}[X,Y]$ without dealing with the leaf $\mathbb{Z}[\theta]_{(5+2\theta).2}$. Clearing the denominators, we get:

$$(5+2\theta)f = [-90X^2Y + 45(\theta - 2)X^2 + 54\theta XY^2 + 27(5+2\theta)XY + (6\theta + 15)X$$
$$-4(5+2\theta)Y - 9(5+2\theta)]f_1 + [15(\theta - 2)X^3Y + 9(5+2\theta)X^2Y^2]f_2. \quad (A)$$

It remains to execute the dynamical division algorithm of f by $G_2 = \{f_1 = 3XY + 1, f_7 = -\frac{27}{4+2\theta}X + 1, f_8 = Y + \frac{9}{4+2\theta}\}$ in the ring $\mathbb{Z}[\theta]_{4+2\theta}[X,Y]$. The division is as follows:

q_1	q_7	q_8	p
0	0	$(4\theta - 1)X^2$	$6\theta XY^2 - \frac{81}{4+2\theta}X^2 + 3X - 4Y - 9$
$2\theta Y$	0	$(4\theta - 1)X^2$	$\frac{-81}{4+2\theta}X^2 + 3X - (4+2\theta)Y - 9$
$2\theta Y$	$3X$	$(4\theta - 1)X^2$	$-(4+2\theta)Y - 9$
$2\theta Y$	$3X$	$(4\theta - 1)X^2 - (4+2\theta)$	0

Thus, the answer to this ideal membership problem in the ring $\mathbb{Z}[\theta]_{4+2\theta}[X,Y]$ is positive and one obtains:

$$f = 2\theta Y f_1 + 3X f_7 + ((4\theta - 1)X^2 - (4+2\theta))f_8.$$

But since

$$f_7 = f_1 - \frac{3}{4+2\theta}X f_2, \text{ and}$$

$$f_8 = \left(Y + \frac{9}{4+2\theta}\right)f_1 - \frac{3}{4+2\theta}XY f_2, \text{ one infers that}$$

$$(4+2\theta)f = [(14\theta - 44)X^2Y + 9(4\theta - 1)X^2 - 4(4+2\theta)Y$$
$$+3(4+2\theta)X - 9(4+2\theta)]f_1 + [-9X^2 - 3(4\theta - 1)X^3Y + 3(4+2\theta)XY]f_2. \quad (B)$$

Using the Bézout identity $(5+2\theta) - (4+2\theta) = 1$, (A) $-$ (B) $\Rightarrow$

$$f = [(46 - 14\theta)X^2Y + 9(\theta - 9)X^2 + 54\theta XY^2 + 27(5+2\theta)XY + 3X - 4Y - 9]f_1$$
$$+ [3(9\theta - 11)X^3Y + 9(5+2\theta)X^2Y^2 + 9X^2 - 3(4+2\theta)XX]f_2,$$

a complete positive answer.

Example 61. (Examples 46 and 57 continued)

Let $\mathbf{V}_1 = \mathbb{Z}_{2\mathbb{Z}} = \{\frac{a}{b} \in \mathbb{Q} \mid (a,b) \in \mathbb{Z} \times \mathbb{Z} \text{ and } b \text{ is odd}\}$, $\mathbf{V}_2 = \mathbb{Z}/16\mathbb{Z} = \mathbb{Z}/2^4\mathbb{Z}$, and $\mathbf{T} = \mathbf{V}_1 \times \mathbf{V}_2$. Note that $\mathbf{T}$ is a Gröbner arithmetical ring by Proposition 64.

Consider the ideal

$$I = \langle f_1 = (2,1) + (0,2)Y + (0,8)X + (12,0)XY, \; f_2 = (8,0)Y^2 \rangle \subseteq \mathbf{T}[X,Y],$$

and fix the lexicographic order as monomial order with $X > Y$.

Let us denote by $\mathbf{e_1} = (1,0)$, $\mathbf{e_2} = (0,1)$, and $\mathbf{1} = (1,1) = \mathbf{e_1} + \mathbf{e_2}$ the unit of $\mathbf{R}$. We know in advance that the monoids $S_1 = \mathbf{e_1}^{\mathbb{N}}$ and $S_2 = \mathbf{e_2}^{\mathbb{N}}$ are comaximal with

$$S_1^{-1}\mathbf{T} = \mathbf{T}\left[\frac{1}{\mathbf{e_1}}\right] \cong \mathbf{V}_1 = \mathbb{Z}_{2\mathbb{Z}} \text{ and } S_2^{-1}\mathbf{T} = \mathbf{T}\left[\frac{1}{\mathbf{e_2}}\right] \cong \mathbf{V}_2 = \mathbb{Z}/16\mathbb{Z}.$$

This can be represented as follows:

$$
\begin{array}{ccc}
 & \mathbf{T} & \\
 \swarrow & & \searrow \\
\mathbf{T}[\tfrac{1}{\mathbf{e_1}}] \cong \mathbf{V}_1 & & \mathbf{T}[\tfrac{1}{\mathbf{e_2}}] \cong \mathbf{V}_2
\end{array}
$$

Denoting by π_1 and π_2, the first and second projections, respectively, we have $\pi_1(\mathbf{T}) = \mathbf{V}_1$, $\pi_2(\mathbf{T}) = \mathbf{V}_2$, $\pi_1(I) = \langle 2 + 12XY, \; f_2 = 8Y^2 \rangle$, and $\pi_2(I) = \langle 1 + 2Y + 8X \rangle$.

By Examples 46 and 57, we know that $G_1 = \{2\}$ is a Gröbner basis for $\pi_1(I)$ and $G_2 = \{1\}$ is a Gröbner basis for $\pi_2(I)$. So, denoting by $\mathscr{G}_i = \mathbf{e_i}G_i = \{\mathbf{e_i}g : g \in G_i\}$, $G = \{(\mathscr{G}_1, \mathbf{e_1}^{\mathbb{N}}), (\mathscr{G}_2, \mathbf{e_2}^{\mathbb{N}})\} = \{(\{(2,0)\}, \mathbf{e_1}^{\mathbb{N}}), (\{(0,1)\}, \mathbf{e_2}^{\mathbb{N}})\}$ is a dynamical Gröbner basis for I.

1.3.1   A parallelisable algorithm for computing dynamical Gröbner bases over $\mathbb{Z}/m\mathbb{Z}$

Our objective here is to explain how the method of using the Chinese remainder theorem for computing Gröbner bases over $\mathbb{Z}/m\mathbb{Z}$ can be seen as a particular case of computing dynamical Gröbner bases.

Let $m \in \mathbb{N} \setminus \{0, 1, 2\}$ and suppose that we know the prime factorization $m = p_1^{\alpha_1} \cdots p_\ell^{\alpha_\ell}$ of m, where $\ell, \alpha_i \in \mathbb{N}^*$ and the p_i's are pairwise different prime numbers. The goal is to present a simple way for constructing a dynamical Gröbner basis over the Dedekind ring $\mathbf{R} := \mathbb{Z}/m\mathbb{Z}$ whose leaves (i.e., comaximal localizations) are known in advance by the Chinese remainder theorem.

It is worth pointing out that if the prime factorization of m is not possible then one has to follow the general theory of dynamical Gröbner bases.

By the Chinese remainder theorem, we have the ring isomorphism

$$\mathbb{Z}/m\mathbb{Z} \cong (\mathbb{Z}/p_1^{\alpha_1}\mathbb{Z}) \times (\mathbb{Z}/p_2^{\alpha_2}\mathbb{Z}) \times \cdots \times (\mathbb{Z}/p_\ell^{\alpha_\ell}\mathbb{Z}).$$

So, we can assume that $\mathbf{R} = \prod_{i=1}^{\ell} (\mathbb{Z}/p_i^{\alpha_i}\mathbb{Z})$.

Our objective now is to explain how to construct a dynamical Gröbner basis over $\mathbf{R}$. The advantage of working with the ring $\mathbf{R} = \prod_{i=1}^{\ell}(\mathbb{Z}/p_i^{\alpha_i}\mathbb{Z})$ is that we know in advance that the binary tree we will construct when computing dynamically a Gröbner basis of an ideal of $\mathbf{R}[X_1,\ldots,X_n]$ is formed by only ℓ leaves as follows (we denote by $\mathbf{e_i} = (0,\ldots,0,1,0,\ldots,0)$ where 1 is on the ith position, and $\mathbf{1} = (1,\ldots,1) = \mathbf{e_1} + \cdots + \mathbf{e_\ell}$):

$$
\begin{array}{ccc}
 & \mathbf{R} & \\
\swarrow & \cdots & \searrow \\
\mathbf{R}[\tfrac{1}{\mathbf{e_1}}] & \cdots & \mathbf{R}[\tfrac{1}{\mathbf{e_\ell}}]
\end{array}
$$

Note that, as $\mathbf{R}[\tfrac{1}{\mathbf{e_i}}] \cong \mathbb{Z}/p_i^{\alpha_i}\mathbb{Z}$, in order to compute a dynamical Gröbner basis of an ideal I of $\mathbf{R}[X_1,\ldots,X_n]$, one only has to execute ℓ times (possibly in a parallel way) Buchberger's Algorithm over $\mathbb{Z}/p^{\alpha}\mathbb{Z}$ (Algorithm 38). Denoting by G_i $(1 \le i \le \ell)$ the computed Gröbner basis for $\pi_i(I)$, where π_i is the ith canonical projection, and setting $\mathbf{e_i}G_i = \{\mathbf{e_i}g : g \in G_i\}$, $G = \{(\mathbf{e_1}G_1, \mathbf{e_1}^{\mathbb{N}})\ldots,(\mathbf{e_\ell}G_\ell, \mathbf{e_\ell}^{\mathbb{N}})\}$ is a dynamical Gröbner basis for I.

Example 62. Take $\mathbf{A} = \mathbb{Z}/216\mathbb{Z}$, fix the lexicographic order as monomial order with $X > Y$, and suppose that we want to compute a Gröbner basis for the following ideal of $\mathbf{A}[X,Y]$:

$$J = \langle u_1 = 144, u_2 = X + 162Y - 80, u_3 = 162X^2 + 81, u_4 = -24Y^2, u_5 = -80Y^3 + 40Y^2 - 120\rangle.$$

As $216 = 2^3 \times 3^3$, by the Chinese remainder theorem, we have $\mathbb{Z}/216\mathbb{Z} \overset{\varphi}{\cong} (\mathbb{Z}/8\mathbb{Z}) \times (\mathbb{Z}/27\mathbb{Z})$, where

$$
\begin{array}{rccc}
\varphi: & \mathbb{Z}/216\mathbb{Z} & \longrightarrow & (\mathbb{Z}/8\mathbb{Z}) \times (\mathbb{Z}/27\mathbb{Z}) \\
 & \bar{x} & \longmapsto & (\dot{x}, \tilde{x}),
\end{array}
$$

and $\bar{x}, \dot{x}, \tilde{x}$ denote the classes of $x \in \mathbb{Z}$ modulo 216, 8, 27, respectively. Moreover, we have:

$$
\begin{array}{rccc}
\varphi^{-1}: & (\mathbb{Z}/8\mathbb{Z}) \times (\mathbb{Z}/27\mathbb{Z}) & \longrightarrow & \mathbb{Z}/216\mathbb{Z} \\
 & (\dot{x}, \tilde{y}) & \longmapsto & \overline{81x - 80y}.
\end{array}
$$

So our problem can be translated into the ring $\mathbf{R} := (\mathbb{Z}/2^3\mathbb{Z}) \times (\mathbb{Z}/3^3\mathbb{Z})$ in which the considered ideal becomes $I = \varphi(J) = \langle f_1 = (0,9), f_2 = (1,1)X + (2,0)Y + (0,1), f_3 = (2,0)X^2 + (1,0), f_4 = (0,3)Y^2, f_5 = (0,1)Y^3 + (0,13)Y^2 - (0,12)\rangle$. Let us denote by $I_1 := \pi_1(I) = \langle g_1 = X + 2Y, g_2 = 2X^2 + 1\rangle$, and $I_2 := \pi_2(I) = \langle h_1 = 9, h_2 = X + 1, h_3 = 3Y^2, h_4 = Y^3 + 13Y^2 - 12\rangle$. Using the algorithm given in Section 1.2.1, one finds that $\mathcal{G}_1 = \{1\}$ and $\mathcal{G}_2 = \{h_1, h_2, h_3, h_4\}$ are reduced Gröbner bases for I_1 and I_2 respectively, and thus,

$$\mathcal{G} = \{(\mathbf{e_1}\mathcal{G}_1, \mathbf{e_1}^{\mathbb{N}}), (\mathbf{e_2}\mathcal{G}_2, \mathbf{e_2}^{\mathbb{N}})\}$$

is a dynamical Gröbner basis for I in the ring $\mathbf{R}[X,Y]$, where $\mathbf{e_1} = (1,0)$ and $\mathbf{e_2} = (0,1)$. Going back to the ring $\mathbf{A}[X,Y]$, we conclude that $G = \{81, 72, -80X - 80, -24Y^2, -80Y^3 + 40Y^2 - 120\}$ is a Gröbner basis for J in the ring $\mathbf{A}[X,Y]$ (and thus so is $\{9, -80X - 80, -24Y^2, -80Y^3 + 40Y^2 - 120\}$).

Remark 63. It is worth pointing out that if the prime factorization of m is not possible then instead of using the Chinese remainder theorem one has to follow the general theory of dynamical Gröbner bases: do as if $\mathbb{Z}/m\mathbb{Z}$ were a valuation domain, or also, as if m was a power of a prime number. When, during the computations, one meets two integers s and r ($2 \le s, r < m$) which are not comparable under division modulo m, then compute $d = s \wedge r := \gcd(s,r)$, write $s = d\,s'$, $r = d\,r'$ where $s' \wedge r' = 1$ and s', r' are not invertible modulo m, or also, $s' \wedge m \ne 1$ and $r' \wedge m \ne 1$. Now, the ring $\mathbb{Z}/m\mathbb{Z}$ has to be replaced by the rings $\mathbf{A}_1 = (\mathbb{Z}/m\mathbb{Z})[\frac{1}{s' \wedge m}]$ and $\mathbf{A}_2 = (\mathbb{Z}/m\mathbb{Z})[\frac{1}{r' \wedge m}]$. We have:

- $(s' \wedge m) \wedge (r' \wedge m) = 1$ as $s' \wedge r' = 1$,

- s divides r in $\mathbf{A}_1$, and

- r divides s in $\mathbf{A}_2$.

In fact, this can be rephrased as follows: we found two coprime factors $s' \wedge m$ and $r' \wedge m$ of m which can be used to partially factorize m and then to write $\mathbb{Z}/m\mathbb{Z}$ as a product of simpler rings by the Chinese remainder theorem.

1.3.2   A parallelisable algorithm for computing Gröbner bases over $(\mathbb{Z}/p^\alpha\mathbb{Z}) \times (\mathbb{Z}/p^\alpha\mathbb{Z})$

Many attacks showed that cryptosystems based on Gröbner bases over a field are not secure. The analysis of all known attacks, like for example the linear algebra attack, showed that they use in some step, the solution of a linear system on the underlying field. Hence one solution to avoid such attack is to work with a ring over which linear systems are difficult to solve. Precisely, over a Dedekind ring with many zero-divisors. For this objective, the ring $(\mathbb{Z}/p^\alpha\mathbb{Z}) \times (\mathbb{Z}/p^\alpha\mathbb{Z})$ (where p is a prime number) may be interesting as, in this ring, the probability that an element is a zero-divisor is equal to $\frac{2p-1}{p^2}$ ($= \frac{3}{4}$ if $p = 2$, see Exercise 199).

The following proposition shows that $(\mathbb{Z}/p^\alpha\mathbb{Z}) \times (\mathbb{Z}/p^\alpha\mathbb{Z})$ is a *Dedekind ring*, i.e., a Noetherian arithmetical ring. Moreover, we explicitly code $(\mathbb{Z}/p^\alpha\mathbb{Z}) \times (\mathbb{Z}/p^\alpha\mathbb{Z})$ as the ring $\mathbb{Z}[t]$ modulo an ideal of $\mathbb{Z}[t]$.

Proposition 64.

(1) *If $\mathbf{R}$ and $\mathbf{T}$ are Gröbner valuation rings (resp., coherent Noetherian valuation rings), then $\mathbf{R} \times \mathbf{T}$ is a Gröbner arithmetical ring (resp., a Dedekind ring).*

(2) $\mathbb{Z}[t]/\langle p^{\alpha}, t^2 - t \rangle \overset{\varphi}{\cong} (\mathbb{Z}/p^{\alpha}\mathbb{Z}) \times (\mathbb{Z}/p^{\alpha}\mathbb{Z}).$

Proof.   (1) Denoting by $\mathbf{e} = (1, 0)$ and $\mathbf{1} = (1, 1)$, we have

$$(\mathbf{R} \times \mathbf{T}) \left[\frac{\mathbf{1}}{\mathbf{e}} \right] \cong (\mathbf{R} \times \mathbf{T})/\langle \mathbf{1} - \mathbf{e} \rangle \cong \mathbf{R} \ \& \ (\mathbf{R} \times \mathbf{T}) \left[\frac{\mathbf{1}}{\mathbf{1} - \mathbf{e}} \right] \cong (\mathbf{R} \times \mathbf{T})/\langle \mathbf{e} \rangle \cong \mathbf{T}.$$

(2) This is very classical, take $\varphi(\bar{f}) = (\overline{f(0)}, \overline{f(1)})$ for $f \in \mathbb{Z}[t]$. Here, for $g \in \mathbb{Z}[t]$ (resp., for $c \in \mathbb{Z}$), $\bar{g}$ (resp., for $\bar{c}$) denotes the class of g modulo $\langle p^{\alpha}, t^2 - t \rangle$ (resp., modulo $\langle p^{\alpha} \rangle$). It is worth pointing out that denoting by $f = \sum_{i=0}^{m} a_i t^i \in \mathbb{Z}[t]$, we have

$$\bar{f} = \bar{a}_0 + \left(\sum_{i=1}^{m} \bar{a}_i \right) t, \text{ and}$$

$$\varphi(\bar{f}) = \left(\bar{a}_0, \sum_{i=0}^{m} \bar{a}_i \right), \ \varphi^{-1}(\bar{a}, \bar{b}) = \bar{a} + (\bar{b} - \bar{a})t.$$

How to compute a reduced dynamical Gröbner basis over
$(\mathbb{Z}/p^{\alpha}\mathbb{Z}) \times (\mathbb{Z}/p^{\alpha}\mathbb{Z})[X_1, \dots, X_n]$ **and** $(\mathbb{Z}[t]/\langle p^{\alpha}, t^2 - t \rangle)[X_1, \dots, X_n]$**:**

Our objective now is to explain how to construct a dynamical Gröbner bases over $(\mathbb{Z}/p^{\alpha}\mathbb{Z}) \times (\mathbb{Z}/p^{\alpha}\mathbb{Z})$. The advantage of working with the ring $(\mathbb{Z}/p^{\alpha}\mathbb{Z}) \times (\mathbb{Z}/p^{\alpha}\mathbb{Z})$ is that we know in advance that the binary tree we will construct when computing dynamically a Gröbner basis of an ideal of $(\mathbb{Z}/p^{\alpha}\mathbb{Z}) \times (\mathbb{Z}/p^{\alpha}\mathbb{Z})$ $[X_1, \dots, X_n]$ is formed by only two leaves as follows (we denote by $\mathbf{V} = (\mathbb{Z}/p^{\alpha}\mathbb{Z}) \times (\mathbb{Z}/p^{\alpha}\mathbb{Z})$, $\mathbf{e} = (1, 0)$ and $\mathbf{1} = (1, 1)$):

$$\mathbf{V}$$
$$\swarrow \qquad \searrow$$
$$\mathbf{V}[\tfrac{1}{\mathbf{e}}] \qquad \mathbf{V}[\tfrac{1}{\mathbf{1} - \mathbf{e}}]$$

Note that, as $\mathbf{V}[\tfrac{1}{\mathbf{e}}] \cong \mathbf{V}[\tfrac{1}{\mathbf{1} - \mathbf{e}}] \cong \mathbb{Z}/p^{\alpha}\mathbb{Z}$, in order to compute a dynamical Gröbner basis of an ideal I of $\mathbf{V}[X_1, \dots, X_n]$, one only has to execute twice Buchberger's Algorithm 38 over $\mathbb{Z}/p^{\alpha}\mathbb{Z}$ (possibly in a parallel way). The first one is with $\pi_1(I)$ and the second one is with $\pi_2(I)$, where π_1 and π_2 are the first and second canonical projection of $(\mathbb{Z}/p^{\alpha}\mathbb{Z}) \times (\mathbb{Z}/p^{\alpha}\mathbb{Z})$ over $\mathbb{Z}/p^{\alpha}\mathbb{Z}$. Moreover, if at each ith leave ($i = 1, 2$), the computed Gröbner basis is denoted by G_i, then the dynamical Gröbner basis is $G = \{(\mathscr{G}_1, \mathbf{e}^{\mathbb{N}}), (\mathscr{G}_2, (\mathbf{1} - \mathbf{e})^{\mathbb{N}})\}$ and is also reduced, where $\mathscr{G}_1 = \mathbf{e}G_1 = \{\mathbf{e}g : g \in G_1\}$, and $\mathscr{G}_2 = (\mathbf{1} - \mathbf{e})G_2 = \{(\mathbf{1} - \mathbf{e})g : g \in G_2\}$.

Note that if J is a finitely-generated ideal of $(\mathbb{Z}[t]/\langle p^{\alpha}, t^2 - t \rangle)[X_1, \dots, X_n]$, denoting by G_i the reduced Gröbner basis for $\pi_i(\phi(J))$ ($i = 1, 2$), the reduced dynamical Gröbner basis for J in $(\mathbb{Z}[t]/\langle p^{\alpha}, t^2 - t \rangle)[X_1, \dots, X_n]$ is defined as

$$G = \{(G_1, (1 - t)^{\mathbb{N}}), (G_2, t^{\mathbb{N}})\}.$$

Here ϕ stands for the extension of φ to $(\mathbb{Z}[t]/\langle p^\alpha, t^2 - t\rangle)[X_1, \ldots, X_n]$ by setting $\phi(X_j) = X_j$. Of course, as mentioned above, G_1 and G_2 can be computed in a parallel way.

Moreover, for $h \in (\mathbb{Z}[t]/\langle p^\alpha, t^2 - t\rangle)[X_1, \ldots, X_n]$, its unique remainder on division by G is

$$\overline{h}^G = \phi^{-1}(\overline{\phi(h)}^{G_1}, \overline{\phi(h)}^{G_2}),$$

and we have:

$$h \in J \Leftrightarrow \overline{h}^G = 0 \Leftrightarrow \overline{\phi(h)}^{G_1} = \overline{\phi(h)}^{G_2} = 0.$$

Example 65. Take $p = 2$ and $\alpha = 3$. Let

$$J = \langle P_1 = (2-t)X + (1+t)Y + 1 - t, P_2 = (1+t)X^2 + 1\rangle$$

$$\subseteq (\mathbb{Z}[t]/\langle 8, t^2 - t\rangle)[X, Y].$$

If coded in the ring $(\mathbb{Z}/8\mathbb{Z}) \times (\mathbb{Z}/8\mathbb{Z})[X, Y]$, J becomes

$$I = \langle f_1 = (2,1)X + (1,2)Y + (1,0), f_2 = (1,2)X^2 + (1,1)\rangle.$$

Computing a reduced dynamical Gröbner basis:

Let us denote by $I_1 := \pi_1(I) = \langle g_1 = 2X + Y + 1, g_2 = X^2 + 1\rangle$, and $I_2 := \pi_2(I) = \langle h_1 = X + 2Y, h_2 = 2X^2 + 1\rangle$. Let us fix the lexicographic order as monomial order with $X > Y$. We will give all the details of the dynamical computation:

$S(g_1, g_1) = 4g_1 = 4Y + 4 =: g_3, S(g_3, g_3) = 2g_3 = 0, S(g_2, g_2) = 0,$
$S(g_1, g_2) = Xg_1 - 2g_2 = XY + X - 2 := g_4, S(g_4, g_4) = 0,$
$S(g_1, g_3) = 2Yg_1 - Xg_3 = 2Y^2 + 2Y - 4X \xrightarrow{g_1} 2Y^2 - 2 =: g_5, S(g_5, g_5) = 0,$
$S(g_1, g_5) = Y^2 g_1 - Xg_5 = 2X + Y^3 + Y^2 \xrightarrow{g_1} Y^3 + Y^2 - Y - 1 =: g_6, S(g_6, g_6) = 0,$
$S(g_1, g_6) = Y^3 g_1 - 2Xg_6 = -2XY^2 + 2XY + 2X + Y^4 + Y^3 \xrightarrow{g_1} Y^4 + 2Y^3 - 2Y$
$\quad -1 \xrightarrow{g_1} 0, S(g_2, g_3) = 4Yg_2 - X^2 g_3 = -4X^2 + 4Y \xrightarrow{g_2} 4Y + 4 \xrightarrow{g_2} 0,$
$S(g_2, g_4) = Yg_2 - Xg_4 = -X^2 + 2X + Y \xrightarrow{g_2} 2X + Y + 1 \xrightarrow{g_1} 0,$
$S(g_2, g_5) = 2Y^2 g_2 - X^2 g_5 = 2X^2 + 2Y^2 \xrightarrow{g_2} 2Y^2 - 1 \xrightarrow{g_5} 0,$
$S(g_2, g_6) = Y^3 g_2 - X^2 g_6 = -X^2 Y^2 + X^2 Y + X^2 + Y^3 \xrightarrow{g_1} Y^3 + Y^2 - Y - 1 \xrightarrow{g_6} 0,$
$S(g_3, g_4) = Xg_3 - 4g_4 = 0, S(g_3, g_5) = Yg_3 - 2g_5 = 4Y + 4 \xrightarrow{g_3} 0,$
$S(g_3, g_6) = Y^2 g_3 - 4g_6 = 4Y + 4 \xrightarrow{g_3} 0,$
$S(g_4, g_5) = 2Yg_4 - Xg_5 = 2XY + 2X - 4Y \xrightarrow{g_1} -Y^2 - 6Y - 1 \longrightarrow Y^2 + 6Y + 1$
$\quad =: g_7,$
$S(g_7, g_7) = 0, g_5, g_6 \xrightarrow{g_7, g_3} 0, S(g_4, g_7) = Yg_4 - Xg_7 = -5XY - X - 2Y \xrightarrow{g_4, g_1, g_3} 0.$

Thus, $\mathcal{G}_1 = \{2X + Y + 1, X^2 + 1, 4Y + 4, XY + X - 2, Y^2 + 6Y + 1\}$ is a reduced Gröbner for I_1.

$S(h_1, h_1) = 0, S(h_2, h_2) = 4h_1 = 4 =: h_3, S(h_3, h_3) = 0, S(h_1, h_2) = 2Xh_1 - h_2$
$\quad = 4XY - 1 \xrightarrow{h_3} -1 \longrightarrow 1.$

Thus, $\mathscr{G}_2 = \{1\}$ is a reduced Gröbner for I_2.

As a conclusion, a reduced dynamical Gröbner basis for I in the ring $(\mathbb{Z}/8\mathbb{Z}) \times (\mathbb{Z}/8\mathbb{Z})[X,Y]$ is

$$\mathscr{G} = \{((\{(2,0)X + (1,0)Y + (1,0), (1,0)X^2 + (1,0), (4,0)Y + (4,0), (1,0)XY$$
$$+ (1,0)X - (2,0), (1,0)Y^2 + (6,0)Y + (1,0)\}, e^{\mathbb{N}}), (\{(0,1)\}, (1-e)^{\mathbb{N}})\}.$$

In the ring $(\mathbb{Z}[t]/\langle 8, t^2 - t\rangle)[X,Y]$, a reduced dynamical Gröbner basis for J is

$$G = \{((\{2X + Y + 1, X^2 + 1, 4Y + 4, XY + X - 2, Y^2 + 6Y + 1\}, (1-t)^{\mathbb{N}}), (\{1\}, t^{\mathbb{N}})\}.$$

Denoting by $\mathbf{W} = \mathbb{Z}[t]/\langle 8, t^2 - t\rangle$, this dynamical Gröbner basis corresponds to the binary tree:

$$\mathbf{W}$$
$$\swarrow \qquad \searrow$$
$$\mathbf{W}[\tfrac{1}{1-t}] \qquad \mathbf{W}[\tfrac{1}{t}]$$

Answering the ideal membership problem:

$$h = (1+t)X^2 + (4+3t)Y + 5 - 3t \in ?J.$$

If coded in the ring $(\mathbb{Z}/8\mathbb{Z}) \times (\mathbb{Z}/8\mathbb{Z})[X,Y]$, this problem becomes

$$u = (u_1, u_2) = (1,2)X^2 + (4,7)Y + (5,2) \in ?I, \text{ or also}$$

$$u_1 = X^2 + 4Y + 5 \in ?I_1 \quad \& \quad u_2 = 2X^2 + 7Y + 2 \in ?I_2.$$

We have $u_1 \in I_1$ as $\overline{u_1}^{\mathscr{G}_1} = 0$. More precisely,

$$u_1 = g_2 + g_3 = 4g_1 + g_2, \text{ or also}$$

$$(1,0)u = (4,0)f_1 + (1,0)f_2. \tag{1.5}$$

On the other hand, of course $u_2 \in I_2 = \langle 1 \rangle$, and we have

$$u_2 = -2X(2X^2 + 7Y + 2)h_1 + (2X^2 + 7Y + 2)h_2, \text{ or also}$$

$$(0,1)u = -2X(2X^2 + 7Y + 2)(0,1)f_1 + (2X^2 + 7Y + 2)(0,1)f_2. \tag{1.6}$$

$$(1.5) + (1.6) \quad \Rightarrow \quad u = (4, -2X(2X^2 + 7Y + 2))f_1 + (1, 2X^2 + 7Y + 2)f_2 \in I,$$
and thus, $h = (4 - (4X^3 + 14XY + 4X - 4)t)P_1 + (1 + (2X^2 + 7Y + 1)t)P_2 \in J$.

1.3.3 Dynamical Gröbner bases over Boolean rings

One main feature of the use of dynamical Gröbner bases is that it enables to easily resolve the delicate problem caused by zero-divisors appearing as leading coefficients. Cai and Kapur concluded their paper [12] by mentioning the open question of how to generalize Buchbergers's algorithm for Boolean rings. As a typical example of a problematical situation, they studied the case where the base ring is $\mathbf{A} = \mathbb{F}_2[a,b]/\langle a^2 - a, b^2 - b\rangle$. In that case, the method they proposed in [12] does not work due to the fact that an annihilator of $ab + a + b + 1 \in \mathbf{A}$ can be either a or b and thus there may exist noncomparable multiannihilators for an element in $\mathbf{A}$. Dynamical Gröbner bases allow to fairly overcome this difficulty. As a matter of fact, in this precise case, as will be explained below, we will be led to compute a dynamical Gröbner bases made up of four Gröbner bases on localizations of $\mathbf{A}$. We will see that at each leaf of the constructed binary tree, the problem Cai and Kapur pointed disappears completely. Thus, by systemizing the dynamical construction above, it is straightforward that dynamical Gröbner bases over Dedekind rings could be a satisfactory solution to this open problem.

It is folklore that the presence of a nontrivial idempotent $\mathbf{e}$ in a ring $\mathbf{B}$ (i.e., $\mathbf{e}^2 = \mathbf{e}$ and $\mathbf{e} \notin \{0, 1\}$) splits $\mathbf{B}$ into two subrings as follows:

$$\mathbf{B} = \mathbf{e}\mathbf{B} + (1 - \mathbf{e})\mathbf{B} \cong \mathbf{e}\mathbf{B} \times (1 - \mathbf{e})\mathbf{B} \cong \mathbf{B}/\langle 1 - \mathbf{e}\rangle \times \mathbf{B}/\langle \mathbf{e}\rangle \cong \mathbf{B}\left[\frac{1}{\mathbf{e}}\right] \times \mathbf{B}\left[\frac{1}{1-\mathbf{e}}\right].$$

Let us consider the ring $\mathbf{A} = \mathbb{F}_2[a,b]\langle a^2 - a, b^2 - b\rangle = \mathbb{F}_2 + \mathbb{F}_2 a + \mathbb{F}_2 b + \mathbb{F}_2 ab$ with the relations $a^2 = a$ and $b^2 = b$. When working with the ring $\mathbf{A}$, we know in advance that the binary tree we will construct when computing dynamically a Gröbner basis of an ideal of $\mathbf{A}[X_1, \ldots, X_n]$ is formed by only four leaves as follows:

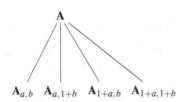

$$\mathbf{A}_{a,b} \quad \mathbf{A}_{a,1+b} \quad \mathbf{A}_{1+a,b} \quad \mathbf{A}_{1+a,1+b}$$

where

$$\mathbf{A}_{a,b} := \mathbf{A}\left[\tfrac{1}{a}, \tfrac{1}{b}\right] \cong \mathbf{A} \text{ with } \begin{cases} a = 1 \\ b = 1 \end{cases} \cong \mathbb{F}_2,$$

$$\mathbf{A}_{a,1+b} := \mathbf{A}\left[\tfrac{1}{a}, \tfrac{1}{1+b}\right] \cong \mathbf{A} \text{ with } \begin{cases} a = 1 \\ b = 0 \end{cases} \cong \mathbb{F}_2,$$

$$\mathbf{A}_{1+a,b} := \mathbf{A}\left[\tfrac{1}{1+a}, \tfrac{1}{b}\right] \cong \mathbf{A} \text{ with } \begin{cases} a = 0 \\ b = 1 \end{cases} \cong \mathbb{F}_2,$$

$$\mathbf{A}_{1+a,1+b} := \mathbf{A}\left[\tfrac{1}{1+a}, \tfrac{1}{1+b}\right] \cong \mathbf{A} \text{ with } \begin{cases} a = 0 \\ b = 0 \end{cases} \cong \mathbb{F}_2.$$

So, computing a dynamical Gröbner basis over $\mathbf{A}$ amounts to computing four (classical) Gröbner bases over $\mathbb{F}_2$ (possibly, in a parallel way). Moreover, we can define a "reduced dynamical Gröbner basis" as a set $\{(G_1, a^{\mathbb{N}} b^{\mathbb{N}}), (G_2, a^{\mathbb{N}}(1 + $

$b)^{\mathbb{N}}), (G_3, (1+a)^{\mathbb{N}} b^{\mathbb{N}}), (G_4, (1+a)^{\mathbb{N}} (1+b)^{\mathbb{N}})\}$ where each G_i is the reduced Gröbner basis over $\mathbb{F}_2$.

More generally, computing a dynamical Gröbner basis over a Boolean ring $\mathbb{F}_2[a_1, \dots, a_m]/\langle a_1^2 - a_1, \dots, a_m^2 - a_m \rangle$ amounts to computing 2^m (classical) Gröbner bases over $\mathbb{F}_2$ (possibly, in a parallel way).

Example 66. Let us consider the ideal $I = \langle f = 1 + (1+a+b+ab)X \rangle$ of $\mathbf{A}[X]$ where $\mathbf{A} = \mathbb{F}_2[a, b]$ with $a^2 = a$ and $b^2 = b$. The reduced dynamical Gröbner basis of I is

$$\{(\{1\}, a^{\mathbb{N}} b^{\mathbb{N}}), (\{1\}, a^{\mathbb{N}} (1+b)^{\mathbb{N}}), (\{1\}, (1+a)^{\mathbb{N}} b^{\mathbb{N}}), (\{1+X\}, (1+a)^{\mathbb{N}} (1+b)^{\mathbb{N}})\}.$$

Now, suppose that we want to answer to the ideal membership problem

$$g = 1 + (a+b+ab)X + X^2 \in? I.$$

Obviously, over the first three localizations, the answer is yes. On the other hand, taking $a = b = 0$, g becomes $1 + X^2 = (1+X)^2 \in \langle 1+X \rangle$, and thus the global answer to the ideal membership problem is positive. Moreover, as the relation $g = (1+X)f$ holds at the four localizations (i.e., by successively taking $a = 1$, $b = 1$; $a = 1$, $b = 0$; $a = 0$, $b = 1$; $a = 0$, $b = 0$), it holds globally, i.e., over $\mathbf{A}$.

1.4   Gröbner bases over coherent Bézout rings

The goal of this section is to extend what we did in the previous section for coherent valuation rings to coherent Bézout rings ($\mathbb{Z}$ and $\mathbb{Z}/N\mathbb{Z}$ as main examples).

Definition 67. A ring $\mathbf{R}$ is a *Bézout* ring if every finitely-generated ideal is principal (i.e. of the form $\langle a \rangle = \mathbf{R}a$ with $a \in \mathbf{R}$), and for all $a, b \in \mathbf{R}$, we can find $g, a_1, b_1, c, d \in \mathbf{R}$ such that $a = a_1 g$, $b = b_1 g$, and $ca_1 + db_1 = 1$. A valuation ring is the same thing as a Bézout local ring. A quotient or a localisation of a Bézout ring is again a Bézout ring. A typical example of a Bézout domain is $\mathbb{Z}$. A typical example of a Bézout ring with nonzero zero-divisors is $\mathbb{Z}/N\mathbb{Z}$, where $N \in \mathbb{N}$ is not prime.

A Bézout domain is a Prüfer domain but the converse does not hold (the ring $\mathbb{Z}[\sqrt{-5}]$ is Prüfer but not Bézout).

Remark 68. In some cases, e.g. Euclidean domains ($\mathbb{Z}$ for example) or polynomial rings over a field or $\mathbb{Z}/N\mathbb{Z}$, the ring is equipped with a *division algorithm* which, for arbitrary $a, b_1, \dots, b_n \in \mathbf{R}$, provides an expression $a = b_1 c_1 + \cdots + b_n c_n + r$ with *quotients* $c_1, \dots, c_n$ and a *remainder* r, where $r = 0$ iff $a \in \langle b_1, \dots, b_n \rangle$. In the case of Bézout rings, dividing a by $[b_1, \dots, b_n]$ amounts to dividing a by the gcd d of $[b_1, \dots, b_n]$, since $a = dc + r$ can be read as $a = (c\alpha_1)b_1 + \cdots + (c\alpha_n)b_n + r$, where $d = \alpha_1 b_1 + \cdots + \alpha_n b_n$.

Algorithm 69. (The division algorithm over a Bézout ring)

Consider a Bézout ring **R**, take n indeterminates $X_1, \ldots, X_n$, and consider a free **R**$[\underline{X}]$-module $\mathbf{H}_m$ with basis $(e_1, \ldots, e_m)$. We consider a monomial order $>$ on $\mathbf{H}_m$. Like the classical division algorithm for $\mathbf{F}[\underline{X}]^m$ with **F** a field, this algorithm has the following goal.

Input: $h \in \mathbf{H}_m$, $h_1, \ldots, h_s \in \mathbf{H}_m \setminus \{0\}$.

Output: $q_1, \ldots, q_s \in \mathbf{R}[\underline{X}]$ and $r \in \mathbf{H}_m$ such that $h = q_1 h_1 + \cdots + q_s h_s + r$, $\mathrm{LM}(h) \geq \mathrm{LM}(q_\ell h_\ell)$ whenever $q_\ell h_\ell \neq 0$, $f \notin \langle \mathrm{LT}(h_1), \ldots, \mathrm{LT}(h_s) \rangle$ for each term f of r. The vector r is called *a remainder of h on division by $H = [h_1, \ldots, h_s]$* and is denoted by $r = \overline{h}^H$.

```
 1  local variables  i:{1,…,s} ,  D: subset of {1,…,s} ,  a,c,d,b_i : R ,  p: H_m ;
 2  q_1 ← 0 ;  … ;  q_s ← 0 ;  r ← 0 ;  p ← h ;
 3  while  p ≠ 0  do
 4      D ← {i ; LM(h_i) | LM(p)} ;
 5      find  d,b_i (i ∈ D)  such that
 6          d = gcd(LC(h_i))_{i∈D} = Σ_{i∈D} b_i LC(h_i) ;
 7      find  a,c  such that
 8          LC(p) = ad + c  (with c = 0 iff d divides LC(p), see Remark 68) ;
 9      for  i  in  D  do
10          q_i ← q_i + ab_i(LM(p)/LM(h_i))
11      od ;
12      r ← r + cLM(p) ;
13      p ← p − Σ_{i∈D} ab_i(LM(p)/LM(h_i))h_i − cLM(p)
14  od
```

By convention, if D is empty, then $d = 0$. At each step of the algorithm, the equality $h = q_1 h_1 + \cdots + q_s h_s + p + r$ holds while $\mathrm{mdeg}(p)$ decreases.

Note that in the case of a valuation ring, the gcd d is an $\mathrm{LC}(h_{i_0})$ dividing all the $\mathrm{LC}(h_i)$, and the Bézout identity may be given by setting $b_{i_0} = 1$ and $b_i = 0$ for $i \neq i_0$.

Algorithm 70. (The S-polynomial algorithm over coherent Bézout rings)

Consider a coherent Bézout ring **R**, take n indeterminates $X_1, \ldots, X_n$, and consider a free **R**$[\underline{X}]$-module $\mathbf{H}_m$ with basis $(e_1, \ldots, e_m)$. We consider a monomial order $>$ on $\mathbf{H}_m$. The algorithm has the following goal.

Input: $f, g \in \mathbf{H}_m \setminus \{0\}$.

Output: the S-polynomial given by $b\underline{X}^\beta$ and $a\underline{X}^\alpha$ as $S(f,g) = b\underline{X}^\beta f - a\underline{X}^\alpha g$: if $f = g$, then $S(f,f) = bf$ with b a generator of $\mathrm{Ann}(\mathrm{LC}(f))$; otherwise if $\mathrm{LM}(f) = \underline{X}^\mu e_i$ and $\mathrm{LM}(g) = \underline{X}^\nu e_i$, then $S(f,g) = b\underline{X}^{(\nu-\mu)^+} f - a\underline{X}^{(\mu-\nu)^+} g$ with $b\mathrm{LC}(f) = a\mathrm{LC}(g)$, $\gcd(a,b) = 1$; otherwise, $S(f,g) = 0$. Here $\alpha^+ = (\max(\alpha_1, 0), \ldots, \max(\alpha_n, 0))$ is the *positive part* of $\alpha \in \mathbb{Z}^n$.

```
1  local variables a,b:R, μ,ν:ℕⁿ;
2  if f = g then
3    find b such that Ann(LC(f)) = ⟨b⟩ ;
4    S(f,f) ← bf
5  else
6    if LPos(f) ≠ LPos(g) then
7      S(f,g) ← 0
8    else
9      μ ← mdeg(f) ;  ν ← mdeg(g) ;
10     find a,b such that
11       a gcd(LC(f),LC(g)) = LC(f),
12       b gcd(LC(f),LC(g)) = LC(g) ;
13     S(f,g) ← bX^(ν−μ)⁺ f − aX^(μ−ν)⁺ g
14   fi
15 fi
```

Note the following important properties of $S(f,g)$:

- If $\mathrm{LM}(f) = \underline{X}^\mu e_i$ and $\mathrm{LM}(g) = \underline{X}^\nu e_i$, then either $S(f,g) = 0$ or $\mathrm{LM}(S(f,g)) < \underline{X}^{\sup(\mu,\nu)} e_i$; otherwise, $S(f,g) = 0$;

- $S(\underline{X}^\delta f, \underline{X}^\delta g) = \underline{X}^\delta S(f,g)$ for all $\delta \in \mathbb{N}^n$.

$S(f,f)$ is called the *auto-S-polynomial* of f. It is designed to produce cancellation of the leading term of f by multiplying f with an element of the annihilator of $\mathrm{LC}(f)$. Note that we need this annihilator to be principal. If the leading coefficient of f is regular, then automatically $S(f,f) = 0$ (as in the field case): in case **R** is a domain, we can replace lines 2–5 by "**if** $f \neq g$ **then**".

The S-polynomial $S(f,g)$ is designed to produce cancellation of the leading terms of f and g. It is worth pointing out that $S(f,g)$ is not uniquely determined (up to a unit) when **R** has nonzero zero-divisors. Also $S(g,f)$ is generally not equal (up to a unit) to $S(f,g)$. These issues are repaired[1] through the consideration of the auto-S-polynomials $S(f,f)$ and $S(g,g)$.

Note that in the case of a coherent valuation ring, the computation of the coefficients a,b is particularly easy: see Definition 32.

Algorithm 71. (Buchberger's algorithm over coherent Bézout rings)

Consider a coherent Bézout ring **R**, take n indeterminates $X_1, \ldots, X_n$, and consider a free $\mathbf{R}[\underline{X}]$-module $\mathbf{H}_m$ with basis $(e_1, \ldots, e_m)$. We consider a monomial order $>$ on $\mathbf{H}_m$. The algorithm has the following goal.

Input: $g_1, \ldots, g_s \in \mathbf{H}_m \setminus \{0\}$.

Output: a Gröbner basis $[g_1, \ldots, g_s, \ldots, g_t]$ for $\langle g_1, \ldots, g_s \rangle$.

---

[1]In the field case, the second ambiguity is taken care of by making the S-polynomial monic.

```
 1  local variables S : H_m , i, j, u : ℕ;
 2  t ← s;
 3  repeat
 4      u ← t;
 5      for i from 1 to u do
 6          for j from i to u do
 7              S ← S̄(g_i,g_j)^[g_1,...,g_u]  by Algorithms 70 and 69;
 8              if S ≠ 0 then
 9                  t ← t + 1;
10                  g_t ← S
11              fi
12          od
13      od
14  until t = u
```

This algorithm is almost the same algorithm as in the case where the base ring is a field. The modifications are in the definition of S-polynomials, in the consideration of the auto-S-polynomials, and in the division of terms. In line 7, the algorithm may be sped up by computing the remainder w.r.t. $[g_1, \ldots, g_t]$ instead of $[g_1, \ldots, g_u]$ only.

Remark 72. The correctness of Algorithm 71 is ensured by Theorem 77 below. Its termination is guaranteed in case of the lexicographic monomial order if **R** is an archimedean valuation ring or a Bézout domain of Krull dimension ≤ 1 (see [66, Theorem 272 and Section 3.3.11]). If the algorithm terminates, then we can transform the obtained Gröbner basis into a Gröbner basis $[h_1, \ldots, h_r]$ such that no term of an element h_ℓ lies in $\langle \mathrm{LT}(h_k) ; k \neq \ell \rangle$ by replacing each element of the Gröbner basis with a remainder of it on division by the other nonzero elements and by repeating this process until it stabilises. Such a Gröbner basis is called a *pseudo-reduced Gröbner basis* (it is not unique).

Example 73. Let us consider the ideal $I = \langle f_1 = 10X + 2Y, f_2 = 6X + 3 \rangle$ of $\mathbb{Z}[X,Y]$ and fix the lexicographic order as monomial order with $X > Y$. Then:

$$S(f_1, f_2) = 3f_1 - 5f_2 = 6Y - 15 =: f_3,$$
$$S(f_1, f_3) = 3Yf_1 - 5Xf_3 = 6Y^2 + 75X \xrightarrow{f_3, f_2} 3X + 3Y - 6 =: f_4,$$
$$S(f_1, f_4) = 3f_1 - 10f_4 = -24Y + 60 \xrightarrow{f_3} 0,$$
$$S(f_2, f_3) = Yf - Xf_3 = 3Y + 15X \xrightarrow{f_4, f_3} 0,$$
$$S(f_2, f_4) = f_2 - 2f_4 = -6Y + 15 \xrightarrow{f_3} 0,$$
$$S(f_3, f_4) = Xf_3 - 2Yf_4 = -15X - 6Y^2 + 12Y \xrightarrow{f_4, f_3} 0.$$

Thus, $G = \{f_1, f_2, f_3, f_4\}$ is a Gröbner basis of I. We transform G into a reduced Gröbner basis as follows:

$$f_1 \xrightarrow{f_2, f_4} X - Y + 3 =: f_1,$$
$$f_2 \xrightarrow{f_1, f_3} 0 =: f_2,$$
$$f_4 \xrightarrow{f_1, f_3} 0 =: f_4.$$

Thus, $\{X - Y + 3, 6Y - 15\}$ is the reduced Gröbner basis of I.

The following lemma provides a necessary and sufficient condition for a term to belong to a module generated by terms over a coherent Bézout ring. It generalizes Lemma 30.

Lemma 74. *Let* $\mathbf{R}$ *be a coherent Bézout ring,* $\mathbf{H}_m$ *a free* $\mathbf{R}[\underline{X}]$-*module with basis* $e_1, \ldots, e_m$, *and* $>$ *a monomial order on* $\mathbf{H}_m$. *Let* U *be a submodule of* $\mathbf{H}_m$ *generated by a finite collection of terms* $a_\alpha \underline{X}^\alpha e_{i_\alpha}$ *with* $\alpha \in A$. *A term* $b\underline{X}^\beta e_r$ *lies in* U *iff there is a nonempty subset* A' *of* A *such that* $\underline{X}^\alpha e_{i_\alpha}$ *divides* $\underline{X}^\beta e_r$ *for every* $\alpha \in A'$ *(i.e.* $i_\alpha = r$ *and* $\underline{X}^\alpha \mid \underline{X}^\beta$*) and* $\gcd_{\alpha \in A'}(a_\alpha)$ *divides* b.

Proof. The condition is clearly sufficient. For the necessity, write

$$b\underline{X}^\beta e_r = \sum_{\alpha \in \tilde{A}} c_\alpha a_\alpha \underline{X}^{\gamma_\alpha} \underline{X}^\alpha e_{i_\alpha}$$

with $\tilde{A} \subseteq A$, $c_\alpha \in \mathbf{R} \setminus \{0\}$, and $\underline{X}^{\gamma_\alpha} \in \mathbb{M}_n$. Then $b = \sum_{\alpha \in A'} c_\alpha a_\alpha$, where A' is the set of those α such that $\gamma_\alpha + \alpha = \beta$ and $i_\alpha = r$. For each $\alpha \in A'$, $\underline{X}^\alpha$ divides $\underline{X}^\beta$. Since the gcd of the a_α's with $\alpha \in A'$ divides every a_α, it also divides b.

The following lemma is a key result for the characterisation of Gröbner bases by means of S-polynomials. It generalizes Lemma 36.

Lemma 75. *Let* $\mathbf{R}$ *be a coherent Bézout ring,* $\mathbf{H}_m$ *a free* $\mathbf{R}[\underline{X}]$-*module with basis* $e_1, \ldots, e_m$, $>$ *a monomial order, and* $f_1, \ldots, f_s \in \mathbf{H}_m \setminus \{0\}$ *with the same leading monomial* M. *Let* $c_1, \ldots, c_s \in \mathbf{R}$. *If* $c_1 f_1 + \cdots + c_s f_s$ *vanishes or has leading monomial* $< M$, *then* $c_1 f_1 + \cdots + c_s f_s$ *is a linear combination with coefficients in* $\mathbf{R}$ *of the* S-*polynomials* $S(f_i, f_j)$ *with* $1 \leq i \leq j \leq s$.

Proof. Let us write, for $j \neq i$, $\mathrm{LC}(f_j) = d_{i,j} a_{i,j}$ with $d_{i,j} = \gcd(\mathrm{LC}(f_i), \mathrm{LC}(f_j))$, so that $\gcd(a_{i,j}, a_{j,i}) = 1$ and $S(f_i, f_j) = a_{i,j} f_i - a_{j,i} f_j$. For each permutation $i_1, \ldots, i_s$ of $1, \ldots, s$, we shall transform the sum $a_{i_1,i_2} \cdots a_{i_{s-1},i_s}(c_1 f_1 + \cdots + c_s f_s)$ by replacing successively

$$a_{i_1,i_2} f_{i_1} \quad \text{by } S(f_{i_1}, f_{i_2}) + a_{i_2,i_1} f_{i_2},$$

$$\vdots \qquad\qquad \vdots$$

$$a_{i_{s-1},i_s} f_{i_{s-1}} \quad \text{by } S(f_{i_{s-1}}, f_{i_s}) + a_{i_s,i_{s-1}} f_{i_s}.$$

At the end, the sum will be a linear combination of $S(f_{i_1}, f_{i_2})$, $S(f_{i_2}, f_{i_3})$, ..., $S(f_{i_{s-1}}, f_{i_s})$, and f_{i_s}; let z be the coefficient of f_{i_s} in this combination. The sum as well as each of the S-polynomials vanish or have leading monomial $< M$, so that the hypothesis yields $z\,\mathrm{LC}(f_{i_s}) = 0$; therefore zf_{i_s} is a multiple of $S(f_{i_s}, f_{i_s})$. It remains to obtain a Bézout identity w.r.t. the products $a_{i_1,i_2} \cdots a_{i_{s-1},i_s}$, because it yields an expression of $c_1 f_1 + \cdots + c_s f_s$ as a linear combination of the required form. For this, it is enough to develop the product of the $\binom{s}{2}$ Bézout identities w.r.t. $a_{i,j}$ and $a_{j,i}$, $1 \leq i < j \leq s$: this yields a sum of products of $\binom{s}{2}$ terms, each of which is either $a_{i,j}$ or $a_{j,i}$, $1 \leq i < j \leq s$, so that it is indexed by the tournaments on the vertices $1, \ldots, s$; every such product contains a product of the above form $a_{i_1,i_2} \cdots a_{i_{s-1},i_s}$ because every tournament contains a hamiltonian path (see [44]).

Remark 76. The above proof results from an analysis of that of Lemma 36.

As in the valuation case, Lemmas 74 and 75 enable us to generalise some classical results about the existence and characterisation of Gröbner bases over fields to the case of coherent Bézout rings.

Theorem 77. (Buchberger's criterion for Gröbner bases over coherent Bézout rings)

Let R be a coherent Bézout ring, $U = \langle g_1, \ldots, g_s \rangle$ a submodule of a free $R[\underline{X}]$-module H_m, and $>$ a monomial order on H_m. Then $G = [g_1, \ldots, g_s]$ is a Gröbner basis for U if and only if the remainder of $S(g_i, g_j)$ on division by G vanishes for all pairs $i \leq j$.

The following corollary provides a general explanation for the termination of Buchberger's algorithm 71 for cohérent Bézout rings. It also explains why one can avoid branching when computing a Gröbner basis over a Bézout ring (contrary to the general case of an arithmetical ring).

Theorem 78. *Let R be a coherent Bézout ring, $U = R[\underline{X}]f_1 + \cdots + R[\underline{X}]f_s$ a nonzero submodule of a free $R[\underline{X}]$-module H_m, and $>$ a monomial order on H_m. If $LT(U)$ is finitely generated, then Buchberger's Algorithm 71 computes a Gröbner basis for U.*

Proof. It suffices to prove the result when R is local and $m = 1$, in which case this is Theorem 39. Let us explain in a few words how to pass from the local to the global case. Suppose that we are computing $S(f, g)$ and that the leading coefficients a and b of f and g are incomparable under division. A key fact is that if we write $a = \gcd(a, b)a'$, $b = \gcd(a, b)b'$ with $\gcd(a', b') = 1$, then a divides b in $R[\frac{1}{a'}]$, b divides a in $R[\frac{1}{b'}]$, and the two multiplicative subsets $a'^{\mathbb{N}}$ and $b'^{\mathbb{N}}$ are comaximal because $1 \in \langle a', b' \rangle$. Then R splits into $R[\frac{1}{a'}]$ and $R[\frac{1}{b'}]$, and we can continue as if R were a valuation ring. If $\mathrm{mdeg}(f) = \mu$, $\mathrm{mdeg}(g) = \nu$, then $S(f, g)$ is being computed as follows:

- in the ring $R[\frac{1}{b'}]$, $S(f, g) = X^{(\nu - \mu)^+} f - \frac{a'}{b'} X^{(\mu - \nu)^+} g =: S_1$;

- in the ring $R[\frac{1}{a'}]$, $S(f, g) = \frac{b'}{a'} X^{(\nu - \mu)^+} f - X^{(\mu - \nu)^+} g =: S_2$.

But, letting $S := b' X^{(\nu - \mu)^+} f - a' X^{(\mu - \nu)^+} g$, we have

$$S = b'S_1 = a'S_2.$$

As S is equal to S_1 up to a unit in $R[\frac{1}{b'}]$, and to S_2 in $R[\frac{1}{a'}]$, it can replace both of them, and thus there was no need to open the two branches $R[\frac{1}{a'}]$ and $R[\frac{1}{b'}]$.

1.5   The syzygy theorem over fields, $\mathbb{Z}$, $\mathbb{Z}_{p\mathbb{Z}}$, and $\mathbb{Z}/N\mathbb{Z}$

This section essentially follows [25, 66]. The main goal is to obtain constructive versions of Hilbert's syzygy theorem below for Bézout domains of Krull dimension ≤ 1 and for coherent zero-dimensional Bézout rings with (e.g. for $\mathbb{Z}$, $\mathbb{Z}_{p\mathbb{Z}}$, and $\mathbb{Z}/N\mathbb{Z}$) following Schreyer's clever method [57] which uses Gröbner bases.

Definition and Proposition 79. (Finitely-presented modules)

(1) Let **R** be a ring. A *finitely-presented* **R**-module is an **R**-module given by a finite number of generators and relations. Thus, it is a finitely-generated **R**-module having a finitely-generated relations module. Equivalently, it is an **R**-module isomorphic to the cokernel of a linear application

$$\gamma : \mathbf{R}^m \to \mathbf{R}^q.$$

The matrix $G \in \mathbf{R}^{q \times m}$ of γ has as columns a generating set of the relations module between the generators g_i which are the images of the canonical basis by the epimorphism $\pi : \mathbf{R}^q \to M$. The matrix G is called a presentation matrix of the module M for the generating system $(g_1, \ldots, g_q)$. We have:

- $[g_1 \cdots g_q] \, G = 0$, and

- each relation between the g_i's is a linear combination of the columns of G, i.e., if $[g_1 \cdots g_q] C = 0$ with $C \in \mathbf{R}^{q \times 1}$ then there exists $C' \in \mathbf{R}^{m \times 1}$ such that $C = GC'$.

For example, a free module of rank k (i.e., isomorphic to $\mathbf{R}^k$) is finitely-presented. Its presentation matrix is a column matrix formed by k zeroes. More generally, if P is a finitely-generated projective module then, as it is isomorphic to the image of an idempotent matrix $F \in \mathbf{R}^{n \times n}$ for some $n \in N^*$ and $\mathbf{R}^n = \text{Im}(F) \oplus \text{Im}(I_n - F)$, we get $P \cong \text{Coker}(I_n - F)$ and, thus, P is finitely-presented.

(2) The definition above can be rephrased as follows: An **R**-module M is finitely-presented if there is an epimorphism $\pi : \mathbf{R}^q \to M$ for some $q \in \mathbb{N}^*$ (and thus, $\mathbf{R}^q / \text{Ker}(\pi) \cong M$) whose kernel $\text{Ker}(\pi)$ is finitely-generated. The module M is specified using finitely many generators (the images of the q generators of $\mathbf{R}^q$) and finitely many relations (the generators of $\text{Ker}(\pi)$).

For example, for $a \in \mathbf{R}$, the ideal $a\mathbf{R}$ is finitely-presented if and only if the annihilator $\text{Ann}(a) := \{b \in \mathbf{R} \mid ba = 0\}$ of a is finitely-generated. The epimorphism π corresponds to the multiplication by a and its kernel is $\text{Ann}(a)$. More generally, a finitely-generated ideal $\langle a_1, \ldots, a_n \rangle$ of **R** is finitely-presented if and only if the syzygy module

$$\text{Syz}(a_1, \ldots, a_n) := \{(b_1, \ldots, b_n) \in \mathbf{R}^n \mid b_1 a_1 + \cdots + b_n a_n = 0\}$$

is finitely-generated.

Definition and Proposition 80. (Finite free resolution)

Let **R** be a ring.

(1) A *complex* $\mathscr{F}$ of **R**-modules is a sequence of modules F_i and maps $\varphi_i : F_i \to F_{i-1}$ such that $\varphi_i \circ \varphi_{i+1} = 0$ for all i. The module $H_i := \mathrm{Ker}(\varphi_i)/\mathrm{Im}(\varphi_{i+1})$ is called the *homology* of this complex at F_i.

If $H_i = 0$ for all i, we say the complex $\mathscr{F}$ is *exact*. For example, if U is a submodule of a module M then the complex $0 \to U \xrightarrow{i} M \xrightarrow{\pi} M/U \longrightarrow 0$ is exact, where i is inclusion and π is the canonical projection.

For $a \in \mathbf{R}$, the homology of the complex $0 \longrightarrow \mathbf{R} \xrightarrow{\varphi_a} \mathbf{R}$ (called Koszul complex of length 1), where $\varphi_a(x) = ax$, is the annihilator $\mathrm{Ann}(a) := \{b \in \mathbf{R} \mid ba = 0\}$ of a.

A *finite free resolution* of length n of a module M is a complex

$$0 \longrightarrow \mathbf{R}^{r_n} \xrightarrow{\varphi_n} \cdots \xrightarrow{\varphi_2} \mathbf{R}^{r_1} \xrightarrow{\varphi_1} \mathbf{R}^{r_0} \longrightarrow 0$$

which is exact except at $\mathbf{R}^{r_0}$ and such that $M = \mathrm{Coker}(\varphi_1)$ and $r_i \in \mathbb{N}^*$.

(2) **Hilbert's syzygy theorem:** *If $\mathbf{K}$ is a field then every finitely-generated module over $\mathbf{K}[X_1, \ldots, X_k]$ has a finite free resolution of length $\leq k$.*

Definition 81. (Schreyer's monomial order)

Let **R** be a ring, $n, m \geq 1$, and consider n indeterminates $X_1, \ldots, X_n$. Let $\mathbf{H}_m$ be a free $\mathbf{R}[\underline{X}]$-module with basis $(e_1, \ldots, e_m)$ and a monomial order $>$. Consider $G = [g_1, \ldots, g_s]$ with $g_j \in \mathbf{H}_m \setminus \{0\}$.

Let $(\varepsilon_1, \ldots, \varepsilon_s)$ be the canonical basis of $\mathbf{R}[\underline{X}]^s$. *Schreyer's monomial order* induced by $>$ and $[g_1, \ldots, g_s]$ on $\mathbf{R}[\underline{X}]^s$ is the order denoted by $>_{g_1, \ldots, g_s}$, or again by $>$, defined as follows:

$$\underline{X}^{\alpha} \varepsilon_i > \underline{X}^{\beta} \varepsilon_j \quad \text{if} \quad \left| \begin{array}{l} \text{either } \mathrm{LM}(\underline{X}^{\alpha} g_i) > \mathrm{LM}(\underline{X}^{\beta} g_j) \\ \text{or both } \mathrm{LM}(\underline{X}^{\alpha} g_i) = \mathrm{LM}(\underline{X}^{\beta} g_j) \text{ and } i < j. \end{array} \right.$$

For the definition of monomial order on a finite-rank free $\mathbf{R}[\underline{X}]$-module, see Definition 3.

Algorithm 82. (The syzygy algorithm for terms over a coherent Bézout ring)

Consider a coherent Bézout ring **R**, take n indeterminates $X_1, \ldots, X_n$, and consider a free $\mathbf{R}[\underline{X}]$-module $\mathbf{H}_m$ with basis $(e_1, \ldots, e_m)$. We consider a monomial order $>$ on $\mathbf{H}_m$. The algorithm has the following goal.

Input: terms $T_1, \ldots, T_s \in \mathbf{H}_m$.

Output: a generating system $[S_{i,j}]_{1 \leq i \leq j \leq s, \mathrm{LPos}(T_j) = \mathrm{LPos}(T_i)}$ for $\mathrm{Syz}(T_1, \ldots, T_s)$. In this algorithm, $(\varepsilon_1, \ldots, \varepsilon_s)$ is the canonical basis of $\mathbf{R}[\underline{X}]^s$.

1 **local variables** $i, j : \{1, \ldots, s\}$, J : subset of $\{1, \ldots, s\}$, $a, b : \mathbf{R}$, $\alpha, \beta : \mathbb{N}^n$;
2 **for** i **from** 1 **to** s **do**
3      $J \leftarrow \{ j \geq i ; \mathrm{LPos}(T_j) = \mathrm{LPos}(T_i) \}$;
4   **for** j **in** J **do**
5   **compute** $b\underline{X}^\beta, a\underline{X}^\alpha$ **such that** $S(T_i, T_j) = b\underline{X}^\beta T_i - a\underline{X}^\alpha T_j$ **by** Algorithm 70;
6          $S_{i,j} \leftarrow b\underline{X}^\beta \varepsilon_i - a\underline{X}^\alpha \varepsilon_j$
7   **od**
8 **od**

Proposition 83. *Algorithm 86 is correct.*

Proof. It suffices to prove correctness in the local case (i.e., in case where the base ring is a coherent valuation ring) and in case $m = 1$ (i.e., the ideal case). For this, let $\mathbf{V}$ be a valuation ring, $c_1, \ldots, c_s \in \mathbf{V} \setminus \{0\}$, and $M_1, \ldots, M_s$ be monomials in $\mathbf{V}[\underline{X}]$. Denoting $\mathrm{LCM}(M_i, M_j)$ by $M_{i,j}$, we have to prove that the syzygy module $\mathrm{Syz}(c_1 M_1, \ldots, c_s M_s)$ is generated by $\{S_{i,j} \in \mathbf{V}[\underline{X}]^s \mid 1 \leq i \leq j \leq s\}$, where for $i \neq j$,

$$
S_{i,j} = \begin{cases} \dfrac{M_{i,j}}{M_i} \varepsilon_i - \dfrac{c_i}{c_j} \dfrac{M_{i,j}}{M_j} \varepsilon_j & \text{if } c_j \mid c_i \\[2ex] \dfrac{c_j}{c_i} \dfrac{M_{i,j}}{M_i} \varepsilon_i - \dfrac{M_{i,j}}{M_j} \varepsilon_j & \text{else,} \end{cases}
$$

and $S_{i,i} = d_i \varepsilon_i$, with d_i a generator of the annihilator of c_i in $\mathbf{V}$ ($S_{i,i}$ is defined up to a unit).

It is clear that for all $i \leq j$, $S_{i,j}$ is a syzygy of $M = (c_1 M_1, \ldots, c_s M_s)$. Now, in order to verify that $\{S_{i,j}, 1 \leq i \leq j \leq s\}$ is really a syzygy basis, we need to show that every syzygy H of M can be written as $H = \sum_{1 \leq i \leq j \leq s} u_{i,j} S_{i,j}$ where $u_{i,j} \in \mathbf{V}[\underline{X}]$.

For this, let $H = {}^t(h_1, \ldots, h_s)$ be a syzygy of M, that is, such that $MH = 0$. Letting $\gamma(H) = \max_{1 \leq i \leq s} \mathrm{mdeg}(h_i M_i)$, we have

$$
\sum_{\mathrm{mdeg}(h_i M_i) = \gamma(H)} c_i h_i M_i + \sum_{\mathrm{mdeg}(h_i M_i) < \gamma(H)} c_i h_i M_i = 0.
$$

Thus,

$$
\sum_{\mathrm{mdeg}(h_i M_i) = \gamma(H)} c_i \mathrm{LT}(h_i) M_i + \sum_{\mathrm{mdeg}(h_i M_i) = \gamma(H)} c_i (h_i - \mathrm{LT}(h_i)) M_i
$$

$$
+ \sum_{\mathrm{mdeg}(h_i M_i) < \gamma(H)} c_i h_i M_i = 0.
$$

We can write $H = G + \widetilde{G}$, where $G = (g_1, \ldots, g_s)$ with $g_i = \mathrm{LT}(h_i)$ if $\mathrm{mdeg}(h_i M_i) = \gamma(H)$, 0 else; $\widetilde{G} = (\widetilde{g}_1, \ldots, \widetilde{g}_s)$ with $\widetilde{g}_i = h_i - \mathrm{LT}(h_i)$ if $\mathrm{mdeg}(h_i M_i) = \gamma(H)$, h_i else.

Since $\gamma(\widetilde{G}) < \gamma(H)$, it suffices, by induction on $\gamma(H)$ (the induction is legitimated by Corollary 13), to prove the result for G. In particular, we can assume that $h_i = a_i M_i'$ with $a_i \in \mathbf{V}$ (a_i can be zero). Let $i_1 < i_2 < \ldots < i_t$ be the indices corresponding to the nonzero a_i's, and denote $\gamma(H)$ by γ. The facts that $a_1 M_1' c_1 M_1 + \cdots + a_s M_s' c_s M_s = 0$ and $a_i M_i' c_i M_i = a_i c_i X^\gamma$ imply that

$$a_{i_1} c_{i_1} + \cdots + a_{i_t} c_{i_t} = 0. \quad (*)$$

It follows that

$$(h_1, \ldots, h_s) = (a_1 M_1', \ldots, a_s M_s') = a_{i_1} M_{i_1}' \varepsilon_{i_1} + \ldots + a_{i_t} M_{i_t}' \varepsilon_{i_t}$$
$$= a_{i_1} \frac{X^\gamma}{M_{i_1}} \varepsilon_{i_1} + \cdots + a_{i_t} \frac{X^\gamma}{M_{i_t}} \varepsilon_{i_t}.$$

As $\mathbf{V}$ is a valuation ring, there exists an integer $q \in \{1, \ldots, t\}$ such that c_{i_q} divides all the c_{i_j}'s. So, the previous expression can be written as

$$a_{i_1} \frac{X^\gamma}{M_{i_1}} \varepsilon_{i_1} + \cdots + a_{i_t} \frac{X^\gamma}{M_{i_t}} \varepsilon_{i_t} = \sum_{1 \le j \le q-1} a_{ij} \frac{X^\gamma}{M_{ij,iq}} \left(\frac{M_{ij,iq}}{M_{ij}} \varepsilon_{ij} - \frac{c_{ij}}{c_{iq}} \frac{M_{ij,iq}}{M_{iq}} \varepsilon_{iq} \right)$$

$$- \sum_{q+1 \le j \le t} a_{ij} \frac{X^\gamma}{M_{ij,iq}} \left(\frac{c_{ij}}{c_{iq}} \frac{M_{ij,iq}}{M_{iq}} \varepsilon_{iq} - \frac{M_{ij,iq}}{M_{ij}} \varepsilon_{ij} \right) + \left(\sum_{j \ne q} a_{ij} \frac{c_{ij}}{c_{iq}} + a_{iq} \right) \frac{X^\gamma}{M_{iq}} \varepsilon_{iq}. \quad (**)$$

Note that we have $\left(\sum_{j \ne q} a_{ij} \frac{c_{ij}}{c_{iq}} + a_{iq} \right) c_{iq} = 0$ by virtue of $(*)$, and thus, $\left(\sum_{j \ne q} a_{ij} \frac{c_{ij}}{c_{iq}} + a_{iq} \right) \varepsilon_{iq} \in \langle d_{iq} \rangle \varepsilon_{iq}$. Note also that, for $1 \le j \le q-1$, $\frac{M_{ij,iq}}{M_{ij}} \varepsilon_{ij} - \frac{c_{ij}}{c_{iq}} \frac{M_{ij,iq}}{M_{iq}} \varepsilon_{iq} = S_{ij,iq}$, and for $q+1 \le j \le t$,

$$\frac{c_{ij}}{c_{iq}} \frac{M_{ij,iq}}{M_{iq}} \varepsilon_{iq} - \frac{M_{ij,iq}}{M_{ij}} \varepsilon_{ij} = \begin{cases} \frac{c_{ij}}{c_{iq}} S_{iq,ij} & \text{if } c_{ij} \mid c_{iq} \\ S_{iq,ij} & \text{else.} \end{cases}$$

Thus, $\mathrm{Syz}(c_1 M_1, \ldots, c_s M_s) \subseteq \langle S_{ij}, 1 \le i \le j \le s \rangle$.

So, in case of a coherent valuation ring, Proposition 83 can be rephrased as follows:

Proposition 84. (Generating set for the syzygy module of a list of terms for a coherent valuation ring)

Let $\mathbf{V}$ be a coherent valuation ring, H_m a free $\mathbf{V}[\underline{X}]$-module with basis $(e_1, \ldots, e_m)$, and terms $T_1, \ldots, T_s$ in H_m. Considering the canonical basis $(\varepsilon_1, \ldots, \varepsilon_s)$ of $\mathbf{V}[\underline{X}]^s$, the syzygy module $\mathrm{Syz}(T_1, \ldots, T_s)$ is generated by the

$$S_{i,j} \in \mathbf{V}[\underline{X}]^s \text{ with } 1 \le i \le j \le s \text{ and } \mathrm{LPos}(T_i) = \mathrm{LPos}(T_j),$$

as computed by the syzygy algorithm for terms 86.

Note that in the syzygy algorithm for terms 86, the a, b will be found as in the S-polynomial algorithm 32, so that we get

$$
S_{i,j} = \begin{cases} b\varepsilon_i & \text{if } i = j, \text{ where } \langle b \rangle = \text{Ann}(\text{LC}(T_i)), \\ \underline{X}^\beta \varepsilon_i - a\underline{X}^\alpha \varepsilon_j & \text{if } i < j \text{ and } \text{LC}(T_j) \mid \text{LC}(T_i), \text{ where } \text{LC}(T_i) = a\text{LC}(T_j), \\ b\underline{X}^\beta \varepsilon_i - \underline{X}^\alpha \varepsilon_j & \text{otherwise, where } b\text{LC}(T_i) = \text{LC}(T_j). \end{cases}
$$

$$(1.7)$$

Here $\beta = (\text{mdeg}(T_j) - \text{mdeg}(T_i))^+$ and $\alpha = (\text{mdeg}(T_i) - \text{mdeg}(T_j))^+$.

Example 85. Let $\mathbf{V} = \mathbb{Z}/8\mathbb{Z}$, $f_1 = 4X^2$, $f_2 = 2XY^3$, $f_3 = 6Y$, $f_4 = 5$ in $\mathbf{V}[X, Y]$. With the previous notation, we have $S_{1,1} = (2, 0, 0, 0)$, $S_{2,2} = (0, 4, 0, 0)$, $S_{3,3} = (0, 0, 4, 0)$, $S_{4,4} = (0, 0, 0, 0)$. In addition, since $c_4 \mid c_3 \mid c_2 \mid c_1$, the syzygy module $\text{Syz}(f_1, \ldots, f_4)$ is generated by $\{ S_{i,j} = \frac{M_{i,j}}{M_i} \varepsilon_i - \frac{c_i}{c_j} \frac{M_{i,j}}{M_j} \varepsilon_j \mid 1 \leq i < j \leq 4 \} \cup \{ S_{1,1}, S_{2,2}, S_{3,3} \}$, that is,

$$
\text{Syz}(f_1, \ldots, f_4) = \langle {}^{\mathsf{t}}(Y^3, 6X, 0, 0), {}^{\mathsf{t}}(Y, 0, 2X^2, 0), {}^{\mathsf{t}}(1, 0, 0, 4X^2), {}^{\mathsf{t}}(0, 1, 5XY^2, 0),
$$
$$
{}^{\mathsf{t}}(0, 1, 0, 6XY^3), {}^{\mathsf{t}}(0, 0, 1, 2Y), {}^{\mathsf{t}}(2, 0, 0, 0), {}^{\mathsf{t}}(0, 4, 0, 0), {}^{\mathsf{t}}(0, 0, 4, 0) \rangle.
$$

Algorithm 86. (Schreyer's syzygy algorithm over a coherent Bézout ring)

Consider a coherent Bézout ring $\mathbf{R}$, take n indeterminates $X_1, \ldots, X_n$, and consider a free $\mathbf{R}[\underline{X}]$-module $\mathbf{H}_m$ with basis $(e_1, \ldots, e_m)$. We consider a monomial order $>$ on $\mathbf{H}_m$. The algorithm has the following goal.

Input: $[g_1, \ldots, g_s]$ for a submodule of $\mathbf{H}_m$.

Output: a Gröbner basis $[u_{i,j}]_{1 \leq i \leq j \leq s, \text{LPos}(g_j) = \text{LPos}(g_i)}$ for $\text{Syz}(g_1, \ldots, g_s)$ w.r.t. Schreyer's monomial order induced by $>$ and $[g_1, \ldots, g_s]$. In this algorithm, $(\varepsilon_1, \ldots, \varepsilon_s)$ is the canonical basis of $\mathbf{R}[\underline{X}]^s$.

```
1  local variables i, j : {1,...,s} ,  J : subset of {1,...,s} ,  a, b : R ,  α, β : ℕⁿ ,  qℓ : R[X] ;
2  for i from 1 to s do
3      J ← { j ≥ i ; LPos(gⱼ) = LPos(gᵢ) } ;
4      for j in J do
5          compute bX^β, aX^α such that  S(gᵢ, gⱼ) = bX^β gᵢ − aX^α gⱼ  by Algorithm 70;
6          compute q₁,...,qₛ such that
7              S(gᵢ, gⱼ) = q₁g₁ + ··· + qₛgₛ  by Algorithm 69 (note that
8              LM(S(gᵢ, gⱼ)) ≥ LM(qℓgℓ) whenever qℓgℓ ≠ 0) ;
9          uᵢ,ⱼ ← bX^β εᵢ − aX^α εⱼ − q₁ε₁ − ··· − qₛεₛ
10     od
11 od
```

The polynomials $q_1, \ldots, q_s$ of lines 6–8 may have been computed while constructing the Gröbner basis.

Remark 87. For an arbitrary system of generators $[h_1, \ldots, h_r]$ for a submodule U of $\mathbf{H}_m$, the syzygy module of $[h_1, \ldots, h_r]$ is easily obtained from the syzygy module of a Gröbner basis for U (see [66, Theorem 296]).

1.5.1 The syzygy theorem and Schreyer's algorithm for a coherent valuation ring

In the book *Gröbner bases in commutative algebra*, Ene and Herzog propose the following exercise.

[20, Problem 4.11, p. 81]: Let $>$ be a monomial order on the free S-module $F = \bigoplus_{j=1}^m S e_j$ (where $S = \mathbf{K}[\underline{X}]$ with $\mathbf{K}$ a field), let $U \subseteq F$ be a submodule of F, and suppose that $\mathrm{LT}(U) = \bigoplus_{j=1}^m I_j e_j$. Show that U is a free S-module if and only if I_j is a principal ideal for $j = 1, \ldots, m$.

It is obvious that this condition is sufficient. Unfortunately, it is not necessary as shows the following example, so that the statement of [20, Problem 4.11] is not correct.

Example 88. Let $>$ be a TOP monomial order on $\mathbf{K}[X,Y]^2$ for which $X > Y$, $\mathbf{K}$ being a field, let $e_1 = (1,0)$ and $e_2 = (0,1)$, and consider the free submodule U of $\mathbf{K}[X,Y]^2$ generated by $u_1 = (X,Y)$ and $u_2 = (Y,0)$. Then $\mathrm{LT}(u_1) = X e_1$, $\mathrm{LT}(u_2) = Y e_1$, $\mathrm{S}(u_1, u_2) = Y u_1 - X u_2 = Y^2 e_2 =: u_3$, and $\mathrm{S}(u_1, u_3) = \mathrm{S}(u_2, u_3) = 0$. It follows that $\{u_1, u_2, u_3\}$ is a Gröbner basis for U, and $\mathrm{LT}(U) = \langle X, Y \rangle e_1 \oplus \langle Y^2 \rangle e_2$. One can see that $\langle X, Y \rangle$ is not principal, $\mathrm{LT}(U)$ is not free while U is free.

So we content ourselves with the following observation.

Remark 89. Let $>$ be a monomial order on the free S-module $F = \bigoplus_{j=1}^m S e_j$, where $S = \mathbf{V}[\underline{X}]$ and $\mathbf{V}$ is a valuation domain. Let U be a submodule of F and suppose that $\mathrm{LT}(U) = \bigoplus_{j=1}^m I_j e_j$, where I_j is a principal ideal for $j = 1, \ldots, m$. Then $\mathrm{LT}(U)$ and U are free S-modules. Of course, this is not true anymore if $\mathbf{V}$ is a valuation ring with nonzero zero-divisors. For example, consider the ideal $U = \langle 8X + 2 \rangle$ in $(\mathbb{Z}/16\mathbb{Z})[X]$: we have $\mathrm{LT}(U) = \langle 2 \rangle$ (so that it is principal), but U is not free since $8U = (0)$.

Now we shall follow closely Schreyer's ingenious proof [57] of Hilbert's syzygy theorem via Gröbner bases, but with a valuation ring instead of a field. Schreyer's proof is very well explained in [20, §§ 4.4.1–4.4.3].

Theorem 90. (Schreyer's algorithm for a coherent valuation ring)

Let $\mathbf{V}$ *be a coherent valuation ring,* $\mathbf{H}_m$ *a free* $\mathbf{V}[\underline{X}]$-*module with basis* $(e_1, \ldots, e_m)$, *and* $>$ *a monomial order on* $\mathbf{H}_m$. *Let* U *be a submodule of* $\mathbf{H}_m$ *with Gröbner basis* $[g_1, \ldots, g_s]$. *Then the relations* $u_{i,j}$ *computed by Schreyer's syzygy algorithm 86 form a Gröbner basis for the syzygy module* $\mathrm{Syz}(g_1, \ldots, g_s)$ *w.r.t. Schreyer's monomial order induced by* $>$ *and* $[g_1, \ldots, g_s]$. *Moreover, for* $1 \leq i \leq j \leq s$ *such that* $\mathrm{LPos}(g_i) = \mathrm{LPos}(g_j)$,

$$
\mathrm{LT}(u_{i,j}) = \begin{cases} b\varepsilon_i & \text{if } i = j, \text{ where } \langle b \rangle = \mathrm{Ann}(\mathrm{LC}(g_i)), \\ \underline{X}^{(\mathrm{mdeg}(g_j)-\mathrm{mdeg}(g_i))^+}\varepsilon_i & \text{if } i < j \text{ and } \mathrm{LC}(g_j) \mid \mathrm{LC}(g_i), \\ b\underline{X}^{(\mathrm{mdeg}(g_j)-\mathrm{mdeg}(g_i))^+}\varepsilon_i & \text{otherwise, where } b\mathrm{LC}(g_i) = \mathrm{LC}(g_j). \end{cases}
$$
$$(1.8)$$

Proof (a slight modification of the proof of [20, Theorem 4.16]). Let us use the notation of Schreyer's syzygy algorithm 86.

Let $1 \leq i = j \leq s$. As $\mathrm{LM}(q_\ell g_\ell) \leq \mathrm{LM}(\mathrm{S}(g_i, g_i)) < \mathrm{LM}(g_i)$ whenever $q_\ell g_\ell \neq 0$, we infer that $\mathrm{LT}(u_{i,i}) = b\varepsilon_i$ with $\langle b \rangle = \mathrm{Ann}(\mathrm{LC}(g_i))$.

Let $1 \leq i < j \leq s$ such that $\mathrm{LPos}(g_i) = \mathrm{LPos}(g_j)$. Suppose that $\mathrm{LC}(g_i) = a\mathrm{LC}(g_j)$ for some a: as $\mathrm{LM}(\underline{X}^\beta g_i) = \mathrm{LM}(a\underline{X}^\alpha g_j)$ and $i < j$, we have $\mathrm{LT}(\underline{X}^\beta \varepsilon_i - a\underline{X}^\alpha \varepsilon_j) = \underline{X}^\beta \varepsilon_i$, and because $\mathrm{LM}(q_\ell g_\ell) \leq \mathrm{LM}(\mathrm{S}(g_i, g_j)) < \mathrm{LM}(\underline{X}^\beta g_i)$ whenever $q_\ell g_\ell \neq 0$, we infer that $\mathrm{LT}(u_{i,j}) = \underline{X}^\beta \varepsilon_i$; otherwise, with b such that $b\mathrm{LC}(g_i) = \mathrm{LC}(g_j)$, we obtain similarly $\mathrm{LT}(u_{i,j}) = b\underline{X}^\beta \varepsilon_i$.

Thus, considering Equation (1.7) with $T_\ell = \mathrm{LT}(g_\ell)$, $\mathrm{LT}(u_{i,j}) = \mathrm{LT}(\mathrm{S}_{i,j})$ holds for all $1 \leq i \leq j \leq s$.

Let us show now that the relations $u_{i,j}$ form a Gröbner basis for the syzygy module $\mathrm{Syz}(g_1, \ldots, g_s)$. For this, let $v = \sum_{\ell=1}^s v_\ell \varepsilon_\ell \in \mathrm{Syz}(g_1, \ldots, g_s)$ and let us show that there exist $1 \leq i \leq j \leq s$ with $\mathrm{LPos}(g_i) = \mathrm{LPos}(g_j)$ such that $\mathrm{LT}(u_{i,j})$ divides $\mathrm{LT}(v)$.

Let us write $\mathrm{LM}(v_\ell \varepsilon_\ell) = N_\ell \varepsilon_\ell$ and $\mathrm{LC}(v_\ell \varepsilon_\ell) = c_\ell$ for $1 \leq \ell \leq s$. Then $\mathrm{LM}(v) = N_i \varepsilon_i$ for some $1 \leq i \leq s$. Now let $v' = \sum_{\ell \in \mathscr{S}} c_\ell N_\ell \varepsilon_\ell$, where $\mathscr{S}$ is the set of those ℓ for which $N_\ell \mathrm{LM}(g_\ell) = N_i \mathrm{LM}(g_i)$. By definition of Schreyer's monomial order, we have $\ell \geq i$ for all $\ell \in \mathscr{S}$. Substituting each ε_ℓ in v' by T_ℓ, the sum becomes zero.

Therefore v' is a relation of the terms T_ℓ with $\ell \in \mathscr{S}$. By virtue of Proposition 84, v' is a linear combination of elements of the form $\mathrm{S}_{\ell,k}$ with $\ell \leq k$ in $\mathscr{S}$. Since $\ell > i$ for all $\ell \in \mathscr{S}$ with $\ell \neq i$, we infer, by virtue of Lemma 30, that $\mathrm{LT}(v')$ is a multiple of $\mathrm{LT}(\mathrm{S}_{i,j})$ for some $j \in \mathscr{S}$. The desired result follows since $\mathrm{LT}(v) = \mathrm{LT}(v')$. $\square$

As a consequence of Theorem 90, we obtain the following constructive versions of Hilbert's syzygy theorem for a valuation domain.

Theorem 91. (Syzygy theorem for a valuation domain)

Let $M = \boldsymbol{H}_m/U$ be a finitely-presented $\mathbf{V}[X_1, \ldots, X_n]$-module, where $\mathbf{V}$ is a valuation domain. Assume that $\mathrm{LT}(U)$ is finitely-generated according to a lexicographic monomial order. Then M admits a free $\mathbf{V}[X_1, \ldots, X_n]$-resolution

$$
0 \to F_p \to F_{p-1} \to \cdots \to F_1 \to F_0 \to M \to 0
$$

of length $p \leq n+1$.

Proof. It suffices to prove that U has a free $\mathbf{V}[\underline{X}]$-resolution of length $p \leq n$. Let us put the lexicographic monomial order with $X_n > X_{n-1} > \cdots > X_1$ on $\mathbf{V}[\underline{X}]$.

Let $(g_1, \ldots, g_s)$ be a Gröbner basis for U w.r.t. the corresponding TOP order. We can w.l.o.g. suppose that whenever $\mathrm{LM}(g_i)$ and $\mathrm{LM}(g_j)$ involve the same basis element for some $i < j$, say $\mathrm{LM}(g_i) = N_i \varepsilon_k$ and $\mathrm{LM}(g_j) = N_j \varepsilon_k$, then $N_i > N_j$. More precisely, whenever $N_i = N_j$, one of $\mathrm{LC}(g_i)$ and $\mathrm{LC}(g_j)$ divides the other, say $\mathrm{LC}(g_j) = b\,\mathrm{LC}(g_i)$, and the corresponding g_j may be reduced into $g_j - bg_i$. In a nutshell, all the possible reductions between the $\mathrm{LT}(g_k)$'s are exhausted. Now, since we have used the lexicographic order with $X_n > X_{n-1} > \cdots > X_1$, it turns out that the indeterminate X_n cannot appear in the leading terms of the $u_{i,j}$'s in (1.8). Thus, after at most n computations of the iterated syzygies, we reach a situation where none of the indeterminates $X_n, \ldots, X_1$ appears in the leading terms of the computed Gröbner basis of the iterated syzygy module. This implies that the iterated syzygy module is free (as noted in Remark 89). $\qquad \blacksquare$

Remark 92. In the proof of the above theorem, we need to work with the TOP lexicographic monomial order. We do not know what happens for other monomial orders. This applies also for Theorems 96 and 100 below.

Corollary 93. (Syzygy theorem for one-dimensional valuation domains) *Let $M = \mathbf{H}_m / U$ be a finitely-presented $\mathbf{V}[X_1, \ldots, X_n]$-module, where $\mathbf{V}$ is a one-dimensional valuation domain. Then M admits a finite free $\mathbf{V}[X_1, \ldots, X_n]$-resolution*

$$0 \to F_p \to F_{p-1} \to \cdots \to F_1 \to F_0 \to M \to 0$$

of length $p \le n+1$ as described in the previous theorem.

Example 94. Let $g_1 = X^4 - X$, $g_2 = 2X$, $g_3 = Y^3 - 1 \in \mathbb{Z}_{2\mathbb{Z}}[X,Y]$, and let us consider the lexicographic order $>_1$ for which $X >_1 Y$ as monomial order. We have

$$\mathrm{S}(g_1, g_2) = 2g_1 - X^3 g_2 = -2X = -g_2,$$
$$\mathrm{S}(g_1, g_3) = Y^3 g_1 - X^4 g_3 = -XY^3 + X^4 = g_1 - Xg_3,$$
$$\mathrm{S}(g_2, g_3) = Y^3 g_2 - 2Xg_3 = 2X = g_2.$$

Thus, (g_1, g_2, g_3) is a (pseudo-reduced) Gröbner basis for $I = \langle g_1, g_2, g_3 \rangle$ and $\mathrm{LT}(I) = \langle X^4, 2X, Y^3 \rangle$. By Theorem 90, $u_{1,3} = [Y^3 - 1, 0, X - X^4]$, $u_{1,2} = [2, 1 - X^3, 0]$, $u_{2,3} = [0, Y^3 - 1, -2X]$ form a (pseudo-reduced) Gröbner basis for the syzygy module $\mathrm{Syz}(g_1, g_2, g_3)$ w.r.t. Schreyer's monomial order $>_2$ induced by $>_1$ and $[g_1, g_2, g_3]$. In particular,

$$\mathrm{LT}(\mathrm{Syz}(g_1, g_2, g_3)) = \langle \mathrm{LT}(u_{1,3}), \mathrm{LT}(u_{1,2}), \mathrm{LT}(u_{2,3}) \rangle$$
$$= \langle Y^3 e_1, 2e_1, Y^3 e_2 \rangle = \langle 2, Y^3 \rangle e_1 \oplus \langle Y^3 \rangle e_2,$$

where (e_1, e_2, e_3) stands for the canonical basis of $\mathbb{Z}_{2\mathbb{Z}}[X,Y]^3$. We have

$$S(u_{1,3}, u_{1,2}) = 2u_{1,3} - Y^3 u_{1,2} = (-2, -Y^3 + X^3 Y^3, 2X - 2X^4)$$
$$= -u_{1,2} + (X^3 - 1)u_{2,3},$$
$$S(u_{1,3}, u_{2,3}) = S(u_{1,2}, u_{2,3}) = 0.$$

We recover that $[u_{1,3}, u_{1,2}, u_{2,3}]$ is a Gröbner basis for $\mathrm{Syz}(g_1, g_2, g_3)$. By Theorem 90, the elements $u_{1,3;1,2} = [2, 1 - Y^3, 1 - X^3]$ forms a (pseudo-reduced) Gröbner basis for the syzygy module $\mathrm{Syz}(u_{1,3}, u_{1,2}, u_{2,3})$ w.r.t. Schreyer's monomial order $>_3$ induced by $>_2$ and $[u_{1,3}, u_{1,2}, u_{2,3}]$. In particular, $\mathrm{LT}(\mathrm{Syz}(u_{1,3}, u_{1,2}, u_{2,3})) = \langle \mathrm{LT}(u_{1,3;1,2}) \rangle = \langle 2 \rangle \varepsilon_1$, where $(\varepsilon_1, \varepsilon_2, \varepsilon_3)$ stands for the canonical basis of $\mathbb{Z}_{2\mathbb{Z}}[X,Y]^3$. By Remark 89, $\mathrm{Syz}(u_{1,3}, u_{1,2}, u_{2,3})$ is free. We conclude that I admits the following length 2 free $\mathbb{Z}_{2\mathbb{Z}}[X,Y]$-resolution:

$$0 \longrightarrow \mathbb{Z}_{2\mathbb{Z}}[X,Y] \xrightarrow{u_{1,3;1,2}} \mathbb{Z}_{2\mathbb{Z}}[X,Y]^3 \xrightarrow{\begin{pmatrix} u_{1,3} \\ u_{1,2} \\ u_{2,3} \end{pmatrix}} \mathbb{Z}_{2\mathbb{Z}}[X,Y]^3 \xrightarrow{\begin{pmatrix} g_1 \\ g_2 \\ g_3 \end{pmatrix}} I \to 0.$$

It follows that $\mathbb{Z}_{2\mathbb{Z}}[X,Y]/I$ admits the following length 3 free $\mathbb{Z}_{2\mathbb{Z}}[X,Y]$-resolution:

$$0 \to \mathbb{Z}_{2\mathbb{Z}}[X,Y] \to \mathbb{Z}_{2\mathbb{Z}}[X,Y]^3 \to \mathbb{Z}_{2\mathbb{Z}}[X,Y]^3 \to \mathbb{Z}_{2\mathbb{Z}}[X,Y] \xrightarrow{\pi} \mathbb{Z}_{2\mathbb{Z}}[X,Y]/I \to 0.$$

Obviously, when resolving a finitely-presented $\mathbf{K}[\underline{X}]$-module over a field $\mathbf{K}$ (a field is a valuation domain of Krull dimension 0), we have one step less to do (since $\mathbf{K}[X_1]$ is a PID and by virtue of Remark 89).

Corollary 95. (Hilbert's syzygy theorem)

Let $M = H_m/U$ be a finitely-presented $\mathbf{K}[X_1, \ldots, X_n]$-module, where $\mathbf{K}$ is a field. Then M admits a finite free $\mathbf{K}[X_1, \ldots, X_n]$-resolution

$$0 \to F_p \to F_{p-1} \to \cdots \to F_1 \to F_0 \to M \to 0$$

of length $p \leq n$ as described in the previous theorem.

Another consequence of Theorem 90 is the following result.

Theorem 96. *Let $M = H_m/U$ be a finitely-presented $\mathbf{V}[X_1, \ldots, X_n]$-module, where $\mathbf{V}$ is a coherent valuation ring with nonzero zero-divisors. Assume that $\mathrm{LT}(U)$ is finitely-generated according to a lexicographic monomial order. Then M admits a resolution by finite free $\mathbf{V}[X_1, \ldots, X_n]$-modules*

$$\cdots \xrightarrow{\varphi_{p+3}} F_p \xrightarrow{\varphi_{p+2}} F_p \xrightarrow{\varphi_{p+1}} F_p \xrightarrow{\varphi_p} F_{p-1} \xrightarrow{\varphi_{p-1}} \cdots \xrightarrow{\varphi_2} F_1 \xrightarrow{\varphi_1} F_0 \xrightarrow{\varphi_0} M \longrightarrow 0$$

such that for some $p \leq n + 1$,

- $\mathrm{LT}(\mathrm{Ker}(\varphi_p)) = \bigoplus_{j=1}^{m_p} \langle b_j \rangle e_j$ *with $b_1, \ldots, b_{m_p} \in \mathbf{V}$ and $(e_1, \ldots, e_{m_p})$ a basis of F_p,*

- $\text{LT}(\text{Ker}(\varphi_{p+2k-1})) = \bigoplus_{j=1}^{m_p} \text{Ann}(b_j)e_j$ *for $k \geq 1$,*

- $\text{LT}(\text{Ker}(\varphi_{p+2k})) = \bigoplus_{j=1}^{m_p} \text{Ann}(\text{Ann}(b_j))e_j$ *for $k \geq 1$,*

and at each step where indeterminates remain present, the considered monomial order is Schreyer's monomial order (as in the proof of Theorem 91).

Proof. The part

$$F_p \xrightarrow{\varphi_p} F_{p-1} \xrightarrow{\varphi_{p-1}} \cdots \xrightarrow{\varphi_2} F_1 \xrightarrow{\varphi_1} F_0 \xrightarrow{\varphi_0} M \longrightarrow 0$$

of the free $\mathbf{V}[\underline{X}]$-resolution with $p \leq n+1$ and $\text{LT}(\text{Ker}(\varphi_p)) = \bigoplus_{j=1}^{m_p} \langle b_j \rangle e_j$ follows from the proof of Theorem 91. W.l.o.g., the b_j's are $\neq 0$. Let us denote by $[g_1, \ldots, g_{m_p}]$ a Gröbner basis for $\text{Ker}(\varphi_p)$ such that $\text{LT}(g_j) = b_j e_j$ for $1 \leq j \leq m_p$. So $\text{S}(g_i, g_j) = 0$ for $i < j$. Thus the fact that $\text{LT}(\text{Ker}(\varphi_{p+1})) = \bigoplus_{j=1}^{m_p} \text{Ann}(b_j)e_j$, $\text{LT}(\text{Ker}(\varphi_{p+2})) = \bigoplus_{j=1}^{m_p} \text{Ann}(\text{Ann}(b_j))e_j$, etc. follows immediately from Theorem 90. Finally, let us recall the equality $\text{Ann}(\text{Ann}(\text{Ann}(I))) = \text{Ann}(I)$ for an ideal I.

Let us remark that this shows that the free resolution is in general not a finite one.

Corollary 97. *Let $M = H_m/U$ be a finitely presented $\mathbf{V}[X_1, \ldots, X_n]$-module, where $\mathbf{V}$ is a zero-dimensional coherent valuation ring. Then M admits a free $\mathbf{V}[X_1, \ldots, X_n]$-resolution as described in Theorem 96.*

Example 98. Let $g_1 = X^4 - X$, $g_2 = 2X$, $g_3 = Y^3 - 1 \in (\mathbb{Z}/4\mathbb{Z})[X,Y]$, and let us consider the lexicographic order $>_1$ for which $X >_1 Y$ as monomial order. We have

$$\text{S}(g_1, g_1) = 0g_1 = 0, \quad \text{S}(g_1, g_2) = 2g_1 - X^3 g_2 = -2X = -g_2,$$

$$\text{S}(g_2, g_2) = 2g_2 = 0, \quad \text{S}(g_2, g_3) = Y^3 g_2 - 2Xg_3 = 2X = g_2,$$

$$\text{S}(g_3, g_3) = 0g_3 = 0, \quad \text{S}(g_1, g_3) = Y^3 g_1 - X^4 g_3 = -XY^3 + X^4 = g_1 - Xg_3.$$

Thus, (g_1, g_2, g_3) is a (pseudo-reduced) Gröbner basis for $I = \langle g_1, g_2, g_3 \rangle$, and $\text{LT}(I) = \langle X^4, 2X, Y^3 \rangle$. By Theorem 90, $u_{1,3} = (Y^3 - 1, 0, X - X^4)$, $u_{1,2} = (2, 1 - X^3, 0)$, $u_{2,3} = (0, Y^3 - 1, -2X)$, $u_{2,2} = (0, 2, 0)$ form a (pseudo-reduced) Gröbner basis for the syzygy module $\text{Syz}(g_1, g_2, g_3)$ w.r.t. Schreyer's monomial order $>_2$ induced by $>_1$ and $[g_1, g_2, g_3]$. In particular,

$$\text{LT}(\text{Syz}(g_1, g_2, g_3)) = \langle \text{LT}(u_{1,3}), \ldots, \text{LT}(u_{2,2}) \rangle$$
$$= \langle Y^3 e_1, 2e_1, Y^3 e_2, 2e_2 \rangle = \langle 2, Y^3 \rangle e_1 \oplus \langle 2, Y^3 \rangle e_2,$$

where (e_1, e_2, e_3) stands for the canonical basis of $(\mathbb{Z}/4\mathbb{Z})[X,Y]^3$. We have:

$$S(u_{1,3}, u_{1,3}) = 0u_{1,3} = 0,$$
$$S(u_{1,3}, u_{1,2}) = 2u_{1,3} - Y^3 u_{1,2}$$
$$= (-2, -Y^3 + X^3 Y^3, 2X - 2X^4)$$
$$= -u_{1,2} + (X^3 - 1)u_{2,3},$$
$$S(u_{1,3}, u_{2,3}) = S(u_{1,3}, u_{2,2}) = 0,$$
$$S(u_{1,2}, u_{1,2}) = 2u_{1,2} = (0, 2 - 2X^3, 0) = (1 - X^3)u_{2,2},$$

$$S(u_{1,2}, u_{2,3}) = S(u_{1,2}, u_{2,2}) = 0,$$
$$S(u_{2,3}, u_{2,3}) = 0u_{2,3} = 0,$$
$$S(u_{2,3}, u_{2,2}) = 2u_{2,3} - Y^3 u_{2,2}$$
$$= (0, -2, 0X)$$
$$= (0, -2, 0) = -u_{2,2},$$
$$S(u_{2,2}, u_{2,2}) = 2u_{2,2} = 0.$$

We recover that $(u_{1,3}, u_{1,2}, u_{2,3}, u_{2,2})$ is a Gröbner basis for $\mathrm{Syz}(g_1, g_2, g_3)$. By Theorem 90, $u_{1,3;1,2} = (2, 1 - Y^3, 1 - X^3, 0)$, $u_{1,2;1,2} = (0, 2, 0, X^3 - 1)$, $u_{2,3;2,2} = (0, 0, 2, 1 - Y^3)$, $u_{2,2;2,2} = (0, 0, 0, 2)$ form a (pseudo-reduced) Gröbner basis for the syzygy module $\mathrm{Syz}(u_{1,3}, u_{1,2}, u_{2,3}, u_{2,2})$ w.r.t. Schreyer's monomial order $>_3$ induced by $>_2$ and $[u_{1,3}, u_{1,2}, u_{2,3}, u_{2,2}]$. In particular,

$$\mathrm{LT}(\mathrm{Syz}(u_{1,3}, u_{1,2}, u_{2,3}, u_{2,2})) = \langle \mathrm{LT}(u_{1,3;1,2}), \dots, \mathrm{LT}(u_{2,2;2,2}) \rangle$$
$$= \langle 2\varepsilon_1, \dots, 2\varepsilon_4 \rangle = \langle 2 \rangle \varepsilon_1 \oplus \langle 2 \rangle \varepsilon_2 \oplus \langle 2 \rangle \varepsilon_3 \oplus \langle 2 \rangle \varepsilon_4,$$

where $(\varepsilon_1, \dots, \varepsilon_4)$ stands for the canonical basis of $(\mathbb{Z}/4\mathbb{Z})[X, Y]^4$. By Theorem 90, we find four vectors $u_{(1,3;1,2),(1,3;1,2)}, \dots, u_{(2,2;2,2),(2,2;2,2)} \in (\mathbb{Z}/4\mathbb{Z})[X, Y]^4$ forming a (pseudo-reduced) Gröbner basis for the syzygy module $\mathrm{Syz}(u_{1,3;1,2}, \dots, u_{2,2;2,2})$ w.r.t. Schreyer's monomial order $>_4$ induced by $>_3$ and $[u_{1,3;1,2}, \dots, u_{2,2;2,2}]$. In particular,

$$\mathrm{LT}(\mathrm{Syz}(u_{1,3;1,2}, \dots, u_{2,2;2,2})) = \langle \mathrm{LT}(u_{(1,3;1,2),(1,3;1,2)}), \dots, \mathrm{LT}(u_{(2,2;2,2),(2,2;2,2)}) \rangle$$
$$= \langle 2 \rangle \varepsilon_1 \oplus \langle 2 \rangle \varepsilon_2 \oplus \langle 2 \rangle \varepsilon_3 \oplus \langle 2 \rangle \varepsilon_4,$$

etc. We conclude that I admits the free $(\mathbb{Z}/4\mathbb{Z})[X, Y]$-resolution

$$\cdots \xrightarrow{\varphi_3} (\mathbb{Z}/4\mathbb{Z})[X, Y]^4 \xrightarrow{\varphi_2} (\mathbb{Z}/4\mathbb{Z})[X, Y]^4 \xrightarrow{\varphi_1} (\mathbb{Z}/4\mathbb{Z})[X, Y]^3 \xrightarrow{\varphi_0} I \longrightarrow 0$$

such that $\mathrm{LT}(\mathrm{Ker}(\varphi_i)) = \langle 2 \rangle \varepsilon_1 \oplus \langle 2 \rangle \varepsilon_2 \oplus \langle 2 \rangle \varepsilon_3 \oplus \langle 2 \rangle \varepsilon_4$ for $i \geq 2$.

1.5.2   The syzygy theorem and Schreyer's algorithm for a coherent Bézout ring

As explained in the proof of Theorem 78, one can avoid branching when computing a dynamical Gröbner basis for a Bézout domain of Krull dimension ≤ 1 (e.g. $\mathbb{Z}$ and the ring of all algebraic integers—note that the last one is not a PID) or a zero-dimensional coherent Bézout ring. Note that this is not possible for Prüfer domains of Krull dimension ≤ 1 which are not Bézout domains, or equivalently, which are not gcd domains (for example, $\mathbb{Z}[\sqrt{-5}]$, see Example 60).

Now we generalise these algorithms to the computation of free resolutions. We start with Schreyer's syzygy algorithm.

Theorem 99. (Schreyer's algorithm for coherent Bézout rings)

We consider a coherent Bézout ring **R**. *Let* H_m *be a free* $\mathbf{R}[\underline{X}]$-*module with basis* $(e_1, \ldots, e_m)$ *and* $>$ *a monomial order on* H_m. *Let* $U \subseteq H_m$ *be a submodule of* H_m *with Gröbner basis* $[g_1, \ldots, g_s]$. *Then the relations* $u_{i,j}$ *computed by Algorithm 86 form a Gröbner basis for the syzygy module* $\mathrm{Syz}(g_1, \ldots, g_s)$ *w.r.t. Schreyer's monomial order induced by* $>$ *and* $[g_1, \ldots, g_s]$.

Proof. This follows directly from the local case given by Theorem 90: see the proof of Theorem 78 for an explanation.

Theorem 100. (Syzygy theorem for a Bézout domain)

Let $M = H_m/U$ *be a finitely-presented* $\mathbf{R}[X_1, \ldots, X_n]$-*module, where* **R** *is a Bézout domain and* H_m *a free* $\mathbf{R}[\underline{X}]$-*module. Assume that* $\mathrm{LT}(U)$ *is finitely-generated according to a lexicographic monomial order. Then* M *admits a finite free* $\mathbf{R}[X_1, \ldots, X_n]$-*resolution*

$$0 \to F_p \to F_{p-1} \to \cdots \to F_1 \to F_0 \to M \to 0$$

of length $p \le n + 1$.

Proof. This follows directly from the local case.

Corollary 101. (Syzygy theorem for a one-dimensional Bézout domain)

Let $M = H_m/U$ *be a finitely-presented* $\mathbf{R}[X_1, \ldots, X_n]$-*module, where* **R** *is a one-dimensional Bézout domain. Then* M *admits a finite free* $\mathbf{R}[X_1, \ldots, X_n]$-*resolution*

$$0 \to F_p \to F_{p-1} \to \cdots \to F_1 \to F_0 \to M \to 0$$

of length $p \le n + 1$ *as described in the previous theorem.*

Now we treat the case of zero-dimensional coherent Bézout rings. For a nonzero element a in **R**, we denote by $\mathrm{ann}(a)$ a generator of $\mathrm{Ann}(a)$. We let $\mathrm{ann}(0) = 1$.

Theorem 102. (Syzygy theorem for a zero-dimensional Bézout ring)

Let $M = H_m/U$ *be a finitely-presented* $\mathbf{R}[X_1, \ldots, X_n]$-*module, where* **R** *is a coherent zero-dimensional Bézout ring and* H_m *a free* $\mathbf{R}[X_1, \ldots, X_n]$-*module. Then* M *admits a free* $\mathbf{R}[\underline{X}]$-*resolution*

$$\cdots \xrightarrow{\varphi_{p+3}} F_p \xrightarrow{\varphi_{p+2}} F_p \xrightarrow{\varphi_{p+1}} F_p \xrightarrow{\varphi_p} F_{p-1} \xrightarrow{\varphi_{p-1}} \cdots \xrightarrow{\varphi_2} F_1 \xrightarrow{\varphi_1} F_0 \xrightarrow{\varphi_0} M \longrightarrow 0$$

such that for some $p \le n + 1$,

- $\mathrm{LT}(\mathrm{Ker}(\varphi_p)) = \bigoplus_{j=1}^{m_p} \langle b_j \rangle e_j$ *with* $b_1, \ldots, b_{m_p} \in \mathbf{R}$ *and* $(e_1, \ldots, e_{m_p})$ *a basis of* F_p,

- $\mathrm{LT}(\mathrm{Ker}(\varphi_{p+2k-1})) = \bigoplus_{j=1}^{m_p} \mathrm{Ann}(b_j)e_j \, \text{for } k \geq 1,$

- $\mathrm{LT}(\mathrm{Ker}(\varphi_{p+2k})) = \bigoplus_{j=1}^{m_p} \mathrm{Ann}(\mathrm{Ann}(b_j))e_j \, \text{for } k \geq 1,$

and at each step where indeterminates remain present, the considered monomial order is Schreyer's monomial order.

Proof. Follows directly from the local case.

1.5.3   The case of the integers

The following theorems are particular cases of Theorem 99 and Corollary 101.

Theorem 103. (Schreyer's algorithm for the integers)

Let H_m be a free $\mathbb{Z}[\underline{X}]$-module with basis $(e_1, \dots, e_m)$ and $>$ a monomial order on H_m. Let $U \subseteq H_m$ be a submodule of H_m with Gröbner basis $[g_1, \dots, g_s]$. Then the relations $u_{i,j}$ computed by Algorithm 86 form a Gröbner basis for the syzygy module $\mathrm{Syz}(g_1, \dots, g_s)$ w.r.t. Schreyer's monomial order induced by $>$ and $[g_1, \dots, g_s]$. Moreover, for $1 \leq i < j \leq s$ such that $\mathrm{LPos}(g_i) = \mathrm{LPos}(g_j)$, we have

$$\mathrm{LT}(u_{i,j}) = \frac{\mathrm{LC}(g_j)}{\gcd(\mathrm{LC}(g_i),\mathrm{LC}(g_j))}\underline{X}^{(\mathrm{mdeg}(g_j)-\mathrm{mdeg}(g_i))^+}\varepsilon_i.$$

Theorem 104. (Syzygy theorem for the integers)

Let M be a finitely-generated $\mathbb{Z}[X_1, \dots, X_n]$-module. Then M admits a free $\mathbb{Z}[X_1, \dots, X_n]$-resolution

$$0 \to F_p \to F_{p-1} \to \cdots \to F_1 \to F_0 \to M \to 0$$

of length $p \leq n+1$.

Example 105. Let $g_1 = X^2 - Y + 3$, $g_2 = 4Y^2 - 4$, $g_3 = 6Y + 6 \in \mathbb{Z}[X,Y]$, and let us consider the lexicographic order $>_1$ for which $X >_1 Y$ as monomial order. We have:

$$S(g_1, g_2) = 4Y^2 g_1 - X^2 g_2 = 4g_1 + (-Y+3)g_2,$$
$$S(g_1, g_3) = 6Y g_1 - X^2 g_3 = -6g_1 + (-Y+3)g_3,$$
$$S(g_2, g_3) = 3g_2 - 2Y g_3 = -2g_3.$$

Thus, (g_1, g_2, g_3) is a Gröbner basis for $I = \langle g_1, g_2, g_3 \rangle$, and $\mathrm{LT}(I) = \langle X^2, 4Y^2, 6Y \rangle$. By Theorem 103, $u_{1,2} = (4Y^2 - 4, -X^2 + Y - 3, 0)$, $u_{1,3} = (6Y + 6, 0, -X^2 + Y - 3)$, $u_{2,3} = (0, 3, -2Y + 2)$ form a Gröbner basis for the syzygy module $\mathrm{Syz}(g_1, g_2, g_3)$ w.r.t. Schreyer's monomial order $>_2$ induced by $>_1$ and $[g_1, g_2, g_3]$. In particular,

$$\mathrm{LT}(\mathrm{Syz}(g_1, g_2, g_3)) = \langle \mathrm{LT}(u_{1,2}), \mathrm{LT}(u_{1,3}), \mathrm{LT}(u_{2,3}) \rangle = \langle 4Y^2 e_1, 6Y e_1, 3e_2 \rangle$$
$$= \langle 4Y^2, 6Y \rangle e_1 \oplus \langle 3 \rangle e_2 = 2\langle 2Y^2, 3Y \rangle e_1 \oplus \langle 3 \rangle e_2 = 2\langle Y^2, 3Y \rangle e_1 \oplus \langle 3 \rangle e_2,$$

where (e_1, e_2, e_3) stands for the canonical basis of $\mathbb{Z}[X,Y]^3$. Thus $u'_{1,2} = Yu_{1,3} - u_{1,2} = (2Y^2 + 6Y + 4, X^2 - Y + 3, -X^2Y + Y^2 - 3Y)$, $u_{1,3}$, $u_{2,3}$ form a reduced Gröbner basis for $\mathrm{Syz}(g_1, g_2, g_3)$. We have:

$$S(u'_{1,2}, u_{1,3}) = 3u'_{1,2} - Yu_{1,3} = 2u_{1,3} + (X^2 - Y + 3)u_{2,3},$$
$$S(u'_{1,2}, u_{2,3}) = S(u_{1,3}, u_{2,3}) = 0.$$

We recover that $(u'_{1,2}, u_{1,3}, u_{2,3})$ is a Gröbner basis for $\mathrm{Syz}(g_1, g_2, g_3)$. By Theorem 103, $u_{1,2;1,3} = (3, -Y - 2, -X^2 + Y - 3)$ forms a (pseudo-reduced) Gröbner basis for the syzygy module $\mathrm{Syz}(u'_{1,2}, u_{1,3}, u_{2,3})$ w.r.t. Schreyer's monomial order $>_3$ induced by $>_2$ and $[u'_{1,2}, u_{1,3}, u_{2,3}]$. In particular, $\mathrm{LT}(\mathrm{Syz}(u'_{1,2}, u_{1,3}, u_{2,3})) = \langle \mathrm{LT}(u_{1,2;1,3}) \rangle = \langle 3 \rangle \varepsilon_1$ where $(\varepsilon_1, \varepsilon_2, \varepsilon_3)$ stands for the canonical basis of $\mathbb{Z}[X,Y]^3$. It follows that $\mathrm{Syz}(u'_{1,2}, u_{1,3}, u_{2,3})$ is free. We conclude that I admits the following length 2 free $\mathbb{Z}[X,Y]$-resolution:

$$0 \to \mathbb{Z}[X,Y] \xrightarrow{u_{1,2;1,3}} \mathbb{Z}[X,Y]^3 \xrightarrow{\begin{pmatrix} u_{1,2} \\ u_{1,3} \\ u_{2,3} \end{pmatrix}} \mathbb{Z}[X,Y]^3 \xrightarrow{\begin{pmatrix} g_1 \\ g_2 \\ g_3 \end{pmatrix}} I \to 0.$$

1.5.4   The case of $\mathbb{Z}/N\mathbb{Z}$

The elements of $\mathbb{Z}/N\mathbb{Z}$ are simply written as integers (their representatives in $[\![0, N - 1]\!]$). When talking about the gcd of two nonzero elements in $\mathbb{Z}/N\mathbb{Z}$ we mean the gcd of their representatives in $[\![1, N - 1]\!]$. For a nonzero element a in $\mathbb{Z}/N\mathbb{Z}$, letting $b = \gcd(a, N)$, the class of $\dfrac{N}{b}$ in $\mathbb{Z}/N\mathbb{Z}$ will be denoted by $\mathrm{ann}(a)$; it generates $\mathrm{Ann}(a)$.

- The Division algorithm 69 attains its goal: the gcd and the Bézout relation to be found in line 6 will be computed by finding d, b, b_i $(i \in D)$ in $\mathbb{Z}$ such that $d = \gcd(N, \gcd\{\mathrm{LC}(h_i) ; i \in D\}) = bN + \sum_{i \in D} b_i \mathrm{LC}(h_i)$; the Euclidean division in line 6 will be performed in $\mathbb{Z}$.

- The S-polynomial algorithm 70 attains its goal: note that in this case, the generator of the annihilator of $\mathrm{LC}(f)$ to be found on line 3 may be taken to be $\mathrm{ann}(\mathrm{LC}(f))$, so that the auto-$S$-polynomial of f is

$$S(f, f) = \mathrm{ann}(\mathrm{LC}(f))f;$$

- Buchberger's algorithm 71 attains its goal.

Let $\mathbf{H}_m$ be a free $(\mathbb{Z}/N\mathbb{Z})[\underline{X}]$-module with basis $(e_1, \ldots, e_m)$ and let $>$ be a monomial order on $\mathbf{H}_m$.

The following theorems are particular cases of Theorems 99 and 102.

Theorem 106. (Schreyer's algorithm for $\mathbb{Z}/N\mathbb{Z}$)

Let H_m be a free $(\mathbb{Z}/N\mathbb{Z})[\underline{X}]$-module with basis $(e_1,\ldots,e_m)$, $>$ a monomial order on H_m. Let $U \subseteq H_m$ be a submodule of H_m with Gröbner basis $[g_1,\ldots,g_s]$. Then the relations $u_{i,j}$ computed by Algorithm 86 form a Gröbner basis for the syzygy module $\mathrm{Syz}(g_1,\ldots,g_s)$ w.r.t. Schreyer's monomial order induced by $>$ and $[g_1,\ldots,g_s]$. Moreover, for all $1 \le i \le j \le s$ such that $\mathrm{LPos}(g_i) = \mathrm{LPos}(g_j)$, we have

$$
\mathrm{LT}(u_{i,j}) = \begin{cases} \mathrm{ann}(\mathrm{LC}(g_i))\varepsilon_i & \text{if } i = j, \\[2ex] \dfrac{\mathrm{LC}(g_j)}{\gcd(\mathrm{LC}(g_i),\mathrm{LC}(g_j))}\underline{X}^{(\mathrm{mdeg}(g_j)-\mathrm{mdeg}(g_i))^+}\varepsilon_i & \text{otherwise.} \end{cases}
$$

Theorem 107. (Syzygy theorem for $\mathbb{Z}/N\mathbb{Z}$)

Let M be a finitely-presented $(\mathbb{Z}/N\mathbb{Z})[X_1,\ldots,X_n]$-module. Then M admits a free $(\mathbb{Z}/N\mathbb{Z})[X_1,\ldots,X_n]$-resolution

$$
\cdots \xrightarrow{\varphi_{p+3}} F_p \xrightarrow{\varphi_{p+2}} F_p \xrightarrow{\varphi_{p+1}} F_p \xrightarrow{\varphi_p} F_{p-1} \xrightarrow{\varphi_{p-1}} \cdots \xrightarrow{\varphi_2} F_1 \xrightarrow{\varphi_1} F_0 \xrightarrow{\varphi_0} M \longrightarrow 0
$$

such that for some $p \le n+1$,

$$
\mathrm{LT}(\mathrm{Ker}(\varphi_p)) = \bigoplus_{j=1}^{m_p} \langle b_j \rangle e_j, \quad \mathrm{LT}(\mathrm{Ker}(\varphi_{p+1})) = \bigoplus_{j=1}^{m_p} \frac{N}{\gcd(N,b_j)} e_j,
$$

$$
\mathrm{LT}(\mathrm{Ker}(\varphi_{p+2})) = \bigoplus_{j=1}^{m_p} \langle b_j \rangle e_j, \quad \mathrm{LT}(\mathrm{Ker}(\varphi_{p+3})) = \bigoplus_{j=1}^{m_p} \frac{N}{\gcd(N,b_j)} e_j, \text{ etc.,}
$$

where $(e_1,\ldots,e_{m_p})$ is a basis of F_p, $b_1,\ldots,b_{m_p} \in \mathbb{Z}/N\mathbb{Z}$, and the considered monomial order is Schreyer's monomial order.

Example 108. Let $g_1 = X+1$, $g_2 = Y^3+Y^2+6$, $g_3 = 3Y^2$, $g_4 = 9 \in (\mathbb{Z}/12\mathbb{Z})[X,Y]$, and let us consider the lexicographic order $>_1$ for which $X >_1 Y$ as monomial order. We have

$$S(g_1,g_1) = 0g_1 = 0, \qquad\qquad S(g_2,g_3) = 3g_2 - Yg_3 = g_3 + 2g_4,$$

$$S(g_1,g_2) = Y^3 g_1 - Xg_2 = (-Y^2-6)g_1 + g_2, \quad S(g_2,g_4) = 9g_2 - Y^3 g_3 = (Y^2+6)g_4,$$

$$S(g_1,g_3) = 3Y^2 g_1 - Xg_3 = g_3, \qquad\qquad S(g_3,g_3) = 4g_3 = 0,$$

$$S(g_1,g_4) = 9g_1 - Xg_4 = g_4, \qquad\qquad S(g_3,g_4) = 3g_3 - Y^2 g_4 = 0,$$

$$S(g_2,g_2) = 0g_2 = 0, \qquad\qquad S(g_4,g_4) = 4g_4 = 0.$$

Thus, (g_1,g_2,g_3,g_4) is a (pseudo-reduced) Gröbner basis for $I = \langle g_1,g_2,g_3,g_4 \rangle$, and $\mathrm{LT}(I) = \langle X, Y^3, 3Y^2, 9 \rangle$. By Theorem 106, $u_{1,2} = (Y^3+Y^2+6, -X-1, 0, 0)$, $u_{1,3} = (3Y^2, 0, -X-1, 0)$, $u_{1,4} = (9, 0, 0, -X-1)$, $u_{2,3} = (0, 3, -Y-1, -2)$, $u_{2,4} = (0, 9, -Y^3, -Y^2-6)$, $u_{3,3} = (0,0,4,0)$, $u_{3,4} = (0,0,3,-Y^2)$, $u_{4,4} = (0,0,0,4)$

form a Gröbner basis for the syzygy module $\mathrm{Syz}(g_1, g_2, g_3, g_4)$ w.r.t. Schreyer's monomial order $>_2$ induced by $>_1$ and $[g_1, g_2, g_3, g_4]$. In particular,

$$
\begin{aligned}
\mathrm{LT}(\mathrm{Syz}(g_1, g_2, g_3, g_4)) &= \langle \mathrm{LT}(u_{1,2}), \dots, \mathrm{LT}(u_{4,4}) \rangle \\
&= \langle Y^3, 3Y^2, 9 \rangle e_1 \oplus \langle 3, 9 \rangle e_2 \oplus \langle 4, 3 \rangle e_3 \oplus \langle 4 \rangle e_4 \\
&= \langle Y^3, 3 \rangle e_1 \oplus \langle 3 \rangle e_2 \oplus \langle 1 \rangle e_3 \oplus \langle 4 \rangle e_4,
\end{aligned}
$$

where (e_1, e_2, e_3, e_4) stands for the canonical basis of $(\mathbb{Z}/12\mathbb{Z})[X, Y]^4$. Thus, $u_{1,2}$, $u'_{1,4} = -u_{1,4} = (3, 0, 0, X+1)$, $u_{2,3}$, $u'_{3,3} = u_{3,3} - u_{3,4} = (0, 0, 1, Y^2)$, $u_{4,4}$ form a reduced Gröbner basis for $\mathrm{Syz}(g_1, g_2, g_3, g_4)$. We have:

$$
\begin{aligned}
S(u_{1,2}, u'_{1,4}) = 3u_{1,2} - Y^3 u'_{1,4} &= (Y^2 + 2)u'_{1,4} + (3X+3)u_{2,3} + (3XY + 3X + 3Y + 3)u'_{3,3} \\
&\quad + (2XY^3 + 2XY^2 + 2Y^3 + 2Y^2 + X + 1)u_{4,4}, \\
S(u'_{1,4}, u'_{1,4}) = 4u'_{1,4} &= (X+1)u_{4,4}, \\
S(u_{2,3}, u_{2,3}) = 4u_{2,3} &= (8Y+8)u'_{3,3} + (Y^3 + Y^2 + 1)u_{4,4}, \\
S(u_{4,4}, u_{4,4}) = 3u_{4,4} &= 0.
\end{aligned}
$$

By Theorem 106, the elements $u_{1,2;1,4} = (3, -Y^3 - Y^2 - 2, -3X - 3, -3XY - 3X - 3Y - 3, -2XY^3 - 2XY^2 - 2Y^3 - 2Y^2 - X - 1)$, $u_{1,4;1,4} = (0, 4, 0, 0, -X - 1)$, $u_{2,3;2,3} = (0, 0, 4, -8Y - 8, -Y^3 - Y^2 - 1)$, $u_{4,4;4,4} = (0, 0, 0, 0, 3)$ form a (pseudo-reduced) Gröbner basis for the syzygy module $\mathrm{Syz}(u_{1,2}, u'_{1,4}, u_{2,3}, u'_{3,3}, u_{4,4})$ w.r.t. Schreyer's monomial order $>_3$ induced by $>_2$ and $[u_{1,2}, u'_{1,4}, u_{2,3}, u'_{3,3}, u_{4,4}]$.

In particular, $\mathrm{LT}(\mathrm{Syz}(u_{1,2}, u'_{1,4}, u_{2,3}, u'_{3,3}, u_{4,4})) = \langle 3 \rangle \varepsilon_1 \oplus \langle 4 \rangle \varepsilon_2 \oplus \langle 4 \rangle \varepsilon_3 \oplus \langle 3 \rangle \varepsilon_5$, where $(\varepsilon_1, \dots, \varepsilon_5)$ stands for the canonical basis of $(\mathbb{Z}/12\mathbb{Z})[X, Y]^5$.

We conclude that I admits a free $(\mathbb{Z}/12\mathbb{Z})[X, Y]$-resolution

$$
\cdots \overset{\varphi_4}{\to} (\mathbb{Z}/12\mathbb{Z})[X, Y]^4 \overset{\varphi_3}{\to} (\mathbb{Z}/12\mathbb{Z})[X, Y]^4 \overset{\varphi_2}{\to} (\mathbb{Z}/12\mathbb{Z})[X, Y]^5 \overset{\varphi_1}{\to} (\mathbb{Z}/12\mathbb{Z})[X, Y]^4 \overset{\varphi_0}{\to} I \to 0
$$

such that $\mathrm{LT}(\mathrm{Ker}(\varphi_{2i})) = \langle 4 \rangle \omega_1 \oplus \langle 3 \rangle \omega_2 \oplus \langle 3 \rangle \omega_3 \oplus \langle 4 \rangle \omega_4$ and $\mathrm{LT}(\mathrm{Ker}(\varphi_{2i+1})) = \langle 3 \rangle \omega_1 \oplus \langle 4 \rangle \omega_2 \oplus \langle 4 \rangle \omega_3 \oplus \langle 3 \rangle \omega_4$ for $i \geq 1$, where $(\omega_1, \dots, \omega_4)$ stands for the canonical basis of $(\mathbb{Z}/12\mathbb{Z})[X, Y]^4$.

1.6 Exercises

Exercise 109. Let $\{g_1, \dots, g_s\} \subseteq \mathbf{K}[X_1, \dots, X_n]$ ($\mathbf{K}$ a field), let $h \in \mathbf{K}[X_1, \dots, X_n] \setminus \{0\}$, and let $>$ be a monomial order on $\mathbf{K}[X_1, \dots, X_n]$. Show that $\{g_1, \dots, g_s\}$ is a Gröbner basis for $\langle g_1, \dots, g_s \rangle$ if and only if $\{hg_1, \dots, hg_s\}$ is a Gröbner basis for $\langle hg_1, \dots, hg_s \rangle$.

Exercise 110. Suppose that $F = \{\ell_1, \dots, \ell_s\} \subseteq \mathbf{K}[X_1, \dots, X_n]$ is a set of linear forms (polynomials of total degree 1 with zero constant term), where $\mathbf{K}$ is a field. Let I be the ideal generated by F. Give necessary and sufficient conditions for F to be:

a) A minimal Gröbner basis.

b) A reduced Gröbner basis.

Exercise 111. Let $>$ be a monomial order on $\mathbf{A} = \mathbf{K}[X_1,\ldots,X_n]$, where $\mathbf{K}$ is a field.

1) We consider two nonzero polynomials f and g in $\mathbf{A}$ the leading monomials of which are relatively prime (i.e., $\mathrm{LCM}(\mathrm{LM}(f),\mathrm{LM}(g)) = \mathrm{LM}(f)\,\mathrm{LM}(g)$). Show that the remainder of $S(f,g)$ on division by $\{f,g\}$ is zero.

2) Deduce that if $f_1,\ldots,f_s$ are in $\mathbf{A}\setminus\{0\}$ and their leading monomials are pairwise relatively prime, then $\{f_1,\ldots,f_s\}$ is a Gröbner basis for $\langle f_1,\ldots,f_s\rangle$.

Exercise 112. (Buchberger's first criterion)

Let $>$ be a monomial order on $\mathbf{K}[X_1,\ldots,X_n]$, where $\mathbf{K}$ is a field. Consider two nonzero polynomials f and g in $\mathbf{K}[X_1,\ldots,X_n]$ and let $d := \gcd(f,g)$. Show that $\{f,g\}$ is a Gröbner basis for $\langle f,g\rangle$ if and only if $\mathrm{LM}(\frac{f}{d})$ and $\mathrm{LM}(\frac{g}{d})$ are relatively prime.

Exercise 113. [35, 46]

Let $>$ be a monomial order on $\mathbf{K}[X_1,\ldots,X_n]$, where $\mathbf{K}$ is a field. Consider two nonzero polynomials f and g in $\mathbf{K}[X_1,\ldots,X_n]$.

1) Let $\mathrm{LM}(f) = X_1^{\mu_1}\cdots X_n^{\mu_1}$ and $\mathrm{LM}(g) = X_1^{\nu_1}\cdots X_n^{\nu_1}$. Show that if $\mu_k \geq \nu_k > 0$ for some k, then either $\{f,g\}$ is not a Gröbner basis for $\langle f,g\rangle$ or $\{f+s,g\}$ is not a Gröbner basis for $\langle f+s,g\rangle$ for any $s \in \mathbf{K}^\times$.

2) Deduce that $\{f+s,g+t\}$ is a Gröbner basis for $\langle f+s,g+t\rangle$ for all $s,t \in \mathbf{K}$ if and only if $\mathrm{LM}(f)$ and $\mathrm{LM}(g)$ are relatively prime.

Exercise 114. Let $\mathbf{V}$ be a coherent valuation ring, fix a monomial order on $\mathbf{V}[X_1,\ldots,X_n]$, and consider finitely many terms $a_1 M_1,\ldots,a_s M_s$, with $a_i \in \mathbf{V}\setminus\{0\}$ and M_i is a monomial at $X_1,\ldots,X_n$. Show that $\{a_1 M_1,\ldots,a_s M_s\}$ is a Gröbner basis for $\langle a_1 M_1,\ldots,a_s M_s\rangle$.

Exercise 115. (Eisenbud, Sturmfels)

Let $\mathbf{K}$ be a field. By a binomial (resp. trinomial) in a polynomial ring $\mathbf{K}[X_1,\ldots,X_n]$ we mean a polynomial with at most two terms (resp. three terms), say $aX^\alpha + bX^\beta$ (resp. $aX^\alpha + bX^\beta + cX^\gamma$), where $a,b,c \in \mathbf{K}$ and $\alpha,\beta,\gamma \in \mathbb{N}^n$.

1) Show that every system $f_1(x_1,\ldots,x_n) = \cdots = f_s(x_1,\ldots,x_n) = 0$ of polynomial equations can be seen as a system of trinomial equations.

2) Let $>$ be a monomial order on $\mathbf{K}[X_1,\ldots,X_n]$ and I a binomial ideal, that is, I can be generated by a finite number of binomials in $\mathbf{K}[X_1,\ldots,X_n]$. Show that:

a) The reduced Gröbner basis G of I consists of binomials.

b) The remainder of any term on division by G is again a term.

3) Deduce that an ideal I of $\mathbf{K}[X_1, \ldots, X_n]$ is binomial if and only if some (equivalently, every) reduced Gröbner basis for I consists of binomials.

4) Show that if I is a binomial ideal of $\mathbf{K}[X_1, \ldots, X_n]$, then the elimination ideal $I \cap \mathbf{K}[X_1, \ldots, X_r]$ is a binomial ideal for every $r \leq n$.

5) Show that if $I, I', J_1, \ldots, J_s$ are ideals in $\mathbf{K}[X_1, \ldots, X_n]$ such that I and I' are generated by binomials and $J_1, \ldots, J_s$ are generated by monomials, then

$$(I + I') \cap (I + J_1) \cap (I + J_2) \cap \cdots \cap (I + J_s)$$

is generated by binomials.

Exercise 116. Suppose that I is an ideal of $\mathbf{K}[X_1, \ldots, X_n]$, where $\mathbf{K}$ is field. Let $\bar{\mathbf{K}}$ be an algebraic closure of $\mathbf{K}$.

1) Suppose that $f \in \mathbf{K}[X_1, \ldots, X_n]$ can be written as a linear combination of elements of I with coefficients in $\bar{\mathbf{K}}[X_1, \ldots, X_n]$. Show that f can be written as a linear combination of elements of I with coefficients in $\mathbf{K}[X_1, \ldots, X_n]$.

2) Let G be a Gröbner basis for I. Let $\bar{I}$ be the ideal of $\bar{\mathbf{K}}[X_1, \ldots, X_n]$ generated by the elements of I. Show that G is a Gröbner basis for $\bar{I}$.

Exercise 117. (Lombardi's constructive definition of Krull dimension)

The following gives a constructive characterization of the Krull dimension (equivalent to the classical one). This characterization ensues from the geometrical intuition that a variety (over an algebraically closed field) is of dimension $\leq k$ if and only if any subvariety has a boundary of dimension $< k$. Krull dimension -1 which means that the ring is trivial ($1 = 0$), or, that the variety (over an algebraically closed field) is empty.

For any ring $\mathbf{R}$ and $\ell \in \mathbb{N}$, we say that $\mathbf{R}$ has Krull dimension at most ℓ (in short, Kdim $\mathbf{R} \leq \ell$) if for any for any $x_0, \ldots, x_\ell \in \mathbf{R}$, there exist $a_0, \ldots, a_\ell \in \mathbf{R}$ and $m_0, \ldots, m_\ell \in \mathbb{N}$ such that

$$x_0^{m_0} (x_1^{m_1} \cdots (x_\ell^{m_\ell}(1 + a_\ell x_\ell) + \cdots + a_1 x_1) + a_0 x_0) = 0$$

(such equality is called a *collapse*).

1) Show that if $\mathbf{K}$ is a field, then the Krull dimension of $\mathbf{K}[X_1, \ldots, X_\ell]$ is $\leq \ell$.

2) Show that the Krull dimension of $\mathbb{Z}$ is ≤ 1. Find a collapse between 700 and 6.

3) Show that a finite ring has Krull dimension ≤ 0.

4) Show that a valuation domain has Krull dimension ≤ 1 if and only if its valuation group is archimedean.

Exercise 118. Recall that for a ring $\mathbf{R}$ and $u = (u_1, \ldots, u_m) \in \mathbf{R}^m$, the syzygy module of u is

$$\mathrm{Syz}(u) = \mathrm{Syz}(u_1, \ldots, u_m) := \{(v_1, \ldots, v_m) \in \mathbf{R}^m \mid v_1 u_1 + \cdots + v_m u_m = 0\}.$$

An $s \in \mathbf{R}^m$ ($m \geq 2$) is said to be an *obvious syzygy* of u if $s = u_j e_i - u_i e_j =: s_{i,j}$ with $1 \leq i < j \leq m$, where $(e_1, \ldots, e_m)$ stands for the canonical basis of $\mathbf{R}^m$. Of course there are $\binom{m}{2} = \frac{m(m-1)}{2}$ obvious syzygies for u, and we have:

$$\langle s_{i,j} \mid 1 \leq i < j \leq m \rangle \subseteq \mathrm{Syz}(u).$$

1) Give an example showing that in general the inclusion above can be strict.

2) Show that for $u = (u_1, \ldots, u_m) \in \mathbf{R}^m$, we have

$$\langle u_1, \ldots, u_m \rangle \, \mathrm{Syz}(u) \subseteq \langle s_{i,j} \mid 1 \leq i < j \leq m \rangle \subseteq \mathrm{Syz}(u),$$

where the $s_{i,j}$'s are the obvious syzygies of u.

Deduce that, if u is unimodular, i.e., $1 \in \langle u_1, \ldots, u_m \rangle$, then

$$\mathrm{Syz}(u) = \langle s_{i,j} \mid 1 \leq i < j \leq m \rangle.$$

3) We suppose now that $\mathbf{R}$ be a Bézout domain and we consider $u = (u_1, \ldots, u_m) \in \mathbf{R}^m \setminus \{0\}$ with $m \geq 2$. Find a generating set for $\mathrm{Syz}(u)$.

Exercise 119. Let a be an element of a ring $\mathbf{R}$. Recall that the annihilator of a in $\mathbf{R}$ is the ideal
$$\mathrm{Ann}(a) := \{x \in \mathbf{R} \mid xa = 0\}.$$

As the sequence $(\mathrm{Ann}(a^n))_{n \in \mathbb{N}}$ is nondecreasing,

$$\mathrm{Ann}(a^\infty) := \cup_{n \in \mathbb{N}} \mathrm{Ann}(a^n)$$

is an ideal of $\mathbf{R}$.

1) Show that for any ring $\mathbf{R}$ and $a \in \mathbf{R}$, we have:

$$\langle 1 + aX \rangle \cap \mathbf{R} = \mathrm{Ann}(a^\infty) \quad \& \quad \mathrm{LT}(\langle 1 + aX \rangle) = \mathrm{Ann}(a^\infty)[X] + \langle aX \rangle.$$

In particular, $\mathrm{LT}(\langle 1 + aX \rangle)$ is finitely-generated if and only if so is $\mathrm{Ann}(a^\infty)$.

2) Show that if $\mathbf{K}$ is a field then $\mathbf{K}[X_n : n \geq 0]$ is a non-Noetherian Gröbner ring.

3) Consider the ring $\mathbf{R} := \mathbf{K}[X_n : n \geq 0]/\langle X_0^k X_k : k \geq 1 \rangle$, where $\mathbf{K}$ is a field. What is $\mathrm{Ann}(\bar{X}_0^\infty)$? Deduce that $\mathbf{R}$ is not Gröbner.

Exercise 120. Let $\mathbf{V}$ be a valuation domain of Krull dimension ≥ 2, or equivalently, such that there exist $a, b \in \mathrm{Rad}(\mathbf{V}) \setminus \{0\}$ such that a^n divides b for all $n \geq 0$.

Check that when applied to $\{g_1 = aX + 1, g_2 = b\}$, Buchberger's Algorithm 38 never stops.

Exercise 121. Let $\mathbf{R}$ be a ring and consider a monomial order $>$ on $\mathbf{R}[X, Y]$ such that $X > Y$. Let $a \in \mathbf{R}$, and suppose that $\langle Y + aX \rangle$ has a Gröbner basis $G = \{g_1, \ldots, g_s\}$. Show that the annihilator of a is generated by the set $\{b \in \mathbf{R} \mid bY \in G\}$.

Exercise 122. Let $\mathbf{R}$ be a coherent ring, consider $f_1, \ldots, f_k \in \mathbf{R}[X_1, \ldots, X_n]$, and denote by

$$S := \mathrm{Syz}_{\mathbf{R}[X_1, \ldots, X_n]}(f_1, \ldots, f_k).$$

Show that $S \cap \mathbf{R}^k$ is a finitely-generated $\mathbf{R}$-module.

Exercise 123. (A one-dimensional domain which is not Gröbner)

Let t, u be two independent indeterminates over the field $\mathbb{Q}$ of rationals, denote by $K = \mathbb{Q}(\sqrt{2})(u)$ and consider the following domain

$$\mathbf{A} := \mathbb{Q} + t\mathbf{K}[t] = \{f(t) \in \mathbf{K}[t] \mid f(0) \in \mathbb{Q}\}.$$

1) Show that the ring $\mathbf{A}$ is one-dimensional and shares the ideal $\mathfrak{m} := t\mathbf{K}[t]$ with $\mathbf{K}[t]$.

2) Show that $\mathfrak{m}$ is not finitely-generated as an ideal of $\mathbf{A}$.

3) Show that, in $\mathbf{A}$, we have $\langle t \rangle \cap \langle \sqrt{2}t \rangle = t^2 \mathbf{K}[t] = t\mathfrak{m}$.

4) Deduce that, in $\mathbf{A}$, the ideal $\langle t \rangle \cap \langle \sqrt{2}t \rangle$ is not finitely-generated.

5) Show that $I \cap \mathbf{A} = \langle t \rangle \cap \langle \sqrt{2}t \rangle$.

6) Deduce that $\mathrm{LT}(I)$ is not finitely-generated, and thus, that $\mathbf{A}$ is not Gröbner.

1.7 Solutions to the exercises

Exercise 109:

This is due to the fact that $\mathrm{LT}(\langle hg_1, \ldots, hg_s \rangle) = \mathrm{LT}(h)\,\mathrm{LT}(\langle g_1, \ldots, g_s \rangle)$.

Exercise 110:

Suppose that $X_1 > \cdots > X_n$, write $\ell_j = \sum_{i=1}^s a_{i,j} X_i$, and consider the matrix $A = (a_{i,j})_{\substack{1 \leq i \leq n \\ 1 \leq j \leq s}}$.

a) Computing a minimal Gröbner basis corresponds to putting A in a column echelon form by Gaussian elimination.

b) Computing a reduced Gröbner basis corresponds to putting A in a reduced column echelon form by Gaussian elimination.

Exercise 111:

1) We can without loss of generality suppose that $\mathrm{LC}(f) = \mathrm{LC}(g) = 1$. Denoting by $\mathrm{LM}(f) = X^\alpha$, $\mathrm{LM}(g) = X^\beta$, $\tilde{f} = f - X^\alpha$, and $\tilde{g} = g - X^\beta$, we have

$$S(f,g) = X^\beta f - X^\alpha g = X^\beta \tilde{f} - X^\alpha \tilde{g}.$$

The case where either $\tilde{f}$ or $\tilde{g}$ is zero is clear. Suppose now that $\tilde{f} \neq 0$ and $\tilde{g} \neq 0$. As X^α and X^β are relatively prime, we deduce, by virtue of Gauss's lemma, that $\mathrm{LM}(X^\beta \tilde{f}) \neq \mathrm{LM}(X^\alpha \tilde{g})$. The desired result clearly follows.

2) This follows directly from 1) and Buchberger's Criterion (Theorem 25).

Exercise 112:

Suppose that $\mathrm{LM}(\frac{f}{d})$ and $\mathrm{LM}(\frac{g}{d})$ are relatively prime. By virtue of Exercise 111, $\{\frac{f}{d}, \frac{g}{d}\}$ is a Gröbner basis for $\langle \frac{f}{d}, \frac{g}{d} \rangle$, and, thus, by virtue of Exercise 109, $\{f,g\}$ is a Gröbner basis for $\langle f,g \rangle$.

For the converse, let us first suppose that $\gcd(f,g) = 1$ and that $\{f,g\}$ is a Gröbner basis for $\langle f,g \rangle$, and let us prove that $\mathrm{LM}(f)$ and $\mathrm{LM}(g)$ are relatively prime.

Write $\mathrm{LT}(f) = aDX^\alpha$ and $\mathrm{LT}(g) = bDX^\beta$, where $a,b \in \mathbf{K}^\times$, and D, X^α, X^β are monomials with $\gcd(X^\alpha, X^\beta) = 1$. We have $S(f,g) = \frac{X^\beta}{a}f - \frac{X^\alpha}{b}g$. As $\{f,g\}$ is a Gröbner basis for $\langle f,g \rangle$, we have $\overline{S(f,g)}^{\{f,g\}} = 0$, and hence there exist $u,v \in \mathbf{K}[X_1,\ldots,X_n]$ such that

$$S(f,g) = \frac{X^\beta}{a}f - \frac{X^\alpha}{b}g = uf + vg,$$

with $\mathrm{LM}(uf) \leq \mathrm{LM}(S(f,g))$ and $\mathrm{LM}(vg) \leq \mathrm{LM}(S(f,g))$. From the relation $(\frac{X^\alpha}{b} + v)g = (\frac{X^\beta}{a} - u)f$, we deduce that f divides $\frac{X^\alpha}{b} + v$ and g divides $\frac{X^\beta}{a} - u$ (since $\gcd(f,g) = 1$). Also, as

$$\mathrm{LM}(uf) = \mathrm{LM}(u)DX^\alpha \leq \mathrm{LM}(S(f,g)) < X^\alpha \mathrm{LM}(g) = X^\beta \mathrm{LM}(f) = DX^\alpha X^\beta,$$

we have $\mathrm{LM}(u) < X^\beta$, and hence $\mathrm{LM}(\frac{X^\beta}{a} - u) = X^\beta$. But, as g divides $\frac{X^\beta}{a} - u$, $\mathrm{LM}(g) = DX^\beta$ divides $\mathrm{LM}(\frac{X^\beta}{a} - u) = X^\beta$, and, thus, $D = 1$, that is, $\mathrm{LM}(f)$ and $\mathrm{LM}(g)$ are relatively prime.

Now, let us treat the general case where $d = \gcd(f,g)$ is not necessarily 1. As $\{f,g\}$ is a Gröbner basis for $\langle f,g \rangle$, $\{\frac{f}{d}, \frac{g}{d}\}$ is a Gröbner basis for $\langle \frac{f}{d}, \frac{g}{d} \rangle$, and therefore, by the case $d = 1$, $\mathrm{LM}(\frac{f}{d})$ and $\mathrm{LM}(\frac{g}{d})$ are relatively prime.

Exercise 113:

1) Assume that $\{f,g\}$ is a Gröbner basis for $\langle f,g \rangle$, and let $s \in \mathbf{K}^\times$. We will show that $\{f+s,g\}$ is not a Gröbner basis for $\langle f+s,g \rangle$. Denoting $d = \gcd(f,g)$, $\tilde{f} = \frac{f}{d}$, and $\tilde{g} = \frac{g}{d}$, we know by Exercise 112 that $\mathrm{LM}(\tilde{f})$ and $\mathrm{LM}(\tilde{g})$ are relatively prime. So, as $\mathrm{LM}(f) = \mathrm{LM}(\tilde{f})\,\mathrm{LM}(d)$ and $\mathrm{LM}(g) = \mathrm{LM}(\tilde{g})\,\mathrm{LM}(d)$, we infer that $\deg_{X_k}(d) = v_k$, and, thus, $\mathrm{LM}(\tilde{g})$ is free of X_k.

Let $d' = \gcd(f+s,g)$. Since $f+s$ and f are coprime, so are $f+s$ and d. So, d' divides $\tilde{g}$, and hence $\mathrm{LM}(d')$ is free of X_k. Writing $f+s = f'd'$ and $g = g'd'$, we have $\deg_{X_k}(\mathrm{LM}(f')) = \mu_k$ and $\deg_{X_k}(\mathrm{LM}(g')) = v_k$, and, thus, $\mathrm{LM}(f')$ and $\mathrm{LM}(g')$ are not relatively prime. Using Exercise 112, we deduce that $\{f+s,g\}$ is not a Gröbner basis for $\langle f+s,g \rangle$.

2) The result clearly holds if either f or g is in $\mathbf{K}$. So, we can suppose that both f and g are not constant.

"$\Leftarrow$" Supposing that $\mathrm{LM}(f)$ and $\mathrm{LM}(g)$ are relatively prime, then so are $\mathrm{LM}(f+s)$ and $\mathrm{LM}(g+t)$ for all $s,t \in \mathbf{K}$. By virtue of Exercise 112, we deduce that $\{f+s,g+t\}$ is a Gröbner basis for $\langle f+s,g+t \rangle$ for all $s,t \in \mathbf{K}$.

"$\Rightarrow$" We proceed by contraposition. Suppose that $\mathrm{LM}(f)$ and $\mathrm{LM}(g)$ are not relatively prime, that is, $\mu_k \geq v_k > 0$ for some k, where $\mathrm{LM}(f) = X_1^{\mu_1} \cdots X_n^{\mu_1}$ and $\mathrm{LM}(g) = X_1^{v_1} \cdots X_n^{v_1}$. If $\{f,g\}$ is not a Gröbner basis for $\langle f,g \rangle$ then we are done. Otherwise, $\{f+1,g\}$ is not a Gröbner basis for $\langle f+1,g \rangle$ by virtue of 1).

Exercise 114:

This is an immediate consequence of Theorem 37 since $S(a_i M_i, a_j M_j) = 0$ for all $1 \leq i \leq j \leq m$.

Exercise 115:

1) The trick is to introduce $m-3$ new variables z_i for each equation $a_1 x^{\alpha_1} + \cdots + a_m x^{\alpha_m} = 0$ ($a_j \in \mathbf{K}$, $\alpha_j \in \mathbb{N}^n$) and replace this equation by the system of $m-2$ new equations $z_1 + a_1 x^{\alpha_1} + a_2 x^{\alpha_2} = -z_1 + z_2 + a_3 x^{\alpha_3} = -z_2 + z_3 + a_4 x^{\alpha_4} = \cdots = -z_{m-4} + z_{m-3} + a_{m-2} x^{\alpha_{m-2}} = -z_{m-3} + a_{m-1} x^{\alpha_{m-1}} + a_m x^{\alpha_m} = 0$.

2.a) If we start with a binomial finite generating set for I, then the new Gröbner basis elements produced by a step in Buchberger's algorithm are binomials.

2.b) Each step of the division algorithm modulo a set of binomials takes a term to another term.

3) This follows from 2.a) and the uniqueness of the reduced Gröbner basis.

4) The intersection is generated by a subset of the reduced Gröbner basis of I with respect to the lexicographic monomial order.

5) Suppose first that $s = 1$. In the larger polynomial ring $\mathbf{K}[X_1, \ldots, X_n, t]$, consider the binomial ideal $L = I + tI' + (1-t)J_1$. The claim follows from 4) and the formula

$(I+I')\cap(I+J_1) = L\cap \mathbf{K}[X_1,\ldots,X_n]$. For the general case, induct on s.

Exercise 116:

1) This a direct consequence of the fact that if a linear system with m unknowns over $\mathbf{K}$ is compatible when seen as a linear system over a bigger field then its solutions are in fact in $\mathbf{K}^m$.

2) This follows immediately from Buchberger's Criterion (Theorem 25) since even if we see the elements of G as elements of the ring $\bar{\mathbf{K}}[X_1,\ldots,X_n]$, all the computations ($S$-polynomials, remainders,..) remain in $\mathbf{K}[X_1,\ldots,X_n]$.

Another wording: When executing Buchberger's algorithm on elements of $\mathbf{K}[X_1,\ldots,X_n]$, all the computations ($S$-polynomials, remainders, ...) remain in $\mathbf{K}[X_1,\ldots,X_n]$.

Exercise 117:

1) Let $x_0,\ldots,x_\ell \in \mathbf{K}[X_1,\ldots,X_\ell]$, and $Q(x_0,\ldots,x_\ell) = 0$ be an algebraic relation over $\mathbf{K}$ testifying the dependence between the x_i. Let us order the monomial of Q with nonzero coefficients by the lexicographic order. We can without loss of generality suppose that the first nonzero coefficient of Q is 1. Denoting this monomial by $x_0^{m_0}\cdots x_\ell^{m_\ell}$, it is clear that Q can be written in the form

$$Q=x_0^{m_0}\cdots x_\ell^{m_\ell}+x_0^{m_0}\cdots x_\ell^{1+m_\ell}R_\ell+x_0^{m_0}\cdots x_{\ell-1}^{1+m_{\ell-1}}R_{\ell-1}+\cdots+x_0^{m_0}x_1^{1+m_1}R_1+x_0^{1+m_0}R_0,$$

the desired collapse.

2) Let $a, b \in \mathbb{Z}\setminus\{-1,0,1\}$. Computing successively

$$d_1 = \gcd(a,b),\ d_2 = \gcd\left(\frac{a}{d_1},b\right),\ldots,d_n = \gcd\left(\frac{a}{d_1\cdots d_{n-1}},b\right),$$

we eventually factorize a as

$$a = d_1\cdots d_n a' \text{ with } d_i, a' \in \mathbb{Z},\ d_i\mid b,\text{ and } \gcd(a',b) = 1.$$

Writing a Bézout identity $ca' + db = 1$ for some $c, d \in \mathbb{Z}$, we have

$$b^n(1-db) \in \langle a\rangle,$$

the desired collapse.

Taking $a = 700$ and $b = 6$, we have $d_1 = \gcd(700,6) = 2$, $d_2 = \gcd(350,6) = 2$, with $\gcd(175,6) = 1$, and $175 - 29\times 6 = 1$ as a Bézout identity. We infer that

$$6^2(1+29\times 6) \in \langle 700\rangle.$$

3) Let $\mathbf{R}$ be a finite ring ($\mathbb{Z}/m\mathbb{Z}$ for example). Denoting by $k = \sharp(\mathbf{R})$ ($k \geq 2$; we suppose that $\mathbf{R}$ is not trivial) and considering $x \in \mathbf{R}$, necessarily there exist $0 \leq r < r' \leq k$ such that $x^{r'} = x^r$, the desired collapse.

4) Suppose that $\mathbf{R}$ is a valuation domain (in particular, it is local) with valuation v and valuation group G. Consider $a, b \in \text{Rad}(\mathbf{R}) \setminus \{0\}$. Since $1 + xb \in \mathbf{R}^\times$ and a^n is regular, the desired collapse becomes

$$\forall a, b \in \text{Rad}(\mathbf{R}) \setminus \{0\}, \ \exists n \in \mathbb{N} \text{ such that } a \mid b^n,$$

or also, in other terms,

$$\forall a, b \in \mathbf{R} \text{ with } v(a), v(b) > 0, \ \exists n \in \mathbb{N} \text{ such that } n v(b) \geq v(a),$$

and this is exactly Archimedeanity.

Exercise 118:

1) For example, for $\alpha \notin \mathbf{R}^\times$ and $\alpha \neq 0$, $s = (-\alpha, 1) \in \text{Syz}(\alpha, \alpha^2)$ but $s \notin \langle s_{1,2} \rangle = \mathbf{R}(\alpha^2, -\alpha)$.

2) It suffices to prove that $u_m \text{Syz}(u) \subseteq \langle s_{i,j} \mid 1 \leq i < j \leq m \rangle$. Letting $v = (v_1, \ldots, v_m) \in \text{Syz}(u)$, as $v_m u_m = -(v_1 u_1 + \cdots + v_{m-1} u_{m-1})$, we have

$$u_m v = v_1 s_{1,m} + \cdots + v_{m-1} s_{m-1,m}.$$

3) Denoting by $d = \gcd(u_1 \ldots, u_m)$, $\text{Syz}(u)$ is generated as $\mathbf{R}$-module by the obvious syzygies of $\frac{1}{d} u$. This follows from 2) since $\text{Syz}(u) = \text{Syz}(\frac{1}{d} u)$ and $\frac{1}{d} u$ is unimodular.

Exercise 119:

1) Letting $c \in \langle 1 + aX \rangle \cap \mathbf{R}$, there exists $g = \sum_{i=0}^{m} b_i X^i \in \mathbf{R}[X]$ such that

$$(1 + aX) g = c \in \mathbf{R}.$$

By identification, we have $a b_m = 0$, $b_m + a b_{m-1} = 0, \ldots, b_1 + a b_0 = 0$, $b_0 = c$, and thus $b_k = (-a)^k c \ \forall \, 0 \leq k \leq m$ and $a^{m+1} c = 0$.

Conversely, letting $b \in \text{Ann}(a^\infty)$, there exists $n \in \mathbb{N}$ such that $b a^n = 0$. It follows that

$$b(1 + aX)(1 - aX + \cdots + (-a)^{n-1} X^{n-1}) = b(1 - (-a)^n X^n) = b,$$

and, thus, $b \in \langle 1 + aX \rangle \cap \mathbf{R}$. We conclude that $\langle 1 + aX \rangle \cap \mathbf{R} = \text{Ann}(a^\infty)$ and necessarily $\text{Ann}(a^\infty)[X] + \langle aX \rangle \subseteq \text{LT}(\langle 1 + aX \rangle)$.

Letting $f = c_0 + c_1 X + \cdots + c_n X^n \in \langle 1 + aX \rangle$ (we suppose that $n \geq 1$), there exists $g = \sum_{i=0}^{m} b_i X^i \in \mathbf{R}[X]$ $(m + 1 \geq n)$ such that

$$(1 + aX) g = f.$$

By identification, we have

$$S: \begin{cases} ab_m = 0 \\ b_m + ab_{m-1} = 0 \\ \vdots \\ b_{n+1} + ab_n = 0 \\ b_n + ab_{n-1} = c_n \\ \vdots \\ b_1 + ab_0 = c_1 \\ b_0 = c_0, \end{cases}$$

and, thus, $b_n = c_n - ac_{n-1} + \cdots + (-a)^n c_0$ and $a^{m-n+1} b_n = 0$. It follows that $b_n \in \text{Ann}(a^\infty)$ and $c_n \in \text{Ann}(a^\infty) + \langle a \rangle$, as desired.

The final particular affirmation easily follows by adapting the second members in the equalities of S.

2) $\mathbf{K}[X_n : n \geq 0]$ is Gröbner because a finitely-generated ideal of $\mathbf{K}[X_n : n \geq 0][Y_1, \ldots, Y_m]$ involves in its generators only a finite number of indeterminates among the X_i's, and we know that $\mathbf{K}$ is Gröbner. The ring $\mathbf{K}[X_n : n \geq 0]$ is clearly non-Noetherian since the nondecreasing sequence $(\langle X_0, \ldots, X_n \rangle)_{n \geq 0}$ does not pause.

3) Clearly $\text{Ann}(\bar{X}_0^\infty) = \langle \bar{X}_k : k \geq 1 \rangle$, which is not a finitely-generated ideal of $\mathbf{R}$. It follows, by virtue of 1), that $\mathbf{R}$ is not Gröbner.

Exercise 120:

Since $S(g_1, g_2) = (\frac{b}{a})g_1 - Xg_2 = \frac{b}{a}$ and $\frac{b}{a}$ is not divisible by b, then one must add $g_3 = \frac{b}{a}$ when executing Buchberger's Algorithm 38.

In the same way, $S(g_1, g_3) = (\frac{b}{a^2})g_1 - Xg_3 = \frac{b}{a^2}$ and $\frac{b}{a^2}$ is not divisible by b nor by $\frac{b}{a}$. Thus, one must add $g_4 = \frac{b}{a^2}$, and so on, we observe that Buchberger's Algorithm 38 does not terminate.

Exercise 121:

Let $c \in \text{Ann}(a)$. We have $cY = c(Y + aX) \in \langle Y + aX \rangle$. Thus, $cY \in \langle \text{LT}(g_1), \ldots, \text{LT}(g_s) \rangle$ and, obviously, it belongs to the ideal generated by $\{b \in \mathbf{R} \mid bY \in G\}$.

Conversely, let $b \in \mathbf{R}$ such that $bY \in G$. There exists $h \in \mathbf{R}[X, Y]$ such that $bY = (Y + aX)h$. Necessarily, h has the form $h = b + Yu(Y) + Xv(X) + XYw(X, Y)$. It follows that $abX + aX^2 v(X) \equiv 0 \bmod Y$, and hence $ab = 0$.

Exercise 122:

Write $(f_1, \ldots, f_k) = \sum_{i=1}^{r} X^{\alpha_i}(a_{i,1} \ldots, a_{i,k})$, where the α_i's are pairwise different elements in $\mathbb{N}^n$, $(a_{i,1} \ldots, a_{i,k}) \in \mathbf{R}^k$, and $r \geq 1$. Clearly, we have

$$S \cap \mathbf{R}^k = \cap_{i=1}^{r} \text{Syz}_{\mathbf{R}}(a_{i,1} \ldots, a_{i,k}).$$

So, if $\mathbf{R}$ is coherent, we have a finite generating set for $S \cap \mathbf{R}^k$.

Exercise 123:

1) It is clear that the ring $\mathbf{A}$ shares the ideal $\mathfrak{m} = t\mathbf{K}[t]$ with $\mathbf{K}[t]$.

The fact that $\mathbf{A}$ has Krull dimension 1 follows "classically" from the fact that the prime spectrum of $\mathbf{A}$ is $\{(f\mathbf{K}[t]) \cap \mathbf{A} \mid f \in \mathbf{K}[t]$ and f irreducible$\}$. Let us now give a constructive proof (as in Exercise 117).

Let $f, g \in \mathbf{A}$. We want to find $h, h' \in \mathbf{A}$ such that $f^n(1 - hf) + gh' = 0$ for some $n \in \mathbb{N}$. Two cases may arise:

Case 1: $f(0) \neq 0$. As $\mathbf{K}[t]$ has Krull dimension one, there exists $n \in \mathbb{N}$ such that $f^n(1 - \varphi f) + (tg)\psi = 0$, with $\varphi, \psi \in \mathbf{K}[t]$. Putting $t = 0$, we obtain $\varphi(0)f(0) = 1$, and, thus, $\varphi(0) \in \mathbb{Q}$ and $\varphi \in \mathbf{A}$. It suffices to take $h = \varphi$ and $h' = t\psi$.

Case 2: $f(0) = 0$. Set $f = t^p f_1$, $g = t^q g_1$, with $p > 0$, $q \geq 0$, $f_1(0) \neq 0$, and $g_1(0) \neq 0$. As $\mathbf{K}[t]$ has Krull dimension one, there exists $n \in \mathbb{N}$ such that $(tf)^n(1 - \varphi tf) + g_1\psi = 0$, with $\varphi, \psi \in \mathbf{K}[t]$. We have $t^{n+np}f_1^n(1 - \varphi tf) = -g_1\psi$. As t^{n+np} does not divide g_1, it divides ψ, that is, there exists $\eta \in \mathbf{K}[t]$ such that $\psi = t^{n+np}\eta$. It follows that $f_1^n(1 - \varphi tf) = -g_1\eta$. Let N be a positive integer greater than n and q. Multiplying the equality $f_1^n(1 - \varphi tf) = -g_1\eta$ by $t^{pN}f_1^{N-n}$, we obtain $t^{pN}f_1^N(1 - t\varphi f) = -(t^q g_1)(t^{pN-q}f_1^{N-n}\eta)$, or also, $f^N(1 - t\varphi f) + g(t^{pN-q}f_1^{N-n}\eta) = 0$. It suffices to take $h = t\varphi$ and $h' = t^{pN-q}f_1^{N-n}\eta$.

2) Take $\alpha \in \mathbf{K}$, and suppose that $\alpha t \in th_1\mathbf{A} + \cdots + th_r\mathbf{A} \subseteq \mathfrak{m}$, with $h_1, \ldots, h_r \in \mathbf{K}[t]$. Since the only terms of degree 1 at t in $th_1\mathbf{A} + \cdots + th_r\mathbf{A}$ are of the form $(h_1(0)q_1 + \cdots + h_r(0)q_r)t$ with $q_1, \ldots, q_r \in \mathbb{Q}$, one gets $\alpha = h_1(0)q_1 + \cdots + h_r(0)q_r \in \mathbb{Q} \cdot h_1(0) + \cdots + \mathbb{Q} \cdot h_r(0)$. As $\mathbf{K}$ is not finitely-generated as $\mathbb{Q}$-vector space (because u is transcendental over $\mathbb{Q}$), we infer that $\mathfrak{m}$ is not finitely-generated as an ideal of $\mathbf{A}$.

3) First, $t^2\mathbf{K}[t] = t(t\mathbf{K}[t]) = \sqrt{2}t(t\mathbf{K}[t]) \subseteq \langle t \rangle \cap \langle \sqrt{2}t \rangle$. Second, if $x \in \langle t \rangle \cap \langle \sqrt{2}t \rangle$, then

$$x = t(q_1 + tf(t)) = \sqrt{2}t(q_2 + tg(t)),$$

where $q_1, q_2 \in \mathbb{Q}$ and $f, g \in \mathbf{K}[t]$. This implies that $q_1 - \sqrt{2}q_2 = t(f(t) - g(t))$, and thus $q_1 = q_2 = 0$ (as $\sqrt{2} \notin \mathbb{Q}$), and $x = t^2 f(t) \in t^2\mathbf{K}[t]$.

4) By 3), $\langle t \rangle \cap \langle \sqrt{2}t \rangle$ is isomorphic as $\mathbf{A}$-module to $\mathfrak{m}$, and, thus, it is not finitely-generated by virtue of 2).

5) If $x = ta = \sqrt{2}tb \in \langle t \rangle \cap \langle \sqrt{2}t \rangle$, where $a, b \in \mathbf{A}$, then $x = xX + x(1 - X) = ta(1 - X) + \sqrt{2}tbX \in I \cap \mathbf{A}$. Conversely, if $y = \sqrt{2}tXU(X) + t(1 - X)V(X) \in I \cap \mathbf{A}$, for some $U, V \in \mathbf{A}[X]$, then by successively taking $X = 0$ and $X = 1$, one gets $y \in \langle t \rangle \cap \langle \sqrt{2}t \rangle$.

6) LT(I) is not finitely-generated since so is $I \cap \mathbf{A}$.

Chapter 2

Varieties, Ideals, and Gröbner bases

The richness of algebraic geometry as a mathematical discipline comes from the interplay of algebra and geometry, as its basic objects are both geometrical and algebraic. The strong intuition of geometry translates into manipulation of algebraic objects via symbolic computation. In this chapter, we are going to study this correspondence between ideals and varieties. In Section 2.1, we will prove the celebrated Hilbert's Nullstellensatz that translates (in its weak form) when an algebraic variety is empty in algebraic terms. This is the cornerstone to the algebra-geometry dictionary we will present translating any statement about varieties into a statement about ideals (and vice versa). This correspondence being established, Gröbner bases come to perform the necessary algebraic computations.

2.1 The Ideal–Variety Correspondence

The main reference for this section is [14]. Throughout, let $\mathbf{K}$ be a field and let $\mathbb{A}^n(\mathbf{K}) = \mathbf{K}^n = \{(a_1,\ldots,a_n) \in \mathbf{K}^n \mid a_i \in \mathbf{K}\}$ be the affine space of dimension $n \geq 1$.

The proof of the Hilbert's Nullstellensatz relies heavily on yet another celebrated theorem, namely Noether normalization lemma. It is an important result of commutative algebra, introduced by Emmy Noether in 1926. It says that every finitely-generated $\mathbf{K}$-algebra over a field $\mathbf{K}$ is isomorphic with a module-finite extension of a polynomial ring $\mathbf{K}[z_1,\ldots,z_d]$. Hereafter we follow the same approach as in [34] when visiting Noether normalization lemma.

Lemma 124. (Change of variables "à la Nagata")

Let $\mathbf{K}$ be a field, $f \in \mathbf{K}[X_1,\ldots,X_n] \setminus \mathbf{K}$, and N an integer bounding all the exponents of the variables occurring in the terms of f. Let ϕ be the $\mathbf{K}$-automorphism *of* $\mathbf{K}[X_1,\ldots,X_n]$ *mapping* X_i *to* $X_i + X_n^{N^i}$ *for* $i < n$ *and mapping* X_n *to itself. Then* $\phi(f)$ *is, up to a multiplication by a nonzero scalar of the field, monic when seen as a polynomial at* X_n *with coefficients in* $\mathbf{K}[X_1,\ldots,X_{n-1}]$.

Proof. Let $\alpha = (\alpha_1,\ldots,\alpha_n) \in \mathbb{N}^n$ be an n-tuple of exponents occurring in a nonzero term $T_\alpha = c_\alpha X_1^{\alpha_1} \cdots X_n^{\alpha_n}$ of f. The polynomial

$$c_\alpha^{-1}\phi(T_\alpha) = (X_1 + X_n^N)^{\alpha_1} \cdots (X_{n-1} + X_n^{N^{n-1}})^{\alpha_{n-1}} X_n^{\alpha_n}$$

is monic of degree $d_\alpha = \alpha_n + \alpha_1 N + \cdots + \alpha_{n-1} N^{n-1}$ when seen as a polynomial at X_n with coefficients in $\mathbf{K}[X_1,\ldots,X_{n-1}]$. By uniqueness of representation in basis N, the d_α's are pairwise distinct, and the desired result follows.

Theorem 125. (Noether normalization lemma)

Every finitely-generated $\mathbf{K}$-algebra over a field $\mathbf{K}$ is isomorphic with a module-finite extension of a polynomial ring $\mathbf{K}[z_1,\ldots,z_d]$ (d may be zero).

Proof. Let $\mathbf{R} = \mathbb{Z}[\theta_1,\ldots,\theta_n]$ be a finitely-generated $\mathbf{K}$-algebra. We proceed by induction on n. If $n = 0$ then $\mathbf{R} = \mathbf{K}$. Now suppose $n \geq 1$ and that we know the result for $\mathbf{K}$-algebras generated by $n-1$ or fewer elements. If the θ_i's are algebraically independent over $\mathbf{K}$ then we are done: we may take $d = n$, and $z_i = \theta_i$ for $1 \leq i \leq n$. Therefore we may assume that there is $f \in \mathbf{K}[X_1,\ldots,X_n] \setminus \{0\}$ such that $f(\theta_1,\ldots,\theta_n) = 0$. By a change of variables "à la Nagata" (see Lemma 124), we can suppose that I contains a monic polynomial at the variable X_n. Note that this change of variables amounts to choosing new generators $\theta_1',\ldots,\theta_n' \in \mathbf{R}$ of $\mathbf{R}$ as $\mathbf{K}$-algebra with

$$\theta_1' = \theta_1 - \theta_n^N, \theta_2' = \theta_2 - \theta_n^{N^2},\ldots,\theta_{n-1}' = \theta_{n-1} - \theta_n^{N^{n-1}}, \theta_n' = \theta_n,$$

for some suitable $N \in \mathbb{N}$. As θ_n' is integral over $\mathbf{T} := \mathbf{K}[\theta_1',\ldots,\theta_{n-1}']$, $\mathbf{R} = \mathbf{K}[\theta_1',\ldots,\theta_n']$ is module-finite over $\mathbf{T}$. As $\mathbf{T}$ has $n-1$ generators over $\mathbf{K}$, the desired result follows by the induction hypothesis.

In their paper [1], Abhyankar and Kravitz offered a correction to *Commutative Algebra* [69] by Zariski and Samuel by constructing two counterexamples for two erroneous theorems concerning extensions of Noether normalization lemma to integral domains. More specifically, Zariski and Samuel asserted the following:

Claim ZS: Let $\mathbf{A} = \mathbf{R}[x_1,\ldots,x_n]$ be an integral domain, finitely-generated over an infinite domain $\mathbf{R}$, and let d be the transcendence degree of the field of quotients of $\mathbf{A}$ over the field of quotients $\mathbf{K}$ of $\mathbf{R}$. Then there exist d linear combinations $y_1,\ldots,y_d$ of the x_i's with coefficients in $\mathbf{R}$, such that $\mathbf{A}$ is integral over $\mathbf{R}[y_1,\ldots,y_d]$.

As stated in [1, Example 2.1], the claim ZS is not true. To see this, it suffices to take $\mathbf{R} = \mathbb{Z}$, and $\mathbf{A} = \mathbb{Z}[\frac{1}{2}]$. Note also that Noether normalization theorem over an integral domain $\mathbf{A}$ given by Hochster in [34] is nothing but Noether normalization theorem over the quotient field $\mathbf{K}$ of $\mathbf{A}$ (it is immediate that, when normalizing in $\mathbf{K}$, we will make use of a finite number of denominators $c_1, \ldots, c_m \in \mathbf{A} \setminus \{0\}$ and, thus, we obtain a normalization in $\mathbf{A}[\frac{1}{c}]$ where $c = c_1 \cdots c_m$).

Hereafter a generalization of Noether Normalization theorem to $\mathbb{Z}$ (note that $\mathbb{Z}$ can be replaced with any Bézout domain of Krull dimension one). In the theorem below, the domain $\mathbf{R}$ is supposed to have characteristic 0 as in case it has characteristic p (p a prime number), it is finitely-generated over $\mathbb{F}_p$ and the classical Noether Normalization theorem over fields applies (Theorem 125).

Theorem 126. (Noether normalization lemma over the integers [26, 27])

Let $\mathbf{R} = \mathbb{Z}[\theta_1, \ldots, \theta_n]$ be a domain of characteristic 0, finitely-generated over $\mathbb{Z}$, and let d be the Krull dimension of $\mathbf{R}$. Then, there exist an integer $\delta \in [\![1, m]\!]$, a subring $\mathbf{A}$ of $\mathbf{R}$ which is isomorphic to $\mathbb{Z}[X_1, \ldots, X_{d-\delta}]/\langle h \rangle$ for some nonconstant irreducible polynomial $h \in \mathbb{Z}[X_1, \ldots, X_{d-\delta}]$, and $z_1, \ldots, z_\delta \in \mathbf{R}$ algebraically independent over $\mathbf{A}$, such that $\mathbf{R}$ is module-finite over $\mathbf{A}[z_1, \ldots, z_\delta]$.

Recall that if $\mathbf{K} \subseteq \mathbf{Ł}$ are fields, the fact that $\mathbf{Ł}$ is module-finite over $\mathbf{K}$ is the same as $\mathbf{Ł}$ is a finite-dimensional vector space over $\mathbf{K}$, and this is equivalent to that $\mathbf{Ł}$ is a finite algebraic extension of $\mathbf{K}$.

Corollary 127. (Zariski's lemma)

Let $\mathbf{R}$ be a finitely-generated algebra over a field $\mathbf{K}$, and suppose that $\mathbf{R}$ is a field. Then $\mathbf{R}$ is a finite algebraic extension of $\mathbf{K}$.

Proof. Use Theorem 125 (d must be zero since the Krull dimension of $\mathbf{K}[z_1, \ldots, z_d]$ is d).

Corollary 128. *Let $\mathbf{K}$ be an algebraically closed field, $\mathbf{R}$ a finitely-generated $\mathbf{K}$-algebra, and $\mathfrak{m}$ a maximal ideal of $\mathbf{R}$. Then the composite map $\mathbf{K} \to \mathbf{R} \twoheadrightarrow \mathbf{R}/\mathfrak{m}$ is an isomorphism.*

Proof. The residue field $\mathbf{R}/\mathfrak{m}$ is a finitely-generated $\mathbf{K}$-algebra, and hence $\mathbf{K} \to \mathbf{R}/\mathfrak{m}$ gives an algebraic extension of $\mathbf{K}$. This latter is an isomorphism since $\mathbf{K}$ is algebraically closed.

Definition 129. Let $f_1, \ldots, f_s \in \mathbf{K}[X_1, \ldots, X_n]$. We denote by

$$\langle f_1, \ldots, f_s \rangle = \left\{ \sum_{i=1}^{s} h_i f_i \mid h_i \in \mathbf{K}[X_1, \ldots, X_n] \right\}$$

the ideal of $\mathbf{K}[X_1, \ldots, X_n]$ generated by $f_1, \ldots, f_s$. We set

$$V(f_1,\ldots,f_s) := \{(a_1,\ldots,a_n) \in \mathbb{A}^n(\mathbf{K}) \mid f_i(a_1,\ldots,a_n) = 0 \text{ for all } 1 \leq i \leq s\}$$
$$= \{(a_1,\ldots,a_n) \in \mathbb{A}^n(\mathbf{K}) \mid f(a_1,\ldots,a_n) = 0 \text{ for all } f \in \langle f_1,\ldots,f_s \rangle\}.$$

We call it the algebraic variety defined by $f_1,\ldots,f_s$ or by $\langle f_1,\ldots,f_s \rangle$. The algebraic variety $V(f_1,\ldots,f_s)$ will be denoted also by $V(\langle f_1,\ldots,f_s \rangle)$. It is the set of all solutions of the system of polynomial equations $f_1 = 0,\ldots,f_s = 0$.

Example 130.

1) $\{(1,1),(-1,-1)\} \cup \{(0,\lambda),\ \lambda \in \mathbb{C}\}$ is an algebraic variety which is defined by the polynomials $f_1 = X_1(X_1^2 - 1)$ and $f_2 = X_1(X_1 - X_2)$.

2) $V(1) = \emptyset$ and $V(0) = \mathbb{A}^n(\mathbf{K})$.

3) Consider the affine plane $\mathbb{A}^2(\mathbf{K})$. The set of all points on which a nonzero polynomial with two variables vanishes is an affine algebraic set, and called a *plane curve*. The degree of a curve is determined by the degree of the polynomial. Lines are algebraic sets of degree 1, conics are of degree 2, cubics are of degree 3 (see Figs. 2.1 and 2.2), and so on.

4) The set of four points $\{(-2,-1),(-1,1),(1,-1),(1,2)\}$ in $\mathbb{A}^2(\mathbb{R})$ is the variety $V(X^2 + Y^2 - XY - 3, 3X^2 - Y^2 - XY + 2X + 2Y - 3)$ corresponding to the intersection of the ellipse $V(X^2 + Y^2 - XY - 3)$ and the hyperbola $V(3X^2 - Y^2 - XY + 2X + 2Y - 3)$, see Fig. 2.3.

5) If $a = (a_1,\ldots,a_n) \in \mathbb{A}^n(\mathbf{K})$, then denoting by $\mathscr{I}_a = \langle X_1 - a_1,\ldots,X_n - a_n \rangle$, we have $V(X_1 - a_1,\ldots,X_n - a_n) = V(\mathscr{I}_a) = \{a\}$.

6) In an affine space, every finite set of points $\{a^j,\ 1 \leq j \leq m\}$ is an affine algebraic set $V(\cap_{j=1}^m \mathscr{I}_{a^j}) = V(\prod_{j=1}^m \mathscr{I}_{a^j})$.

The correspondence between varieties and ideals is of central importance. The following result is immediate.

Proposition 131. *Let V be a subset of $\mathbb{A}^n(\mathbf{K})$. Then*

$$\mathscr{I}(V) := \{f \in \mathbf{K}[X_1,\ldots,X_n] \mid f(a_1,\ldots,a_n) = 0 \ \forall\ (a_1,\ldots,a_n) \in V\}$$

is an ideal of $\mathbf{K}[X_1,\ldots,X_n]$.

Hilbert's Nullstellensatz (German for "zero-locus-theorem") is a celebrated theorem discovered by David Hilbert establishing a fundamental relationship between geometry and algebra and, thus, laying the foundations of algebraic geometry in twentieth century mathematics. Hereafter four versions of it.

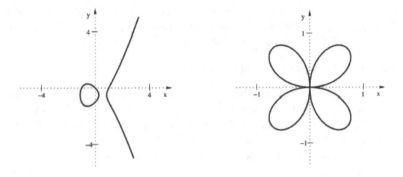

Figure 2.1: $y^2 + y - x^3 + x = 0$ Figure 2.2: $(x^2 + y^2)^3 - 4x^2y^2 = 0$

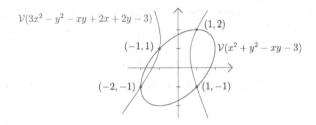

Figure 2.3: The intersection of an ellipse and a hyperbola

Theorem 132. (Hilbert's Nullstellensatz, weak form)

Let $\mathbf{R} = \mathbf{K}[X_1, \ldots, X_n]$ be a polynomial ring over an algebraically closed field $\mathbf{K}$. Then every maximal ideal $\mathfrak{m}$ of $\mathbf{R}$ is the kernel of a $\mathbf{K}$-homomorphism $\mathbf{K}[X_1, \ldots, X_n] \to \mathbf{K}$, and so is determined by the elements $\lambda_1, \ldots, \lambda_n \in \mathbf{K}$ to which $X_1, \ldots, X_n$ map. This maximal ideal is the kernel of the evaluation map $f(X_1, \ldots, X_n) \mapsto f(\lambda_1, \ldots, \lambda_n)$. It may also be described as the ideal $\langle X_1 - \lambda_1, \ldots, X_n - \lambda_n \rangle$ of $\mathbf{R}$.

Proof. Since $\gamma \colon \mathbf{K} \cong \mathbf{R}/\mathfrak{m}$ (by Corollary 128), the $\mathbf{K}$-algebra map $\mathbf{R} \to \mathbf{R}/\mathfrak{m}$ composed with γ^{-1} gives a map $\mathbf{R} \twoheadrightarrow \mathbf{K}$ whose kernel is $\mathfrak{m}$.

Theorem 133. (Hilbert's Nullstellensatz, alternate weak form)

Let $\mathbf{K}$ be an algebraically closed field, and let $f_1, \ldots, f_s \in \mathbf{K}[X_1, \ldots, X_n]$. The algebraic variety $V(f_1, \ldots, f_s)$ is empty if and only if there are $g_1, \ldots, g_s \in \mathbf{K}[X_1, \ldots, X_n]$ such that

$$1 = g_1 f_1 + \cdots + g_s f_s,$$

in other words, we have $V(f_1, \ldots, f_s) = \emptyset$ if and only if $1 \in \langle f_1, \ldots, f_s \rangle$.

Proof. If $1 \in \langle f_1, \ldots, f_s \rangle$ then clearly $V(f_1, \ldots, f_s) = \emptyset$. If $1 \notin \langle f_1, \ldots, f_s \rangle$, then the ideal $\langle f_1, \ldots, f_s \rangle$ is contained in an maximal ideal $\langle X_1 - \lambda_1, \ldots, X_n - \lambda_n \rangle$ (by Theorem 132) of $\mathbf{K}[X_1, \ldots, X_n]$, and, thus, $(\lambda_1, \ldots, \lambda_n) \in V(f_1, \ldots, f_s) \neq \emptyset$.

Theorem 134. (Hilbert's Nullstellensatz)

Let $I = \langle f_1, \ldots, f_s \rangle$ be an ideal of $\mathbf{K}[X_1, \ldots, X_n]$ and $f \in \mathbf{K}[X_1, \ldots, X_n]$, where $\mathbf{K}$ is an algebraically closed field. Then f vanishes on every point of $V(I)$ if and only if there is $k \in \mathbb{N}$ with $f^k \in I$. In other words, f vanishes on $V(I)$ if and only if

$$f \in \sqrt{I} := \{ g \in \mathbf{K}[X_1, \ldots, X_n] \mid \exists\, k \in \mathbb{N}\, (g^k \in I) \}$$

(this is an ideal of $\mathbf{K}[X_1, \ldots, X_n]$ containing I called the radical *of I).*

Proof. "$\Leftarrow$" Immediate.
"$\Rightarrow$" Let U be a new variable. The affine variety defined by $f_1, \ldots, f_s, Uf - 1$ is empty, hence, by Theorem 133, there are $p_1, \ldots, p_{s+1} \in \mathbf{K}[X_1, \ldots, X_n, U]$ such that

$$1 = p_1 f_1 + \cdots + p_s f_s + p_{s+1}(Uf - 1).$$

Thus, by setting $U = \frac{1}{f}$ and multiplying with a suitable power of f, we obtain $f \in \sqrt{I}$.

Theorem 134 can be reformulated as follows:

Corollary 135. (Hilbert's Nullstellensatz)

Let $I = \langle f_1, \ldots, f_s \rangle$ be an ideal of $\mathbf{K}[X_1, \ldots, X_n]$, where $\mathbf{K}$ is an algebraically closed field. Then

$$\mathscr{I}(V(I)) = \sqrt{I}.$$

Note that we always have $\mathscr{I}(V(I)) \supseteq \sqrt{I}$ but this containment may be strict: considering the ideal $I = \langle X^2 + 1 \rangle$ of $\mathbb{R}[X]$, we have $\mathscr{I}(V(I)) = \mathscr{I}(\emptyset) = \mathbb{R}[X] \supsetneq \sqrt{I} = \langle X^2 + 1 \rangle$.

The Ideal–Variety Correspondence 136.

(i) $\mathbf{K}$ *field: affine varieties $\xrightarrow{\mathscr{I}}$ ideals* $\quad v_1 \subseteq v_2 \Rightarrow \mathscr{I}(v_1) \supseteq \mathscr{I}(v_2)$,

$\qquad$ *ideals $\xrightarrow{V}$ affine varieties* $\quad I_1 \subseteq I_2 \Rightarrow V(I_1) \supseteq V(I_2)$.
Moreover, for every variety v, we have $V(\mathscr{I}(v)) = v$.

(ii) *If $\mathbf{K}$ is algebraically closed, then the following assignments are inverse to each other:*

$\qquad$ *affine varieties $\xrightarrow{\mathscr{I}}$ radical ideals (i.e., $I = \sqrt{I}$),*

$\qquad$ *radical ideals $\xrightarrow{V}$ affine varieties.*

Example 137.

1) Let **K** be a field containing the rationals $\mathbb{Q}$. If $f \in \mathbf{K}[X_1, \ldots, X_n] \setminus \{0\}$, then

$$\sqrt{\langle f \rangle} = \left\langle \frac{f}{\mathrm{pgcd}(f, \frac{\partial f}{\partial X_1}, \ldots, \frac{\partial f}{\partial X_n})} \right\rangle$$

(generated by the square-free part of f; recall that a polynomial is said to be *square-free* if it is not divisible by the square of a nonconstant polynomial).

2) $\sqrt{\langle X^m, Y^r \rangle} = \langle X, Y \rangle$.

3) Every prime ideal is a radical ideal.

The Ideal–Variety Correspondence 138. *If I and J are ideals of* $\mathbf{K}[X_1, \ldots, X_n]$, *then*

$$V(I + J) = V(I) \cap V(J),$$

$$V(I \cap J) = V(I.J) = V(I) \cup V(J).$$

Every intersection of affine varietes of $\mathbb{A}^n(\mathbf{K})$ is a variety again, and so are finite unions of affine varieties. This allows to define a topology on $\mathbb{A}^n(\mathbf{K})$, the *Zariski topology*, the closed subsets for which are the affine varieties. This topology is rather different from the usual ones; in particular, it is not separated, i.e., a Hausdorff space. Recall that a topological space is said to be a *Hausdorff space*, if any two distinct points can be separated by disjoint open sets. This is equivalent to the diagonal $\{(x,x) \mid x \in X\}$ being a closed subset of the product space $X \times X$. The reason for an affine space X to be not separated with regard to the Zariski topology is the fact that the Zariski topology on $X \times X$ does not coincide with the product of the Zariski topologies on X. Worse: if **K** is infinite, then any two nonempty open subsets meet (if **K** is finite, then so is $\mathbf{K}^n$, and the Zariski topology coincides with the discrete topology, and hence of no interest). Basically, open subsets are rather big, while the closed ones are rather small. For instance, in $\mathbb{A}^1(\mathbf{K})$ the closed subsets are $\emptyset$, **K**, and all finite subsets of **K**. Every affine subvariety of $\mathbb{A}^n(\mathbf{K})$ can be equipped with the topology induced the Zariski topology.

Definition 139. The Zariski closure of a subset S of the affine space $\mathbb{A}^n(\mathbf{K})$ is the smallest algebraic variety containing S. We denote it with $\bar{S}$. The Zariski closure of S is equal to $V(\mathscr{I}(S))$.

A natural example for a Zariski closure is given by elimination ideals.

The Ideal–Variety Correspondence 140. *Let* **K** *be an algebraically closed field. Let* $V = V(f_1, \ldots, f_s) \subseteq \mathbb{A}^n(\mathbf{K})$ *and for* $1 \leq \ell \leq n$ *consider the projections* $\pi_\ell :$ $\mathbf{K}^n \to \mathbf{K}^{n-\ell}$; $(x_1, \ldots, x_n) \mapsto (x_{\ell+1}, \ldots, x_n)$.

If I_ℓ is the ℓth elimination ideal $I = \langle f_1, \ldots, f_s \rangle$, i.e., $I_\ell = I \cap \mathbf{K}[X_{\ell+1}, \ldots, X_n]$, then $V(I_\ell)$ is the Zariski closure of $\pi_\ell(V)$.

For example, the projection $\pi_1 : (x,y) \mapsto y$ maps the hyperbola $V = V(I)$, with $I = \langle xy - 1 \rangle \subseteq \mathbb{R}[x,y]$, onto the punctured y-axis.[1]

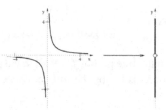

Figure 2.4: Projection

$V(I_1) = V(I \cap \mathbb{R}[x,y]) = V(0)$ is the whole real line while $\pi_1(V) = \{a \in \mathbb{R} \mid a \neq 0\} \subsetneq V(I_1)$. Thus, $\pi_1(V)$ is not a variety (it fails to contain 0).

Definition 141. If I and J are ideals of $\mathbf{K}[X_1,\ldots,X_n]$, then the *ideal quotient* (or *colon ideal*)
$$I : J := \{f \in \mathbf{K}[X_1,\ldots,X_n] \mid fJ \subseteq I\}$$
is an ideal of $\mathbf{K}[X_1,\ldots,X_n]$ containing I.

The Ideal–Variety Correspondence 142.

   (i) *If I and J are ideals of* $\mathbf{K}[X_1,\ldots,X_n]$, *then*
$$\overline{V(I) \setminus V(J)} \subseteq V(I : J).$$

    If $\mathbf{K}$ is algebraically closed and I is a radical ideal, then
$$\overline{V(I) \setminus V(J)} = V(I : J).$$

   (ii) *If V and W are subvarieties of* $\mathbb{A}^n(\mathbf{K})$, *then*

$$\mathscr{I}(V) : \mathscr{I}(W) = \mathscr{I}(V \setminus W).$$

Notice that the set-theoretic difference of two subvarieties in general is not a subvariety. For instance, consider $V = V(xz,yz)$ and $W = V(z)$. Because V is the union of the x,y-plane and the z-axis, $V \setminus W$ is the z-axis without the origin. Since $\langle xz,yz \rangle : \langle z \rangle = \langle x,y \rangle$, we see that $V \setminus W \subsetneq \overline{V \setminus W} = V(x,y)$, which is the z-axis.

---

[1]Figure 2.4 is reproduced from [14].

The affine subvariety of $\mathbb{A}^2(\mathbf{K})$ defined by $xy = 0$ can be decomposed and written as union of the two axes, which in turn (unless $\mathbf{K}$ is finite), cannot be further decomposed as union of affine subvarietes. In the following, we are going to generalize this remark.

Definition 143.

(i) A nonempty topological space E is said to *irreducible* if it can not be expressed as union of two distinct nonempty closed subsets.

(ii) In particular, an affine variety $V \subseteq \mathbb{A}^n(\mathbf{K})$ is said to be irreducible, if, whenever $V = V_1 \cup V_2$, where V_1 and V_2 are affine varieties, we have $V = V_1$ or $V = V_2$.

The following theorem translates the notion of irreducibility for affine varieties into algebraic terms.

Theorem 144. *Let $V \subseteq \mathbb{A}^n(\mathbf{K})$ be an affine variety. Then V is irreducible if and only if $\mathscr{I}(V)$ is a prime ideal.*

For example, if $\mathbf{K}$ is infinite, then $\mathbb{A}^n(\mathbf{K})$ is irreducible because $\mathbf{K}[X_1, \ldots, X_n]$ is integral. Examples for irreducible varieties arise from rational parametric representations.

If V is an algebraic variety, then $\mathbf{K}[V] = \mathbf{K}[X_1, \ldots, X_n]/\mathscr{I}(V)$ is the *coordinate ring* of V. If V is irreducible, then this ring is integral, and we denote by $\mathbf{K}(V)$ its quotient field (called *function field*).

Definition 145.

(1) Roughly speaking, a *rational map* $V_1 \dashrightarrow V_2$ between two irreducible varieties V_1 and V_2 is a function that can be written in coordinates using rational functions. It is defined on the complement of a lower-dimensional subset of V_1.

A birational map from V_1 to V_2 is a rational map $f : V_1 \dashrightarrow V_2$ such that there is a rational map $V_2 \dashrightarrow V_1$ inverse to f. A birational map induces an isomorphism from a nonempty open subset of V_1 to a nonempty open subset of V_2. In this case, V_1 and V_2 are said to be *birationally equivalent*. Algebraically, two varieties over a field $\mathbf{K}$ are birational if and only if their function fields are isomorphic as extension fields of $\mathbf{K}$.

(2) An algebraic subvariety V of $\mathbb{A}^n(\mathbf{K})$ is said to be *unirational* if it can be parametrized as:

$$
\begin{cases}
x_1 &= \dfrac{f_1(t_1,\ldots,t_m)}{g_1(t_1,\ldots,t_m)} \\
\vdots & \quad\vdots \\
x_n &= \dfrac{f_n(t_1,\ldots,t_m)}{g_n(t_1,\ldots,t_m)}
\end{cases}
\tag{2.1}
$$

with $f_i, g_i \in \mathbf{K}[t_1,\ldots,t_m]$.

(3) An irreducible algebraic subvariety V of $\mathbb{A}^n(\mathbf{K})$ is said to be *rational* if $\mathbf{K}(V)$ is isomorphic to $\mathbf{K}(u_1,\ldots,u_d)$, where u_i are algebraically independent over $\mathbf{K}$. This amounts to saying that it is birationally equivalent to an affine space of dimension d over $\mathbf{K}$. Clearly, a rational variety is unirational.

Proposition 146. *If $\mathbf{K}$ is an infinite field and V is an unirational algebraic subvariety of $\mathbb{A}^n(\mathbf{K})$, then V is irreducible.*

A rational parametrization of an algebraic subvariety can be understood as a generic point of the subvariety on which a polynomial vanishes if and only if it does so on the variety. Only irreducible subvarieties allow for a rational parametrization.

As said above, a rational variety is unirational. Conversely, a rational parametrization of V gives rise to a morphism from $\mathbf{K}(V)$ to $\mathbf{K}(u_1,\ldots,u_d)$, which, however, can be nonsurjective. Lüroth's theorem (every field between $\mathbf{K}$ and $\mathbf{K}(X)$ is generated as an extension of $\mathbf{K}$ by an element of $\mathbf{K}(X)$) asserts that every algebraic unirational curve is rational. Furthermore, it is well-known that algebraic curves of degree d which have a rational parametrization are precisely those of genus 0 (see Corollary 304). Note that the genus is at most equal to the arithmetic genus $\frac{(d-1)(d-2)}{2}$ and every singular point decreases it by at least 1 (see Theorem 349 and Proposition 355); so the genus of a nonsingular algebraic curve of degree d is equal to $\frac{(d-1)(d-2)}{2}$ (see Theorem 295). In particular, cubic curves (of arithmetic genus 1) having a rational parametrization are those which have at least one singular point. For instance, the cubic $y^2 = x^3 - x^2$ (see Fig. 2.5) has a polynomial parametrization $x(t) = t^2 - 1$, $y(t) = t^3 - t$, hence is irreducible.

Example 147. Elliptic curves (see Fig. 2.5)[2] over the real numbers are algebraic plane curves defined by an equation of the form $y^2 = x^3 + ax + b$ with $a,b \in \mathbb{R}$ (reduced Weierstrass equation) and do not have singular points (that is to say that the discriminant $\Delta = -16(4a^3 + 27b^2)$ is nonzero), see Chapter 6. According to the above discussion, we infer that elliptic curves are not unirational.

---

[2]The source of Fig. 2.5 is https://en.wikipedia.org/wiki/Elliptic_curve.

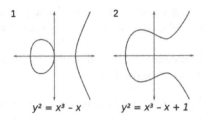

Figure 2.5: Two elliptic curves over the reals

The Ideal–Variety Correspondence 148. *The ring* $\mathbf{K}[X_1, \ldots, X_n]$ *is Noetherian, i.e., every nondecreasing sequence of ideals pauses. Geometrically, every nonincreasing chain of affine subvarieties pauses.*

The Ideal–Variety Correspondence 149. *Every nonempty affine subvariety* V *has a minimal decomposition*

$$V = V_1 \cup \cdots \cup V_r$$

(each V_i *is an irreducible variety and* $V_i \nsubseteq V_j$ *for* $i \neq j$*). This minimal decomposition is unique up to the order in which* $V_1, \ldots, V_r$ *are written. We have*

$$\mathscr{I}(V) = \cap_{i=1}^{r} \mathscr{I}(V_i),$$

where the ideals $\mathscr{I}(V_i)$ *are prime ideals. This is an example of a primary decomposition of an ideal.*

The following correspondence follows from Theorem 132.

The Ideal–Variety Correspondence 150.

(i) *Every point* $(a_1, \ldots, a_n)$ *of* $\mathbb{A}^n(\mathbf{K})$ *corresponds to a maximal ideal* $\langle X_1 - a_1, \ldots, X_n - a_n \rangle$ *of* $\mathbf{K}[X_1, \ldots, X_n]$.

(ii) *If* $\mathbf{K}$ *is algebraically closed, then the maximal ideals of* $\mathbf{K}[X_1, \ldots, X_n]$ *are of the form* $\langle X_1 - a_1, \ldots, X_n - a_n \rangle$ *with* $(a_1, \ldots, a_n) \in \mathbb{A}^n(\mathbf{K})$.

It is instructive to obtain a geometric understanding of the notion of dimension. For instance, a finite variety V does not allow for any "degree of freedom" to move on it. If, on the other hand, we consider the variety $V = V(X - Y)$, then we can move in the direction of $(1, 1)$. Here we are hinting at the notion of dimension of an algebraic variety. In the following we give a formal topological definition of the dimension.

Definition 151. Let X be a topological space. The dimension of X is the supremum of the lengths of all chains $X_0 \subsetneq \cdots \subsetneq X_n$ of irreducible closed subsets $X_0, \ldots, X_n$ of X.

The dimension of X (notation $\dim X$ or $\dim(X)$) is a positive integer, or $+\infty$, or -1 if X is empty (by way of convention).

Proposition 152. *Let X be a topological space and Y a subset of X. We have:*

(i) $\dim Y \le \dim X$.

(ii) *If X is irreducible and has finite dimension, and if Y is a proper closed subset of X, then $\dim Y < \dim X$.*

(iii) *If X is a union of closed subsets $X_1, \ldots, X_N$, then $\dim X = \max \dim X_i$.*

Proposition 153. *Let V be a nonempty variety. Then V has dimension 0 if and only if V is a finite set of points.*

Proof. Let V be a variety of dimension 0. Every irreducible closed subset contains a point, whence coincides with the corresponding singleton set. Thus, the irreducible components of V are its points.

We have seen the correspondence between irreducible subvarieties of an affine variety V and prime ideals of $\mathbf{K}[V]$. We have the following result.

Theorem 154. *The dimension of an affine variety V is equal to the Krull dimension (notation: $\operatorname{Kdim} \mathbf{K}[V]$) of $\mathbf{K}[V]$. In particular, the dimension of an irreducible affine variety V is equal to the transcendence degree of $\mathbf{K}(V)$ over $\mathbf{K}$.*

2.2   Computing on subvarieties of $\mathbb{A}^n(\mathbf{K})$ with Gröbner bases

The goal of this section is give concrete situations where the reliance on "Gröbner bases technology" is of great benefit. We will show how to compute on subvarieties of $\mathbb{A}^n(\mathbf{K})$ using the Algebra-Geometry dictionary already explained. Throughout this section $\mathbf{K}$ is a field.

Application of Gröbner bases 155. Test whether a polynomial $f \in \mathbf{K}[X_1, \ldots, X_n]$ vanishes on an algebraic subvariety $V = V(f_1, \ldots, f_s)$ of $\mathbb{A}^n(\mathbf{K})$. Equivalently, test whether f belongs to $\sqrt{\langle f_1, \ldots, f_s \rangle}$.

Solution: Calculate a reduced Gröbner basis G for the ideal $\langle f_1, \ldots, f_s, Yf - 1 \rangle$ of $\mathbf{K}[X_1, \ldots, X_n, Y]$.

If $G = \{1\}$, then $f \in \sqrt{\langle f_1, \ldots, f_s \rangle}$.
If $G \ne \{1\}$, then $f \notin \sqrt{\langle f_1, \ldots, f_s \rangle}$.

Application of Gröbner bases 156. Test whether an algebraic subvariety $V = V(f_1, \ldots, f_s)$ of $\mathbb{A}^n(\bar{\mathbf{K}})$ is empty.

Solution: Calculate a reduced Gröbner basis G for the ideal $\langle f_1, \ldots, f_s \rangle$ of $\bar{\mathbf{K}}[X_1, \ldots, X_n]$.

If $G = \{1\}$, then $V = \emptyset$.
If $G \neq \{1\}$, then $V \neq \emptyset$.

Application of Gröbner bases 157. Calculate the Zariski closure W of the projection of an algebraic subvariety $V = V(f_1, \ldots, f_s)$ of $\mathbb{A}^n(\mathbf{K})$ ($\mathbf{K}$ an algebraically closed field) under

$$\pi_\ell : \mathbb{A}^n(\mathbf{K}) \to \mathbb{A}^{n-\ell}(\mathbf{K}), \quad (x_1, \ldots, x_n) \mapsto (x_{\ell+1}, \ldots, x_n).$$

Solution: By the Ideal-Variety Correspondence 140, we have $W = V(I_\ell)$, where I_ℓ is the ℓth-elimination ideal of $I = \langle f_1, \ldots, f_s \rangle$, i.e., $I_\ell = I \cap \mathbf{K}[X_{\ell+1}, \ldots, X_n]$. In order to obtain a Gröbner basis for I_ℓ, calculate a Gröbner basis G for the ideal $\langle f_1, \ldots, f_s \rangle$ of $\mathbf{K}[X_1, \ldots, X_n]$ with respect to the lexicographic order with $X_1 > X_2 > \cdots > X_n$. Then $G_\ell = G \cap \mathbf{K}[X_{\ell+1}, \ldots, X_n]$ is a Gröbner basis of I_ℓ with respect to the lexicographic order with $X_{\ell+1} > \cdots > X_n$ (see Theorem 49).

For instance, consider $I = \langle x^2 + yz - 1, \ xz + y^2 - 1, \ xy + z^2 - 1 \rangle \subseteq \mathbb{C}[x, y, z]$. A Gröbner basis G for I with respect to the lexicographic order with $x > y > z$ is given by $G = \{2z^5 - 3z^3 + z, \ yz^3 - z^4 - yz + z^2, \ -2z^4 + y^2 - yz + 3z^2 - 1, \ 2yz^2 + x - y - z\}$.

Thus, $I \cap \mathbb{C}[y, z] = \langle 2z^5 - 3z^3 + z, \ yz^3 - z^4 - yz + z^2, \ -2z^4 + y^2 - yz + 3z^2 - 1 \rangle$, $I \cap \mathbb{C}[z] = \langle 2z^5 - 3z^3 + z \rangle$, and $I \cap \mathbb{C} = \langle 0 \rangle$.

Application of Gröbner bases 158. Calculate the union of two algebraic subvarieties $V = V(f_1, \ldots, f_s)$ and $W = V(g_1, \ldots, g_t)$ of $\mathbb{A}^n(\mathbf{K})$. This amounts to calculating the intersection of the ideals $I = \langle f_1, \ldots, f_s \rangle$ and $J = \langle g_1, \ldots, g_t \rangle$ of $\mathbf{K}[X_1, \ldots, X_n]$.

Solution: Consider the ideal

$$H = \langle U f_1, \ldots, U f_s, (1 - U) g_1, \ldots, (1 - U) g_t \rangle$$

of $\mathbf{K}[X_1, \ldots, X_n, U]$. Then it is immediate that $I \cap J = H \cap \mathbf{K}[X_1, \ldots, X_n]$.

For example, take $I = \langle XY \rangle$ and $J = \langle X^2 - XY \rangle$ in $\mathbb{C}[X, Y]$. Consider the ideal $H = \langle UXY, (1 - U)(X^2 - XY) \rangle$ of $\mathbb{C}[X, Y, U]$. We obtain $G = \{UXY, \ UX^2 + YX - X^2, \ -YX^2 + YX^2\}$ as a Gröbner basis for H with respect to the lexicographic order with $X > Y > U$. Thus, $I \cap J = \langle -YX^2 + YX^2 \rangle$ (as expected).

Application of Gröbner bases 159. Calculate the dimension of an algebraic subvariety $V = V(f_1, \ldots, f_s)$ of $\mathbb{A}^n(\mathbf{K})$. This amounts to calculating the Krull dimension of $\mathbf{K}[V] = \mathbf{K}[X_1, \ldots, X_n]/\langle f_1, \ldots, f_s \rangle$.

Solution: Fix a monomial order compatible with the total degree (which is to say that if $M \leq N$, then $\text{tdeg}(M) \leq \text{tdeg}(N)$). We calculate a Gröbner basis

$\{g_1,\ldots,g_t\}$ of $I = \langle f_1,\ldots,f_s\rangle$ and deduce that $\langle LT(I)\rangle = \langle LT(g_1),\ldots,LT(g_t)\rangle$. We then use the following result (which can be shown by means of Hilbert series, see Proposition 187):

$$\text{Kdim } \mathbf{K}[X_1,\ldots,X_n]/I = \text{Kdim } \mathbf{K}[X_1,\ldots,X_n]/\langle LT(I)\rangle.$$

We thus return to the case of a monomial ideal. For a monomial ideal $J = \langle m_1,\ldots,m_t\rangle$ of $\mathbf{K}[X_1,\ldots,X_n]$ we put

$$d(J,\mathbf{K}[X_1,\ldots,X_n]) = \text{Kdim } \mathbf{K}[X_1,\ldots,X_n]/J.$$

The number $d(J,\mathbf{K}[X_1,\ldots,X_n])$ can be calculated by means of the recursive formula $d(\langle 0\rangle,\mathbf{K}[X_1,\ldots,X_n]) = n$ and

$$d(J,\mathbf{K}[X_1,\ldots,X_n]) = \max\{d(J|_{X_i=0},\mathbf{K}[X_1,\ldots,X_{i-1},X_{i+1},\ldots,X_n]) \mid X_i \text{ divides } m_1\}.$$

For instance, for $J = \langle XZ,YZ\rangle \subseteq \mathbf{K}[X,Y,Z]$, we have

$$d(J,\mathbf{K}[X,Y,Z]) = \max\{d(\langle YZ\rangle,\mathbf{K}[Y,Z]),\ d(\langle 0\rangle,\mathbf{K}[X,Y])\} = \max\{1,1,2\} = 2.$$

Application of Gröbner bases 160. Implicitization of the smallest algebraic variety containing a subset of $\mathbb{A}^n(\mathbf{K})$ defined by a polynomial parametrization

$$F : \mathbf{K}^m \to \mathbf{K}^n, \quad (t_1,\ldots,t_m) \mapsto (f_1(t_1,\ldots,t_m),\ldots,f_n(t_1,\ldots,t_m)),$$

where $f_i \in \mathbf{K}[t_1,\ldots,t_m]$.

Solution: Let I be the ideal $\langle X_1 - f_1,\ldots,X_n - f_n\rangle$ of $\mathbf{K}[t_1,\ldots,t_m,X_1,\ldots,X_n]$. Let $I_m = I \cap \mathbf{K}[X_1,\ldots,X_n]$ be the mth elimination ideal of I. Then $V(I_m)$ is the smallest subvariety of $\mathbb{A}^n(\mathbf{K})$ containing $F(\mathbf{K}^m)$, i.e., its Zariski closure.

For instance, take $F(t) = (t^2 + 3,\ -2t^3 + 1)$. We consider the ideal

$$J = \langle x - t^2 - 3,\ y + 2t^3 - 1\rangle$$

of $\mathbb{R}[t,x,y]$. We calculate the reduced Gröbner basis G of J with regard to the lexicographic order with $t > x > y$ using the computing software MAPLE. We obtain $G = \{-109 + 108x + 2y - 36x^2 - y^2 + 4x^3,\ 2x^2 + 18 + yt - 12x - t,\ 2xt - 1 + y - 6t,\ -x + t^2 + 3\}$. The Zariski closure of $F(\mathbb{R})$ is the algebraic curve defined by the equation $-109 + 108x + 2y - 36x^2 - y^2 + 4x^3 = 0$.

Application of Gröbner bases 161. Calculate the Zariski closure $V \setminus W$ of two algebraic subvarieties $V = V(f_1,\ldots,f_s)$ and $W = V(g_1,\ldots,g_t)$ of $\mathbb{A}^n(\mathbf{K})$, where $\mathbf{K}$ is an algebraically closed field and $\langle f_1,\ldots,f_s\rangle$ is a radical ideal.

Solution: By the Ideal-Variety Correspondence 142, this amounts to calculating the colon ideal $I : J$ of the ideal $J = \langle g_1,\ldots,g_t\rangle$ in the ideal $I = \langle f_1,\ldots,f_s\rangle$ of $\mathbf{K}[X_1,\ldots,X_n]$. It's clear that $I : J = \cap_{i=1}^t I : \langle g_i\rangle$. Thus, it suffices to calculate $I : \langle g\rangle$ for $g \in \mathbf{K}[X_1,\ldots,X_n]$. To this end, we calculate a generating system $\{gp_1,\ldots,gp_r\}$ of $I \cap \langle g\rangle$ with $p_j \in \mathbf{K}[X_1,\ldots,X_n]$. Then obviously $I : \langle g\rangle = \langle p_1,\ldots,p_r\rangle$.

2.3 Singular points of a plane curve

Throughout this section, $\mathbf{K}$ is a field. In this short section, we consider curves in the plane $\mathbb{A}^2(\mathbf{K})$ and study the notion of singularity.

A singular point of an algebraic plane curve is a point where the curve has "nasty" behaviour, such as a cusp (i.e., a point where the tangent is different from the left and from the right of the singularity, see Fig. 2.6) or a node (i.e., a point where the curve intersects itself, see Fig. 2.7) or an acnode (i.e., a point which belongs to the plot of the curve but is isolated from the line that represents the curve, see Fig. 2.8). To make this precise, we will introduce the notion of tangent.

Each of the three curves in Figs. 2.6, 2.7, 2.8 has a point where there is no tangent. Intuitively, a singular point of $V(f)$, where $f \in \mathbf{K}[X,Y] \setminus \mathbf{K}$, is a point where the tangent line fails to exist. In order to make this precise, we need to give an algebraic definition of tangent line, without resorting to concepts from calculus.

Definition 162. Let $f \in \mathbf{K}[X,Y] \setminus \mathbf{K}$ and $P = (a,b) \in C = V(f)$.

1. Let $m \geq 1$, and let L be a line through P. We say that L *meets C with multiplicity m* at P if L can be given parametrically by

$$\begin{cases} x = a + ct \\ y = b + dt \end{cases} \tag{2.2}$$

so that $t = 0$ is a root of multiplicity m of the polynomial $g(t) = f(a + ct, b + dt)$.

It is easy to check that the notion of multiplicity is independent of the parametrization of L (see Exercise 200). Notice further that $t = 0$ is a root of the polynomial g above, i.e., $g(0) = f(a,b) = 0$, since $(a,b) \in V(f)$. Recall also that $t = 0$ is a root of multiplicity m when $g = t^m h$, where $h(0) \neq 0$.

Note that we have $m = I_P(C,L)$, where $I_Q(C,C')$ denotes the intersection multiplicity of two curves C and C' at a point Q (see Section 5.5.2).

2. If for all $i + j < r$, $\frac{\partial^{i+j} f}{\partial X^i \partial Y^j}(P) = 0$ and there is at least one nonzero partial derivative for the case $i + j = r$, then r is defined to be the multiplicity of (a,b) and denoted by $\text{mult}_C(P)$. In such case, r is the minimum of the degrees of the nonzero homogeneous components of $f(X - a, Y - b)$.

Note that $\text{mult}_C(P)$ is nothing but the multiplicity of 0 as a root of the polynomial $f(a + Xt, b + Yt) = f(P) + (\frac{\partial f}{\partial X}(P)X + \frac{\partial f}{\partial Y}(P)Y)t + \cdots$ seen as a polynomial in t with coefficients in $\mathbf{K}[X,Y]$.

Note also that if P is a point on an irreducible curve $C = V(F)$, then for sufficiently large n, $\text{mult}_C(P) = \dim \mathbf{K}(\mathfrak{M}_{C,P}^n / \mathfrak{M}_{C,P}^{n+1})$, where $\mathfrak{M}_{C,P}$ is the maximal ideal of the local ring $\mathcal{O}_{C,P} := \{f \in \bar{\mathbf{K}}(C) \mid f \text{ is regular at } P\}$, $\bar{\mathbf{K}}(C)$ is the quotient field of $\mathbf{K}[C] = \mathbf{K}[X,Y]/\langle F \rangle$, and a rational function $f \in$

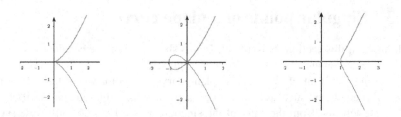

Figure 2.6: $y^2 = x^3$ Figure 2.7: $y^2 = x^3 + x^2$ Figure 2.8: $y^2 = x^3 - x^2$

$\bar{\mathbf{K}}(C)$ is regular at P if it is well-defined at P, i.e., there exist $g, h \in \bar{\mathbf{K}}[C]$ with $h(P) \neq 0$ such that $f = \frac{g}{h}$ (see Definition 282).

Geometrically, $\text{mult}_C(P)$ is the number of tangents to C at P counted with multiplicities (see Definition 345).

In the following, we will need the *gradient vector* of f:

$$\nabla f = \left(\frac{\partial f}{\partial X}, \frac{\partial f}{\partial Y} \right).$$

(Recall that (partial) derivatives can be formally defined for (multivariate) polynomials.)

We will characterize tangents by means of the following.

Proposition 163. *Let $f \in \mathbf{K}[X, Y] \setminus \mathbf{K}$, and let $(a, b) \in V(f)$.*

1. *If $\nabla f(a, b) = (0, 0)$, then every line through (a, b) meets $V(f)$ with multiplicity ≥ 2.*

2. *If $\nabla f(a, b) \neq (0, 0)$, then there is a unique line through (a, b) which meets $V(f)$ with multiplicity ≥ 2.*

Proof. Let a line L through (a, b) be parametrized as in Definition 164 and let $g(t) = f(a + ct, b + dt)$. It can be shown that $t = 0$ is a root of g of multiplicity ≥ 2 if and only if $g'(0) = 0$. Using the chain rule, we calculate

$$g'(0) = \frac{\partial f}{\partial X}(a, b) \cdot c + \frac{\partial f}{\partial Y}(a, b) \cdot d.$$

This shows that if $\nabla f(a, b) = (0, 0)$, then $g'(0)$ always equals 0, whence L always meets $V(f)$ with multiplicity ≥ 2.

As regards the second item, suppose that $\nabla f(a, b) \neq (0, 0)$. We know that $g'(0) = 0$ if and only if

$$\frac{\partial f}{\partial X}(a, b) \cdot c + \frac{\partial f}{\partial Y}(a, b) \cdot d = 0,$$

which is a linear equation in the unknowns c and d, all solutions of which parametrize the same line L.

Definition 164. Let $f \in \mathbf{K}[X, Y] \setminus \mathbf{K}$ and let $(a, b) \in C = V(f)$.

1. If $\nabla f(a, b) = (0, 0)$, then we say that (a, b) is a *singular point* of C. The set of singular points of C is denoted by $\text{Sing}(C)$. It is a finite set (see Lemma 273 or Weak Bézout's Theorem 336).

2. If $\nabla f(a, b) \neq (0, 0)$ (in other terms, (a, b) has multiplicity 1), then the *tangent line* of C at (a, b) is the unique line L through (a, b) which meets C with multiplicity ≥ 2. We say that (a, b) is a *nonsingular point* or a *regular point* of C. Note that the equation of L is $(x - a)\frac{\partial f}{\partial x}(a, b) + (y - b)\frac{\partial f}{\partial y}(a, b) = 0$, whose left-hand side is the term of degree one of the Taylor expansion of f at (a, b). A curve without singular points is called *nonsingular* or *smooth*.

Example 165. The Fermat curve $x^n + y^n = z^n$ ($n \geq 2$), defined over a field of characteristic 0 or coprime with n, is smooth.

Remark 166.

- Let $f \in \mathbf{K}[X, Y] \setminus \mathbf{K}$. We can compute the singular points of $V(f)$ as follows. The gradient ∇f is zero if and only if $\frac{\partial f}{\partial X}$ and $\frac{\partial f}{\partial Y}$ vanish simultaneously. We must further be on $V(f)$, i.e., we need $f = 0$. Thus, the singular points of $V(f)$ are determined by the following equations:

$$f = \frac{\partial f}{\partial X} = \frac{\partial f}{\partial Y} = 0.$$

To give an example, consider the curve $C : Y^2 = X^2(X + 1)$ over a field of characteristic $\neq 2$ (see Fig. 2.7). In order to determine the singular points, we need to solve the following system of equations:

$$\begin{cases} f = Y^2 - X^3 - X^2 &= 0 \\ \frac{\partial f}{\partial X} = -3X^3 - 2X &= 0 \\ \frac{\partial f}{\partial Y} = 2Y &= 0. \end{cases} \tag{2.3}$$

It is now easy to see that $(0, 0)$ is the only singular point of $V(f)$, as expected. It is immediate that both the lines $y = x$ and $y = -x$ meets C at $(0, 0)$ with multiplicity 3. Moreover $(0, 0)$ has multiplicity 2 as a point of C.

- In general, for a hypersurface $f(x_1, \ldots, x_n) = 0$, the singular points are those at which all the partial derivatives simultaneously vanish. For a general algebraic variety $V = V(f_1, \ldots, f_s) \subseteq \mathbb{A}^n(\mathbf{K})$ with $f_i \in \mathbf{K}[X_1, \ldots, X_n]$, the condition on a point P of V to be singular is that the Jacobian matrix $J = \left(\frac{\partial f_i}{\partial X_j}\right)_{\substack{1 \leq i \leq s \\ 1 \leq j \leq n}}$ of the first order partial derivatives of the polynomials

$f_1, \ldots, f_s$ has a rank at P that is lower than the rank at other points of the variety. Points of V that are not singular are called nonsingular or regular. It is always true that almost all points are nonsingular, in the sense that the nonsingular points form a set that is both open and dense in the variety (for the Zariski topology, as well as, in the case of varieties defined over $\mathbb{C}$, for the usual topology). A variety without singular points is called *nonsingular* or *smooth*.

This section will be pursued in Section 5.6, where, among other things, we will separate the singular points in two types, namely, ordinary and nonordinary.

2.4   Solving polynomial systems with Gröbner bases

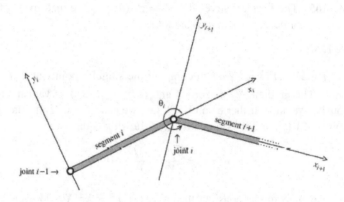

Figure 2.9: Local coordinate systems of a planar robot arm

Polynomial equations appear in several areas such as model geometrical constraints, relations between physical quantities, or properties satisfied by some unknowns (for example, in algebraic cryptanalysis).

A well-known concrete example [14] concerns the space of configurations of a planar robot. Roughly speaking, suppose you want to model a robot arm as a sequence of line segments connected by joints. If (x_i, y_i) is the coordinates of some joint, (x_{i+1}, y_{i+1}) is the coordinates of the next joint (or the hand), and the segment has length ℓ_i, then we get the equation

$$(x_{i+1} - x_i)^2 + (y_{i+1} - y_i)^2 = \ell_i^2,$$

together with the relations $x_{i+1} - x_i = \ell_i \cos \theta_i$, $y_{i+1} - y_i = \ell_i \sin \theta_i$, giving the angle θ_i at the ith joint (see Fig. 2.9).[3] We thus see that we obtain a system of polynomial equations describing all possible positions of the robot arm. To determine whether it is possible to reach a given position, we need to solve a system of polynomial equations to see which configurations are feasible.

Let us give another example about the importance of the resolution polynomial systems in concrete situations. If **K** is a finite field (see Chapter 3 for an overview on finite fields) then all functions from $\mathbf{K}^n$ to **K** can be represented as polynomials in $\mathbf{K}[X_1,\ldots,X_n]$. From this fact stems the interaction between Gröbner bases and coding theory and cryptography. For this reason, Gröbner bases have been successfully used to attack cryptosystems and also have been used to build cryptosystems (this has not been very successful).

Consider a polynomial system

$$\left\{ \begin{array}{c} f_1(x_1,\ldots,x_n) = 0 \\ \vdots \\ f_m(x_1,\ldots,x_n) = 0, \end{array} \right. \tag{2.4}$$

where $f_1,\ldots,f_m$ are polynomials at the variables $x_1,\ldots,x_n$ with coefficients in a field **K** in which one can compute effectively (for example $\mathbb{Q}$ or $\mathbb{F}_p$). As implicitly stated in Theorem 49, when computing a Gröbner basis for the ideal generated by $f_1,\ldots,f_m$ in $\mathbf{K}[x_1,\ldots,x_n]$ accordingly to a lexicographic monomial order, we obtain a Gröbner basis in a "triangular form" and this clearly simplifies the form of the equations considerably. This means that we get equations where the variables are eliminated successively. Also, note that the order of elimination corresponds to the ordering of the variables. A system of equations in this form is easy to solve, especially when the last equation contains only one variable (this corresponds to a zero-dimensional system, see Theorem 174). We can apply one-variable techniques to find its roots, then substitute back into the other equations in the system and solve for the other variables, using a procedure similar to the method of "back-substitution" used to solve a linear system in triangular form by Gaussian elimination. But there is a slight note of caution: in practice, computing a Gröbner basis for the lexicographic order is very expensive. As suggested by Faugère (with other coauthors), in the dimension 0 case, a solution would be to change the monomial order and to use linear algebra (see Section 2.5).

Example 167. Let us try to solve the following polynomial system over the reals:

$$\left\{ \begin{array}{c} x^2 + y + z = 1 \\ x + y^2 + z = 1 \\ x + y + z^2 = 1. \end{array} \right. \tag{2.5}$$

---

[3]Figure 2.9 is reproduced from [14].

A Gröbner basis for the corresponding ideal with respect to the lexicographic order with $x > y > z$ is $G = \{x+y+z^2-1, y^2-y-z^2+z, 2yz^2+z^4-z^2, z^6 - 4z^4+4z^3-z^2\}$. So, System 2.5 boils down to the following triangular system:

$$\begin{cases} x+y+z^2-1 = 0 \\ y^2-y-z^2+z = 0 \\ 2yz^2+z^4-z^2 = 0 \\ z^6-4z^4+4z^3-z^2. \end{cases} \qquad (2.6)$$

Solving the last univariate equation and then substituting back into the others yields to the following solutions $(1,0,0)$, $(0,1,0)$, $(0,0,1)$, $(-1+\sqrt{2},-1+\sqrt{2}, -1+\sqrt{2})$, $(-1-\sqrt{2},-1-\sqrt{2},-1-\sqrt{2})$.

As it is not always possible to express the solutions of a univariate polynomial equation of degree > 5 by radicals (by Galois theory), we cannot, in general, obtain "explicitly" the solutions of a polynomial system. A formal solution of a polynomial system consists in a rewriting of the initial system as an equivalent but simpler system. More precisely, we will see in Section 2.5 that this latter can have the following pleasant form in the dimension 0 case:

$$\begin{cases} P_n(x_n) &=& 0 \\ x_{n-1} &=& P_{n-1}(x_n) \\ &\vdots& \\ x_1 &=& P_1(x_n). \end{cases} \qquad (2.7)$$

2.5   Solving zero-dimensional polynomial systems

In this section we present the approach given in [22] for solving zero-dimensional polynomial systems by only computing ranks of matrices. Let $\mathbf{Q}$ be a field of characteristic zero ($\mathbb{Q}$ for example) and $\mathbf{C}$ an algebraically closed field containing $\mathbf{Q}$ ($\mathbb{C}$ for example). Consider an ideal $I = \langle f_1,\ldots,f_s \rangle$ of $\mathbf{Q}[X_1,\ldots,X_n]$ and denote by $I_{\mathbf{C}}$ the ideal generated by $f_1,\ldots,f_s$ in $\mathbf{C}[X_1,\ldots,X_n]$. The quotient rings $\mathbf{Q}[X_1,\ldots,X_n]/I$ and $\mathbf{C}[X_1,\ldots,X_n]/I_{\mathbf{C}}$ are naturally equipped with a structure of $\mathbf{Q}$-vector space and $\mathbf{C}$-vector space, respectively. To lighten the notation, the class in $\mathbf{Q}[X_1,\ldots,X_n]/I$ of an element $u \in \mathbf{Q}[X_1,\ldots,X_n]$ will also be denoted by u.

Remark 168. It is straightforward that if a linear system with n unknowns over a field $\mathbf{K}$ is compatible when seen as a linear system over a bigger field $Ł$ then its solutions are in fact in $\mathbf{K}^n$.

From Remark 168, we easily infer that $\mathbf{Q}[X_1,\ldots,X_n]/I$ is a subring of $\mathbf{C}[X_1,\ldots,X_n]/I_{\mathbf{C}}$. Moreover, $\mathbf{Q}[X_1,\ldots,X_n]/I$ is finite-dimensional as $\mathbf{Q}$-vector space if and only if

$C[X_1, \ldots, X_n]/I_C$ is finite-dimensional as C-vector space (in such case, they have the same dimension). For example, $\mathbb{Q}[X,Y]/\langle XY \rangle = \mathbb{Q}[X] + \mathbb{Q}[Y]$ is an infinite-dimensional $\mathbb{Q}$-vector space whereas $\mathbb{Q}[X,Y]/\langle X^2, Y^2 \rangle = \mathbb{Q}.1 + \mathbb{Q}.X + \mathbb{Q}.Y + \mathbb{Q}.XY$ is a 4-dimensional $\mathbb{Q}$-vector space.

By a *zero-dimensional ideal* I of $\mathbf{Q}[X_1, \ldots, X_n]$, we mean an ideal I of $\mathbf{Q}[X_1, \ldots, X_n]$ whose corresponding algebraic variety $V(I)$ has (geometrical) dimension zero (i.e., $V(I)$ is finite, or equivalently, $\text{Kdim}(\mathbf{Q}[X_1, \ldots, X_n]/I) < \infty$).

Definition and notation 169. Let $I = \langle f_1, \ldots, f_s \rangle$ be a zero-dimensional ideal of $\mathbf{Q}[X_1, \ldots, X_n]$. We say that a $u \in \mathbf{Q}[X_1, \ldots, X_n]$ is a *separating element* for $V(I)$ if it takes different values when evaluated at the points of $V(I)$.

Example 170. Let $I = \langle X, Y^2 - 1 \rangle \subseteq \mathbb{Q}[X,Y]$. It is clear that I has dimension zero since $V(I) = \{(0,1),(0,-1)\}$. We see that Y is separating for $V(I)$ while X is not.

Notation 171. For $i, j \in \mathbb{N}$ with $i \geq j$, we denote by $\binom{i}{j} = \frac{i!}{j!(i-j)!}$, with the convention $0! = 1$.

The following lemma explains how to find separating elements.

Lemma 172. *Let $I = \langle f_1, \ldots, f_s \rangle$ be a zero-dimensional ideal of $\mathbf{Q}[X_1, \ldots, X_n]$ and denote $D := \sharp V(I)$. Then there exists $i \in \{0, \ldots, (n-1)\binom{D}{2}\}$ such that the linear form*

$$u_i = X_1 + iX_2 + \cdots + i^{n-1}X_n$$

is a separating element for $V(I)$.

Proof. For different points $x = (x_1, \ldots, x_n), y = (y_1, \ldots, y_n)$ in $V(I)$, we denote by $\ell(x,y)$ the number of integers $i \in \{0, \ldots, (n-1)\binom{D}{2}\}$ such that $u_i(x) = u_i(y)$. As the polynomial $(x_1 - y_1) + (x_2 - y_2)T + \cdots + (x_n - y_n)T^{n-1}$ is nonzero (because $x \neq y$), it cannot have more that $n-1$ roots, and, thus, $\ell(x,y) \leq n-1$. The desired result clearly follows. ∎

That is good, but not enough. We want to know how to test whether some linear form u_i is a separating element for $V(I)$. This can be done with linear algebra as shown by the following lemma.

Lemma 173. *Let $I = \langle f_1, \ldots, f_s \rangle$ be a zero-dimensional ideal of $\mathbf{Q}[X_1, \ldots, X_n]$, denote $D := \sharp V(I)$, and consider $u \in \mathbf{Q}[X_1, \ldots, X_n]$. Then u is a separating element for $V(I)$ if and only if $1, u, \ldots, u^{D-1}$ are $\mathbf{Q}$-linearly independent in $\mathbf{Q}[X_1, \ldots, X_n]/I$.*

Proof. "⇒" By way of contraposition, suppose that $1, u, \ldots, u^{D-1}$ are $\mathbf{Q}$-linearly dependent in $\mathbf{Q}[X_1, \ldots, X_n]/I$, that is, $\sum_{i=0}^{D-1} c_i u^i \in I$ for some $(c_0, \ldots, c_{D-1}) \in \mathbf{Q}^D \setminus \{0\}$. As the nonzero polynomial $\sum_{i=0}^{D-1} c_i T^i$ cannot have more that $D-1$ roots, we infer that u is not a separating element for $V(I)$.

"$\Leftarrow$" We can suppose that I is radical (as $V(I) = V(\sqrt{I})$). By way of contraposition, denote by $z_1,\ldots,z_D$ the elements of $V(I)$ and suppose that u is not a separating element for $V(I)$. Then the following Vandermonde determinant

$$\text{VDM}(u(z_1),\ldots,u(z_D)) = \begin{vmatrix} 1 & u(z_1) & \cdots & u(z_1)^{D-1} \\ \vdots & \vdots & \cdots & \vdots \\ \vdots & \vdots & \cdots & \vdots \\ 1 & u(z_D) & \cdots & u(z_D)^{D-1} \end{vmatrix}$$

is zero, and, thus, there exists $(\alpha_1,\ldots,\alpha_D) \in \mathbf{C}^D \setminus \{0\}$ such that

$$\alpha_1 \begin{pmatrix} 1 \\ \vdots \\ 1 \end{pmatrix} + \alpha_2 \begin{pmatrix} u(z_1) \\ \vdots \\ u(z_D) \end{pmatrix} + \cdots + \alpha_D \begin{pmatrix} u(z_1)^{D-1} \\ \vdots \\ u(z_D)^{D-1} \end{pmatrix} = \begin{pmatrix} 0 \\ \vdots \\ 0 \end{pmatrix}.$$

Thus, the polynomial $P(u) = \alpha_1 + \alpha_2 u + \cdots + \alpha_D u^{D-1}$ vanishes over $V(I)$ and $\alpha_1 + \alpha_2 u + \cdots + \alpha_D u^{D-1} = 0$ in $\mathbf{C}[X_1,\ldots,X_n]/I_{\mathbf{C}}$. We deduce that $1, u, \ldots, u^{D-1}$ are $\mathbf{Q}$-linearly dependent in $\mathbf{Q}[X_1,\ldots,X_n]/I$ (see Remark 168).

Now we are in position to give a linear-algebra characterization of zero-dimensional ideals.

Theorem 174. *Let $I = \langle f_1,\ldots,f_s \rangle$ be an ideal of $\mathbf{Q}[X_1,\ldots,X_n]$. Then, I has dimension zero if and only if the dimension of $\mathbf{Q}[X_1,\ldots,X_n]/I$ as $\mathbf{Q}$-vector space is finite. In such case, $\sharp V(I) \leq \dim_{\mathbf{Q}}(\mathbf{Q}[X_1,\ldots,X_n]/I)$. Moreover, for each $1 \leq i \leq n$, $I \cap \mathbf{Q}[X_i]$ is a nonzero principal ideal.*

Proof. "$\Rightarrow$" Suppose that $\sharp V(I) < \infty$. For $1 \leq i \leq n$, denote by π_i the projection from $\mathbf{C}^n$ to $\mathbf{C}$ sending $(x_1,\ldots,x_n)$ onto x_i. Setting $\pi_i(V(I)) = \{\xi_1,\ldots,\xi_d\}$, we can find a nonzero polynomial $g_i \in \mathbf{C}[X_i]$ vanishing at $\pi_i(V(I))$ and, thus, also at $V(I)$. It follows that there exists $k_i \in \mathbb{N}$ such that $g_i^{k_i} \in I_{\mathbf{C}} \cap \mathbf{C}[X_i]$ with $g_i^{k_i} \neq 0$. As a consequence, $\mathbf{C}[X_1,\ldots,X_n]/I_{\mathbf{C}}$ is generated as $\mathbf{C}$-vector space by the classes of monomials $X_1^{\alpha_1} \cdots X_n^{\alpha_n}$ with $0 \leq \alpha_i < k_i \deg(g_i)$. Since $\mathbf{C}[X_1,\ldots,X_n]/I_{\mathbf{C}}$ is finite-dimensional as $\mathbf{C}$-vector space, $\mathbf{Q}[X_1,\ldots,X_n]/I$ is finite-dimensional as $\mathbf{Q}$-vector space.

"$\Leftarrow$" Suppose that $\dim_{\mathbf{Q}}(\mathbf{Q}[X_1,\ldots,X_n]/I) = D < \infty$. For $1 \leq i \leq n$, as $1, X_i, \ldots, X_i^D$ are $\mathbf{Q}$-linearly dependent, we can find a nonzero polynomial $p_i \in I \cap \mathbf{Q}[X_i]$. It follows that, for each $1 \leq i \leq n$, $\pi_i(V(I))$ is finite, and, thus, $V(I)$ is finite. The fact that $\sharp V(I) \leq \dim_{\mathbf{Q}}(\mathbf{Q}[X_1,\ldots,X_n]/I)$ ensues from Lemmas 172 and 173.

Remark 175. Let $I = \langle f_1, \ldots, f_s \rangle$ be a zero-dimensional ideal of $\mathbf{Q}[X_1, \ldots, X_n]$, and denote by $\mathbf{A} := \mathbf{Q}[X_1, \ldots, X_n]/I$ and $\overline{\mathbf{A}} := \mathbf{C}[X_1, \ldots, X_n]/I_\mathbf{C}$. Fix a monomial order $>$, compute a Gröbner basis G for I, and denote by $w_1 < w_2 < \cdots < w_\delta$ the monomials at $X_1, \ldots, X_n$ which are not in $\langle \mathrm{LT}(G) \rangle$. Then $\mathscr{B} := (w_1, \ldots, w_\delta)$ is a (canonical) basis of $\mathbf{A}$ as $\mathbf{Q}$-vector space and for $\overline{\mathbf{A}}$ as $\mathbf{C}$-vector space.

Definition 176. Let $I = \langle f_1, \ldots, f_s \rangle$ be a zero-dimensional ideal of $\mathbf{Q}[X_1, \ldots, X_n]$, and denote $\overline{\mathbf{A}} := \mathbf{C}[X_1, \ldots, X_n]/I_\mathbf{C}$.

The integer $\dim_{\mathbf{Q}}(\mathbf{Q}[X_1, \ldots, X_n]/I)$ is called the *degree* of I (or the degree of $V(I)$) and is denoted by $\deg(I)$. Recall that $\sharp V(I) \leq \deg(I)$ (Theorem 174).

Proposition 177. *Let* $I = \langle f_1, \ldots, f_s \rangle$ *be a zero-dimensional ideal of* $\mathbf{Q}[X_1, \ldots, X_n]$, *and denote* $\overline{\mathbf{A}} := \mathbf{C}[X_1, \ldots, X_n]/I_\mathbf{C}$. *Then, for each* $x \in V(I)$, *there exists* $e_x \in \overline{\mathbf{A}}$ *such that:*

(i) $\displaystyle\sum_{x \in V(I)} e_x = 1$.

(ii) $e_x e_y = 0$ *for* $x \neq y$ *in* $V(I)$.

(iii) $e_x^2 = e_x \ \forall \, x \in V(I)$.

(iv) $e_x(x) = 1 \ \forall \, x \in V(I)$.

(v) $e_x(y) = 1 \ \forall \, y \in V(I) \setminus \{x\}$.

Proof. By virtue of Lemma 172, we can suppose that X_1 is a separating element for $V(I)$. For $x \in V(I)$, we denote by x_1 its first coordinate. By Lagrange's interpolation formula, the polynomial

$$\ell_x := \prod_{y \in V(I) \setminus \{x\}} \frac{X_1 - y_1}{x_1 - y_1}$$

satisfies $\ell_x(x) = 1$ and $\ell_x(y) = 0$ for $y \in V(I) \setminus \{x\}$. Thus, for $x \neq y$ in $V(I)$, $\ell_x \ell_y$ vanishes at $V(I)$. By Hilbert's Nullstellensatz Theorem 134, for each $x \in V(I)$, there exists $n_x \in \mathbb{N}$ such that $\ell_x^{n_x} \ell_y^{n_y} = 0$ in $\overline{\mathbf{A}}$. So, setting $p_x = \ell_x^{n_x}$, we have $p_x p_y = 0$ in $\overline{\mathbf{A}}$ for $x \neq y$ in $V(I)$, and $p_x(x) = 1$ for $x \in V(I)$. By this latter property, the sets $V(I)$ and $\{\text{common roots of the } p_x\text{'s}\}$ are disjoint, and, thus, by Hilbert's Nullstellensatz Theorem 133, there exist polynomials q_x $(x \in V(I))$ such that

$$\sum_{x \in V(I)} q_x p_x = 1 \ \text{ in } \overline{\mathbf{A}}.$$

Setting $e_x = q_x p_x$, all the desired properties are clearly fulfilled.

Definition and Proposition 178. Let $I = \langle f_1, \ldots, f_s \rangle$ be a zero-dimensional ideal of $\mathbf{Q}[X_1, \ldots, X_n]$, and denote $\overline{\mathbf{A}} := \mathbf{C}[X_1, \ldots, X_n]/I_{\mathbf{C}}$.

For $x \in V(I)$, we denote by S_x the monoid formed by the elements in $\overline{\mathbf{A}}$ which don't vanish at x, and $\overline{\mathbf{A}}_x := S_x^{-1}\overline{\mathbf{A}}$. It is a local ring since for $\frac{a}{b} \in \overline{\mathbf{A}}_x$, either $a(x) \neq 0$ and hence $\frac{a}{b} \in \overline{\mathbf{A}}_x^{\times}$, or $a(x) = 0$ and hence $1 - \frac{a}{b} = \frac{b-a}{a} \in \overline{\mathbf{A}}_x^{\times}$ since $(b-a)(x) = b(x) \neq 0$.

The nonnegative integer $\mu(x) := \dim_{\mathbf{C}} \overline{\mathbf{A}}_x$ is called *the multiplicity of x as an element of the variety $V(I)$*. By Proposition 177 (items (i), (ii), and (iii)), we have *a fundamental system of orthogonal idempotents* $(e_x, x \in V(I))$. The element e_x is called *the idempotent associated to x*. Within this system of orthogonal idempotents, the ring $\overline{\mathbf{A}}$ splits as

$$\overline{\mathbf{A}} \cong \prod_{x \in V(I)} \overline{\mathbf{A}}_x,$$

where $\overline{\mathbf{A}}_x \cong e_x \overline{\mathbf{A}}$. In particular, we have

$$\deg(I) = \sum_{x \in V(I)} \mu(x).$$

When I is radical, we have $\mu(x) = 1 \; \forall x \in V(I)$, and, thus,

$$\deg(I) = \sharp V(I).$$

Definition 179. Let $I = \langle f_1, \ldots, f_s \rangle$ be a zero-dimensional ideal of $\mathbf{Q}[X_1, \ldots, X_n]$, and denote $\overline{\mathbf{A}} := \mathbf{C}[X_1, \ldots, X_n]/I_{\mathbf{C}}$. For $f \in \overline{\mathbf{A}}$, the map

$$
\begin{aligned}
M_f : \quad \overline{\mathbf{A}} \quad &\longrightarrow \quad \overline{\mathbf{A}} \\
p \quad &\longmapsto \quad f p
\end{aligned}
$$

is $\mathbf{C}$-linear.

Keeping the above notation:

Theorem 180. (Stickelberger)

For $f \in \overline{\mathbf{A}}$, the map M_f satisfies the following properties:

(i) *The characteristic polynomial of M_f is $P_{M_f}(T) = \prod_{x \in V(I)} (T - f(x))^{\mu(x)}$.*

(ii) *The determinant of M_f is $\prod_{x \in V(I)} f(x)^{\mu(x)}$.*

(iii) *The trace of M_f is $\mathrm{Tr}(M_f) = \sum_{x \in V(I)} \mu(x) f(x)$.*

Proof. For $x \in V(I)$, we denote by e_x the associated idempotent and by $M_{f,x} = M_{e_x f}$ the component of M_f along $\overline{\mathbf{A}}_x$, that is,

$$M_{f,x}: \begin{array}{ccc} \overline{\mathbf{A}}_x & \longrightarrow & \overline{\mathbf{A}}_x \\ \frac{p}{q} & \longmapsto & f\frac{p}{q} \end{array}$$

Since $e_x(f - f(x))$ vanishes at $V(I)$ then, by Hilbert's Nullstellensatz Theorem 134, there exists $k_x \in \mathbb{N}$ such that $(e_x(f - f(x))^{k_x} = 0$ in $\overline{\mathbf{A}}$. It follows that $M_{e_x(f - f(x))}$ is nilpotent (with a unique eigenvalue 0), and, thus, $M_{f,x}$ has a unique eigenvalue $f(x)$ with multiplicity $\mu(x)$.

Example 181. $I = \langle XY + X, X^2, Y^2 - 1 \rangle \subseteq \mathbb{Q}[x,y]$. Fixing the lexicographic order as monomial order with $X > Y$, we find $G = \{XY + X, X^2, Y^2 - 1\}$ as the reduced Gröbner basis for I. Thus, $\mathscr{B} = (1, Y, X)$ is a basis for $\mathbb{Q}[X,Y]/I$ as $\mathbb{Q}$-vector space. We have:

$$\mathscr{M}_X := \mathrm{Mat}(M_X, \mathscr{B}) = \begin{pmatrix} 0 & 0 & 0 \\ 0 & 0 & 0 \\ 1 & -1 & 0 \end{pmatrix}, \quad \mathscr{M}_Y := \mathrm{Mat}(M_Y, \mathscr{B}) = \begin{pmatrix} 0 & 1 & 0 \\ 1 & 0 & 0 \\ 0 & 0 & -1 \end{pmatrix},$$

$$\mathscr{M}_{2XY+1} := \mathrm{Mat}(M_{2XY+1}, \mathscr{B}) = 2\mathscr{M}_X \mathscr{M}_Y + \mathbf{I}_3 = \begin{pmatrix} 1 & 0 & 0 \\ 0 & 1 & 0 \\ -2 & 2 & 1 \end{pmatrix},$$

$P_{M_X}(T) = \det(T\mathbf{I}_3 - \mathscr{M}_X) = T^3$ (X is not separating for $V(I)$),

$P_{M_Y}(T) = (T - 1)(T + 1)^2 = (T - Y(0,1))^{\mu(0,1)}(T - Y(0,-1))^{\mu(0,-1)}$ (Y is separating for $V(I)$).

Stickelberger's theorem will allow us to deduce an algorithm testing whether a zero-dimensional ideal I of $\mathbf{Q}[X_1, \ldots, X_n]$ is radical. We consider the so-called *Hermite quadratic form* associated to I:

$$\mathrm{QuadHerm}_I: \begin{array}{ccc} \overline{\mathbf{A}} & \longrightarrow & \mathbf{Q} \\ p & \longmapsto & \mathrm{Tr}(M_{p^2}). \end{array}$$

It is associated to the bilinear form

$$\text{Herm}_I: \quad \mathbf{A} \times \mathbf{A} \quad \longrightarrow \quad \mathbf{Q}$$
$$(p,q) \quad \longmapsto \quad \text{Tr}(M_{pq}).$$

Keeping the above notation:

Theorem 182. (Rank theorem)

$$\text{rk}(\text{QuadHerm}_I) = \sharp V(I).$$

Proof. Denote by $D = \sharp V(I)$ and $\delta = \deg(I) \geq D$. Considering a separating element u for $V(I)$, we know that $w_1 = 1$, $w_2 = u, \ldots, w_D = u^{D-1}$ are $\mathbf{Q}$-linearly independent in $\mathbf{A}$. These vectors can be completed, by adding $\delta - D$ vectors $w_{D+1}, \ldots, w_\delta$, into a basis of $\mathbf{A}$ as $\mathbf{Q}$-vector space.

Let $g = \sum_{i=1}^{\delta} g_i w_i \in \mathbf{A}$. By Stickelberger's Theorem 180,

$$\text{QuadHerm}_I(g) = \sum_{x \in V(I)} \mu(x) g^2(x).$$

Denoting the elements of $V(I)$ by $\alpha_1, \ldots, \alpha_D$, we have:

$$\text{QuadHerm}_I(g) = \sum_{i=1}^{D} \mu(\alpha_i) g^2(\alpha_i)$$
$$= (g(\alpha_1), \ldots, g(\alpha_D)) \cdot \text{Diag}(\mu(\alpha_1), \ldots, \mu(\alpha_D)) \cdot {}^t(g(\alpha_1), \ldots, g(\alpha_D)).$$

As ${}^t(g(\alpha_1), \ldots, g(\alpha_D)) = M \cdot \tilde{g}$ where $\tilde{g} = {}^t(g_1, \ldots, g_\delta)$ and

$$M = \begin{pmatrix} 1 & u(\alpha_1) & \cdots & u(\alpha_1)^{D-1} & w_{D+1}(\alpha_1) & \cdots & w_\delta(\alpha_1) \\ \vdots & \vdots & \cdots & \vdots & \vdots & \cdots & \vdots \\ 1 & u(\alpha_D) & \cdots & u(\alpha_D)^{D-1} & w_{D+1}(\alpha_D) & \cdots & w_\delta(\alpha_D) \end{pmatrix},$$

we infer that

$$\text{QuadHerm}_I(g) = {}^t(\tilde{g}) \, {}^t(M) \cdot \text{Diag}(\mu(\alpha_1), \ldots, \mu(\alpha_D)) \cdot M \, \tilde{g}.$$

As u is separating for $V(I)$, we deduce that

$$\text{VDM}(u(\alpha_1), \ldots, u(\alpha_D)) = \begin{vmatrix} 1 & u(\alpha_1) & \cdots & u(\alpha_1)^{D-1} \\ \vdots & \vdots & \cdots & \vdots \\ \vdots & \vdots & \cdots & \vdots \\ 1 & u(\alpha_D) & \cdots & u(\alpha_D)^{D-1} \end{vmatrix} \neq 0,$$

and $\text{rk}(\text{QuadHerm}_I) = \text{rk}(M) = D$.

Algorithm 183. (Testing whether a zero-dimensional ideal is radical)

Let $I = \langle f_1, \ldots, f_s \rangle$ be a zero-dimensional ideal of $\mathbf{Q}[X_1, \ldots, X_n]$, and denote $\mathbf{A} = \mathbf{Q}[X_1, \ldots, X_n]/I$.

- First compute the degree δ of I. This is nothing but the cardinality of a monomial basis $\mathscr{B}$ of $\mathbf{A}$.

- Compute $D = \mathrm{rk}(\mathrm{QuadHerm}_I)$ (by computing the rank of its matrix in $\mathscr{B}$).

- I is radical if and only if $D = \delta$.

Now, we have all the ingredients needed to give an algorithm computing a rational parametrization of a zero-dimensional algebraic variety.

Algorithm 184. (Computing a rational parametrization of a zero-dimensional algebraic variety)

Let $V = V(I)$ a zero-dimensional algebraic variety where $I = \langle f_1, \ldots, f_s \rangle$ is an ideal of $\mathbf{Q}[X_1, \ldots, X_n]$, and denote $\mathbf{A} = \mathbf{Q}[X_1, \ldots, X_n]/I$.

- First compute a separating element u for $V(I)$ (follow Lemmas 172 and 173).

- Test with Algorithm 183 whether I is radical. If I is not radical then replace it with the radical ideal $J = \langle f_1, \ldots, f_s, m_u \rangle$, where m_u is the square-free part of $P_{M_u}(u)$ ($P_{M_u}(T)$ being the characteristic polynomial of the multiplication M_u in $\mathbf{A}$ by u). Note that $J = \sqrt{I}$ and, thus, $V(I) = V(J)$.

- We can now suppose that I is radical with degree D. By Stickelberger's Theorem 180, we know that the characteristic polynomial $q(T)$ of the multiplication M_u in $\mathbf{A}$ by u is $q(T) = \prod_{\alpha_i \in V(I)} (T - u(\alpha_i))$.

  Moreover, as $1, u, \ldots, u^{D-1}$ is a basis for $\mathbf{A}$ as $\mathbf{Q}$-vector space, for all $1 \le i \le n$, there exists $(\lambda_{1,i}, \ldots, \lambda_{D,i}) \in \mathbf{Q}^D$ such that

  $$X_i = \sum_{j=1}^{D} \lambda_{j,i} u^{j-1}$$

  in $\mathbf{A}$. Denoting by $q_i(T) := \sum_{j=1}^{D} \lambda_{j,i} T^{j-1}$, we obtain the following parametrization of $V(I)$:

  $$\left\{ \begin{array}{rcl} x_n & = & q_n(T) \\ \vdots & & \vdots \\ x_1 & = & q_1(T) \\ q(T) & = & 0. \end{array} \right. \tag{2.8}$$

For an example, see Exercise 203.

2.6   Complexity of computing a Gröbner basis

We give here a very short discussion of the complexity of Buchberger's algorithm. The general problem of computing a reduced Gröbner basis and the Ideal Membership Problem are EXSPACE-complete (for coefficients in $\mathbb{Q}$), that is, any problem that can be solved with exponential space can be reduced to them (since Gröbner bases can be used to solve polynomial systems and it is not difficult to encode NP-complete problems into polynomial systems).

From the numerous complexity results for polynomial ideals, we mention the following result due to Dubé [17]:

Theorem 185. *Let $I = \langle f_1, \ldots, f_s \rangle \subseteq \mathbb{Q}[X_1, \ldots, X_n]$ be an ideal, let d be the maximal total degree of the f_i's, and fix a monomial order on $\mathbb{Q}[X_1, \ldots, X_n]$. The reduced Gröbner basis for I consists of polynomials whose total degree is bounded by*

$$2 \left(\frac{d^2}{2} + d \right)^{2^{n-1}}.$$

On the other hand, there are examples [48] showing that the above doubly exponential upper bound is optimal. This created widespread pessimism in the field of computer algebra about the practicality of computing with Gröbner bases. Borrowing words from [6], this was quite surprising since the rates of growth familiar to algebraic geometers (for instance Bézout's theorem on the number of points in the intersection of n hypersurfaces in n variables, a generalization of Bézout's Theorem 340) had all been single exponential and many large concrete real-life problems had been efficiently solved with Gröbner bases. It is rather more interesting to try to find generic complexity bounds, i.e., those valid for randomly chosen systems. For more discussion on this issue and on how to reconcile these different points of view with the concept of regularity (see Section 2.7), the interested reader can consult the nice paper [6].

2.7   Hilbert series

The goal of this section is to provide a rapid overview of Hilbert series and their computation with Gröbner bases. For more details, the reader can refer to Kemper's nice book [37]. We will restrict ourselves to the case where the base ring is a field. The extension of Hilbert series to polynomials over (arithmetical) rings is still an unexplored avenue. For sake of simplicity, we will deal with the ideal case.

Definition and notation 186. Let $I = \langle f_1, \ldots, f_s \rangle$ be an ideal in $\mathbf{K}[X_1, \ldots, X_k]$, where $\mathbf{K}$ is a field, and denote by $\mathbf{A} := \mathbf{K}[X_1, \ldots, X_k]/I$.

1. For $i \in \mathbb{N}$, we denote by E_i the **K**-vector subspace of **A** generated by the $\bar{M}$'s, where M is a monomial at $X_1, \ldots, X_k$ of total degree at most i and $\bar{M}$ denotes its class modulo I. Note that E_i is a finite-dimensional **K**-vector space of dimension $\binom{k+i}{i}$. Also, set $F_0 = E_0$, and for $i \geq 1$, decompose $E_i = E_{i-1} \oplus F_i$.

   The function $h_I : \mathbb{N} \to \mathbb{N}$ defined by $h_I(i) := \dim_{\mathbf{K}} E_i$ is called the *Hilbert function* of I.

2. The formal power series

$$\mathrm{HS}_I(t) := \sum_{i \geq 0} (\dim_{\mathbf{K}} E_i) t^i,$$

is called the *Hilbert series* of I. Note that, denoting by $\mathrm{H}_I(t) := \sum_{i \geq 0} (\dim_{\mathbf{K}} F_i) t^i$, we obviously have the relation $(1-t)\, \mathrm{HS}_I(t) = \mathrm{H}_I(t)$.

   For example,

$$\mathrm{HS}_{\langle 0 \rangle}(t) = \sum_{i \geq 0} \binom{k+i}{i} t^i = \frac{1}{(1-t)^{k+1}},$$

$$\mathrm{HS}_{\langle X_1, \ldots, X_k \rangle}(t) = \sum_{i \geq 0} t^i = \frac{1}{1-t},$$

$$\mathrm{H}_{\langle X_r X_\ell;\, 1 \leq r < \ell \leq k \rangle}(t) = 1 + \sum_{i \geq 1} k t^i = 1 + \frac{kt}{(1-t)} = \frac{1 + (k-1)t}{(1-t)}.$$

The following gives a substantial simplification of the computation of Hilbert series.

Proposition 187. *Let* **K** *be a field, and consider a total order* $>$ *on* $\mathbf{K}[X_1, \ldots, X_k]$, *i.e., a monomial order on* $\mathbf{K}[X_1, \ldots, X_k]$ *such that* $M > N$ *whenever* $\mathrm{tdeg}(M) > \mathrm{tdeg}(N)$. *Then* $\mathrm{HS}_I(t) = \mathrm{HS}_{\mathrm{LT}(I)}(t)$.

Proof. Let G be a Gröbner basis of I for the total order.
Denote $\mathbf{A} := \mathbf{K}[X_1, \ldots, X_k]/I$ and consider the one-to-one map

$$\varphi : \quad \mathbf{A} \quad \longrightarrow \quad \mathbf{K}[X_1, \ldots, X_k]$$
$$p \quad \longmapsto \quad \bar{p}^G.$$

For $d \geq 0$, we consider $\varphi_d : \mathbf{A}_{\leq d} \to \mathbf{K}[X_1, \ldots, X_k]$, the restriction of φ to $\mathbf{A}_{\leq d}$. We denote by V_d the **K**-vector space generated by the monomials at $X_1, \ldots, X_k$ of total degree $\leq d$ and which are not in $\mathrm{LT}(I)$. For $f \in V_d$, we have $f = \bar{f}^G = \varphi_d(f + I)$, and, thus, $V_d \subseteq \mathrm{Im}(\varphi_d)$. The reverse inclusion clearly holds.

The following two lemmas are at the heart of the algorithm we will present for computing Hilbert series.

Lemma 188. *Let $I = \langle f \rangle$ be a principal ideal of $\mathbf{K}[X_1,\ldots,X_k]$, where $\mathbf{K}$ is a field. Then*

$$H_I(t) = \begin{cases} \frac{1-t^{\text{tdeg}(f)}}{(1-t)^{k+1}} & \text{if } f \neq 0 \\ \frac{1}{(1-t)^{k+1}} & \text{if } f = 0. \end{cases}$$

Proof. Suppose that $f = 0$ and let us prove the result by induction on k. We denote by $H_k(t)$ the Hilbert series of $\langle 0 \rangle$ in $\mathbf{K}[X_1,\ldots,X_k]$. For $k = 0$, $h_I(d) = 1$ for all $d \in \mathbb{N}$, and, thus, $H_0(t) = \frac{1}{1-t}$. Now, as $\mathbf{K}[X_1,\ldots,X_k]_{\leq d} = \bigoplus_{i+j=d} \mathbf{K}[X_1,\ldots,$

$X_{k-1}]_{\leq i} X_k^j$, $H_k(t) = H_{k-1}(t)\frac{1}{1-t} = \frac{1}{(1-t)^{k+1}}$.

Case $f \neq 0$. We have the exact sequence

$$0 \to \mathbf{K}[X_1,\ldots,X_k]^{[\text{tdeg}(f)]} \xrightarrow{\text{multiplication by } f} \mathbf{K}[X_1,\ldots,X_k] \to \mathbf{K}[X_1,\ldots,X_k]/\langle f \rangle \to 0,$$

where $\mathbf{K}[X_1,\ldots,X_k]^{[\text{tdeg}(f)]}$ is the graded $\mathbf{K}$-algebra $\mathbf{K}[X_1,\ldots,X_k]$ with degrees shifted by $\text{tdeg}(f)$. It follows that $H_I(t) = \frac{1}{(1-t)^{k+1}} - \frac{t^{\text{tdeg}(f)}}{(1-t)^{k+1}} = \frac{1-t^{\text{tdeg}(f)}}{(1-t)^{k+1}}$.

Lemma 189. (Addition formula for Hilbert series of homogeneous ideals)

Let $\mathbf{K}$ be a field and consider two finitely-generated homogeneous ideals I and J in $\mathbf{K}[X_1,\ldots,X_k]$. Then

$$\text{HS}_{I+J}(t) + \text{HS}_{I \cap J}(t) = \text{HS}_I(t) + \text{HS}_J(t).$$

Proof. As the ideals I, J and $I+J$ are homogeneous, the map $I_{\leq d} \to (I+J)_{\leq d}/J_{\leq d}$, $f \mapsto f + J_{\leq d}$, is surjective with kernel $(I \cap J)_{\leq d}$.

Lemma 189 is no longer true if we drop the homogeneity hypothesis. To see this, taking $I = \langle X_1 - X_2^2 \rangle$, $J = \langle X_1^2 \rangle$, we have $I + J = \langle X_1 - X_2^2, X_1^2 \rangle$, $I \cap J = \langle X_1^3 - X_2^2 X_1^2 \rangle$, $\text{HS}_{I+J}(t) + \text{HS}_{I \cap J}(t) = 1 + \sum_{i \geq 1} 4t^i + \frac{1-t^4}{(1-t)^3} = \frac{2+t-5t^2+3t^3-t^4}{(1-t)^3} \neq \text{HS}_I(t) + \text{HS}_J = \frac{2(1-t^2)}{(1-t)^3}$.

The previous two lemmas and proposition yield to the following algorithm and result.

Proposition 190. (Computing Hilbert series and Krull dimension with Gröbner bases)

Let $\mathbf{K}$ be a field, and consider a total order $>$ on $\mathbf{K}[X_1,\ldots,X_k]$. The following holds:

(1) *If $I = \langle f_1, \ldots, f_s \rangle$ is an ideal in $\mathbf{K}[X_1, \ldots, X_k]$, then $\mathrm{HS}_I(t)$ can be computed in a finite number of steps with the following algorithm:*

Input: An ideal $I = \langle f_1, \ldots, f_s \rangle$ in $\mathbf{K}[X_1, \ldots, X_k]$.

Output: The Hilbert series $\mathrm{HS}_I(t)$.

    (i) Fix a total order $>$ on $\mathbf{K}[X_1, \ldots, X_k]$, compute a Gröbner basis G of I with respect to $>$, and denote by $M_1, \ldots, M_r$ the leading monomials of the nonzero elements of G.

    (ii) If $r = 0$, then return $\mathrm{HS}_I(t) = \frac{1}{(1-t)^{k+1}}$.

    (iii) Set $J := (M_2, \ldots, M_r)$ and $\tilde{J} := (\mathrm{LCM}(M_1, M_2), \ldots, \mathrm{LCM}(M_1, M_r))$.

    (iv) Compute the Hilbert series $\mathrm{HS}_J(t)$ and $\mathrm{HS}_{\tilde{J}}(t)$ by a recursive call of the algorithm.

    (v) Return $\mathrm{HS}_I(t) := \frac{1 - t^{\deg(m_1)}}{(1-t)^{k+1}} + \mathrm{HS}_J(t) - \mathrm{HS}_{\tilde{J}}(t)$.

(2) *If $I = \langle f_1, \ldots, f_s \rangle$ is an ideal in $\mathbf{K}[X_1, \ldots, X_k]$, then, as a consequence of the algorithm given in (1), $\mathrm{HS}_I(t)$ has the form*

$$\mathrm{HS}_I(t) = \frac{a_0 + a_1 t + \cdots + a_n t^n}{(1-t)^{k+1}} = \frac{Q(t)}{(1-t)^{\delta+1}},$$

with $n \in \mathbb{N}$, $\delta \in [\![0, k]\!]$, $a_i \in \mathbb{Z}$, and $Q(t) \in \mathbb{Z}[t]$ with $Q(1) \neq 0$. Moreover, the Hilbert function is ultimately polynomial. More precisely, the polynomial (called the Hilbert polynomial *of I)*

$$p_I(X) := \sum_{j=0}^{n} a_j \binom{X + k - j}{k} \in \mathbb{Q}[X],$$

satisfies $h_I(d) = p_I(d)$ for sufficiently large integer d. The least d_0 such that $h_I(d) = p_I(d)$ for $d \geq d_0$ is called the Hilbert regularity. *Moreover, the leading coefficient of p_I is $\frac{Q(1)}{\delta!}$ and*

$$\deg(p_I) = \delta = \mathrm{Kdim}(\mathbf{K}[X_1, \ldots, X_k]/I),$$

and, therefore, $\mathrm{Kdim}(\mathbf{K}[X_1, \ldots, X_k]/I) = \mathrm{Kdim}(\mathbf{K}[X_1, \ldots, X_k]/\mathrm{LT}(I))$. From this fact ensues the following algorithm for computing $\mathrm{Kdim}(\mathbf{K}[X_1, \ldots, X_k]/I)$.

Input: An ideal $I = \langle f_1, \ldots, f_s \rangle$ in $\mathbf{K}[X_1, \ldots, X_k]$.

Output: $\mathrm{Kdim}(\mathbf{A})$ with $\mathbf{A} = \mathbf{K}[X_1, \ldots, X_k]/I$.

   (i) Fix a total order $>$ on $\mathbf{K}[X_1,\ldots,X_k]$, compute a Gröbner basis G of I with respect to $>$, and denote by $M_1,\ldots,M_r$ the leading monomials of the nonzero elements of G.

  (ii) If $M_j = 1$ for some j, return $\mathrm{Kdim}(\mathbf{A}) = -1$.

 (iii) By an exhaustive search, find a set $E \subseteq \{X_1,\ldots,X_k\}$ of minimal size such that every M_j involves at least one indeterminate from E.

 (iv) Return $\mathrm{Kdim}(\mathbf{A}) = k - \sharp(E)$.

Example 191. Consider the ideal $I = \langle XY+X, X^2, Y^2 - 1\rangle \subseteq \mathbb{Q}[X,Y]$. Choosing the graded lexicographic order with $X > Y$, $G = \{XY+X, X^2, Y^2 - 1\}$ is a Gröbner basis for I and $\mathrm{LT}(I) = \langle XY, X^2, Y^2\rangle$. We see that $\mathrm{Kdim}(\mathbb{Q}[X,Y]/I) = 2 - \sharp(\{X,Y\}) = 2 - 2 = 0$. Setting $J = \langle X^2, Y^2\rangle$ and $\tilde{J} = \langle X^2 Y, XY^2\rangle$, we have:

$$
\begin{aligned}
\mathrm{HS}_I(t) = \mathrm{HS}_{\mathrm{LT}(I)}(t) &= \frac{1-t^2}{(1-t)^3} + \mathrm{HS}_J(t) - \mathrm{HS}_{\tilde{J}}(t) \\
&= \frac{1-t^2}{(1-t)^3} + \mathrm{HS}_{\langle X^2\rangle}(t) + \mathrm{HS}_{\langle Y^2\rangle}(t) - \mathrm{HS}_{\langle X^2 Y^2\rangle}(t) - \mathrm{HS}_{\langle X^2 Y\rangle}(t) \\
&\quad - \mathrm{HS}_{\langle XY^2\rangle}(t) + \mathrm{HS}_{\langle X^2 Y^2\rangle}(t) \\
&= \frac{1 - 3t^2 + 2t^3}{(1-t)^3} = \frac{1+2t}{1-t} = \sum_{d \geq 0} t^d + \sum_{d \geq 1} 2t^d = 1 + \sum_{d \geq 1} 3t^d.
\end{aligned}
$$

Thus, $h_I(0) = 1$, and for $d \geq 1$, $h_I(d) = 3$ (as expected).

2.8   Exercises

Exercise 192. Let $I = \langle XY - 4, Y^2 - X^3 + 1\rangle \subseteq \mathbb{R}[X,Y]$. Compute $I \cap \mathbb{R}[Y]$ and $\mathrm{Kdim}(\mathbb{R}[X,Y]/I)$.

Exercise 193. Let I be a zero-dimensional ideal of $\mathbf{K}[X_1,\ldots,X_n]$ ($\mathbf{K}$ a field) and consider a Gröbner basis $G = \{g_1,\ldots,g_s\}$ for I accordingly to a monomial order $>$. We denote by $\omega_1 = 1 < \omega_2 < \cdots < \omega_\delta$ the monomials at $X_1,\ldots,X_n$ which are not in $\mathrm{LT}(I) = \langle \mathrm{LT}(g_1),\ldots,\mathrm{LT}(g_s)\rangle$.

$\mathscr{S}(G) := (\omega_1 = 1, \omega_2,\ldots,\omega_\delta)$ or $\{\omega_1 = 1, \omega_2,\ldots,\omega_\delta\}$ (depending on the context) is called the *staircase of the Gröbner basis* G. Recall that $\mathscr{S}(G)$ is a basis of the $\mathbf{K}$-vector space $\mathbf{A} := \mathbf{K}[X_1,\ldots,X_n]/I$.

We also define the *frontier* of G as $\mathscr{F}(G) := \{X_i e \mid 1 \leq i \leq n,\ e \in \varepsilon(G),\ X_i e \notin \varepsilon(G)\}$.

1) Show that for $e \in \mathscr{S}(G)$, if $X_i \mid e$ then $\frac{e}{X_i} \in \mathscr{S}(G)$.

2) Take $G = \{X^2, Y^2\} \subseteq \mathbf{K}[X,Y]$. Check that $\langle G \rangle$ has dimension zero. Give $\mathscr{S}(G)$ and $\mathscr{F}(G)$.

3) Take $I = \langle X^2, XY + Y^2 \rangle \subseteq \mathbb{Q}[X,Y]$. Compute a Gröbner basis G for I accordingly to the lexicographic order with $X > Y$. Check that I has dimension zero. Give $\mathscr{S}(G)$ and $\mathscr{F}(G)$.

4) Show that if I is a zero-dimensional ideal of $\mathbf{K}[X_1, \ldots, X_n]$ and G is the reduced Gröbner basis for I accordingly to a monomial order $>$, then for every $t \in \mathscr{F}(G)$ one of the following holds:

(i) $\exists\, g \in G$ such that $\mathrm{LM}(g) = t$.

(ii) $\exists\, g' \in \mathscr{F}(G)$ and $1 \leq i \leq n$ such that $t = X_i g'$.

5) Show that if I is a zero-dimensional ideal of $\mathbf{K}[X_1, \ldots, X_n]$ of degree δ and G is a reduced Gröbner basis for I accordingly to a monomial order $>$, then

$$\sharp(G) \leq n\,\delta.$$

Exercise 194. Let $\mathscr{C}$ be the cubic curve given by all points $(x,y) \in \mathbb{A}^2(\mathbb{R})$ such that $y^2 - x(x^2 - 1) = 0$. The cubic $\mathscr{C}$ is shown in Fig. 2.10.

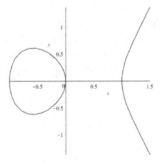

Figure 2.10: An irreducible, disconnected, non-rational cubic

1) Show that $\mathscr{C}$ is disconnected.

2) Show that $\mathscr{C}$ is irreducible.

3) Show that $\mathscr{C}$ is not rational.

Exercise 195. Let $\mathscr{E}$ be the cubic curve (shown in Fig. 2.11) given by all points $(x,y) \in \mathbb{A}^2(\mathbb{R})$ such that $y^2 - x^2(x+1) = 0$. Show that $\mathscr{E}$ is irreducible (by way of obtaining a parametrization).

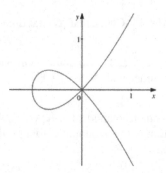

Figure 2.11: A rational irreducible cubic

Exercise 196. Let $\mathbf{K}$ be a field.

1) Let f, g be two polynomials in $\mathbf{K}[X_1, \ldots, X_n]$ and $h = \mathrm{lcm}(f, g)$. Show that

$$\langle f \rangle \cap \langle g \rangle = \langle h \rangle.$$

2) Deduce an algorithm for computing the gcd of two polynomials $f, g \in \mathbf{K}[X_1, \ldots, X_n] \setminus \mathbf{K}$ without factorizing f and g.

Exercise 197. Let $\mathscr{S}$ be the affine algebraic variety given by all points $(x, y, z) \in \mathbb{A}^3(\mathbb{R})$ such that $x^2 = y^2 z$. This variety (see Fig. 2.12) is called the "*Whitney umbrella*". As the picture below indicates, the Whitney umbrella has two parts: the umbrella's handle (the z-axis, in the following: $\mathscr{A}$) along with a 2-dimensional surface.

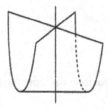

Figure 2.12: Whitney umbrella

1) Show that $\mathscr{A}$ is an irreducible algebraic variety.

2) At first sight it seems interesting to try to isolate the surface of the Whitney umbrella.

a) Calculate $\langle X^2 - Y^2 Z \rangle : \langle X, Y \rangle$.

b) Show that $\overline{\mathscr{S} \setminus \mathscr{A}} = \mathscr{S}$.

c) Verify that $\mathscr{S}$ is irreducible (whence attempting to separate the handle from the surface fails).

Exercise 198. Compute the Zariski closure V of the image of the map

$$\varphi : \mathbb{A}^2(\mathbb{R}) \to \mathbb{A}^3(\mathbb{R}), \; (s,t) \mapsto (st, t, s^2).$$

Exercise 199. For a ring $\mathbf{R}$ in which one can test whether an element is a zero-divisor, we denote by $P_{(\mathbf{R})}$ the probability that an element in $\mathbf{R}$ is a zero-divisor (including zero).

(1) Show that if $\mathbf{R}$ is a finite local ring with n elements then

(a) $P_{(\mathbf{R})} \leq \frac{1}{2}$.

(b) $P_{(\mathbf{R} \times \mathbf{R})} = 2P_{(\mathbf{R})} - P_{(\mathbf{R})}^2$.

(c) If $P_{(\mathbf{R})} = \frac{1}{2}$ (that is, maximal), then $P_{(\mathbf{R} \times \mathbf{R})} = \frac{3}{4}$.

(2) Show that if p, q are prime numbers and $\alpha, \beta \in \mathbb{N}^*$, then

$$P_{(\mathbb{Z}/p^\alpha \mathbb{Z})} = \frac{1}{p} \; \& \; P_{((\mathbb{Z}/p^\alpha \mathbb{Z}) \times (\mathbb{Z}/q^\beta \mathbb{Z}))} = \frac{1}{p} + \frac{1}{q} - \frac{1}{pq}.$$

(3) Deduce $\sup_{\{p, q \text{ prime numbers}; \; \alpha, \beta \geq 1\}} P_{((\mathbb{Z}/p^\alpha \mathbb{Z}) \times (\mathbb{Z}/q^\beta \mathbb{Z}))}$.

Exercise 200. Show that the notion of multiplicity from Definition 164 is independent of how the line is parametrized.

Exercise 201. 1) Show that a line L is tangent to a plane curve C at a nonsingular point P if and only if L meets C at P with multiplicity ≥ 2.

2) Show that if P is a point of a plane curve C of multiplicity 2, and a line L is tangent to C at P with multiplicity 2 (see Definition 345), then L meets C at P with multiplicity ≥ 3.

Exercise 202. Let $f(X,Y) = X^4 + 2XY^2 + Y^3 \in \mathbb{R}[X,Y]$. Show that all lines through the origin meet $V(f)$ (see Fig. 2.13) with multiplicity ≥ 3. Two lines meet $V(f)$ with multiplicity 4. Can you determine them?

Exercise 203. Let $I = \langle f_1 = X^2 - Y^2 - 1, \; f_2 = XY^2 - X^2Y \rangle \subseteq \mathbb{Q}[X,Y]$.

1) Calculate a Gröbner basis of I with respect to the lexicographic order with $X > Y$. Determine $\langle \mathrm{LT}(I) \rangle$ and give a monomial basis $\mathscr{B}$ for $\mathbf{A} = \mathbb{Q}[X,Y]/I$ as $\mathbb{Q}$-vector space.

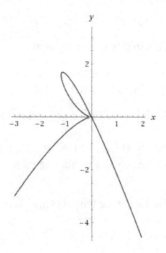

Figure 2.13: $x^4 + 2xy^2 + y^3 = 0$

What is the dimension of the variety $V(I)$?
What is the degree of the ideal I?

2) Is I radical?

3) Find a linear form u which is separating for $V(I)$ (without trying to know the points of $V(I)$).

4) Give the matrix M_u of the multiplication by u in A written in the basis $\mathscr{B}$. Calculate its characteristic polynomial.

5) Deduce a parametrization of the variety $V(I)$ (we are not trying to determine the points of $V(I)$).

Exercise 204. (Seidenberg's Lemma)

I) Let $\mathbf{K}$ be a field, and consider a zero-dimensional ideal I of $\mathbf{K}[X_1, \ldots, X_n]$ such that, for every $1 \leq i \leq n$, there exists a nonzero polynomial $g_i \in \mathbf{K}[X_i]$ such that $g_i \wedge g_i' = 1$. The goal of this exercise is to prove that I is a radical ideal. This is the so-called "Seidenberg's Lemma".

1) Show that if $n = 1$ then I is a radical ideal.

2) Write $g_1 = h_1 \cdots h_s$, where the h_j's are distinct irreducible polynomials in $\mathbf{K}[X_1]$. Show that $I = \cap_{j=1}^{s}(I + \langle h_j \rangle)$.

3) Show that I is a radical ideal for $n \geq 1$.

II) Let $\mathbf{K}$ be a perfect field (see Definition 223) equipped with an algorithm computing the square-free part sqfree(f) of a nonconstant univariate polynomial f

with coefficients in $\mathbf{K}$ (such algorithm ensues from Exercise 238, see [41] for more details). From I), deduce an algorithm computing the radical of a zero-dimensional ideal of $\mathbf{K}[X_1,\ldots,X_n]$.

Exercise 205. Let $\mathbf{K}$ be a field and consider the ideal $I = \langle x_1^2 x_2^2, x_1^2 x_3^2, x_2^2 x_3^2 \rangle$ of $\mathbf{K}[x_1,x_2,x_3]$. Calculate the Hilbert series, the Hilbert function and the Hilbert polynomial of I.

For which d the Hilbert function and the Hilbert polynomial coincide?

Give $\mathrm{Kdim}(\mathbf{K}[x_1,x_2,x_3]/I)$.

Exercise 206. Let I, J be two ideals of $\mathbf{K}[X_1,\ldots,X_k]$ ($\mathbf{K}$ a field) such that $I \subseteq J$ and $H_I = H_J$. Show that $I = J$.

2.9 Solutions to the exercises

Exercise 192:

We find $G = \{-64 + Y^3 + Y^5, -Y^2 - Y^4 + 16X\}$ as the reduced Gröbner basis for I accordingly to the lexicographic order with $X > Y$. It follows that $I \cap \mathbb{R}[Y] = \langle -64 + Y^3 + Y^5 \rangle$, $\mathrm{LT}(I) = \langle Y^5, X \rangle$, and $\mathrm{Kdim}(\mathbb{R}[X,Y]/I) = \mathrm{Kdim}(\mathbb{R}[X,Y]/\mathrm{LT}(I)) = \mathrm{Kdim}(\mathbb{R}[X,Y]/\langle Y^5, X \rangle) = \mathrm{Kdim}(\mathbb{R}[Y]/\langle Y^5 \rangle) = 0$.

Exercise 193:

1) $\frac{e}{X_i} \in \mathrm{LT}(I) \Rightarrow e \in \mathrm{LT}(I)$.

2) $\langle G \rangle$ has dimension zero since $\mathbf{K}[X,Y]/\langle G \rangle$ has a finite dimension ($= 4$) as a $\mathbf{K}$-vector space.
$\mathscr{S}(G) = \{1, X, Y, XY\}$ and $\mathscr{F}(G) = \{X^2, Y^2, XY^2, X^2Y\}$.

3) $G = \{X^2, XY + Y^2, Y^3\}$ is a reduced Gröbner basis G for I.
I has dimension zero since $\mathbb{Q}[X,Y]/I$ has a finite dimension ($= 4$) as a $\mathbb{Q}$-vector space.

$\mathscr{S}(G) = \{1, Y, Y^2, X\}$ and $\mathscr{F}(G) = \{X^2, XY, XY^2, Y^3\}$.

4) Let $t \in \mathscr{F}(G)$ and set $A_t := \{1 \le j \le n$ such that $X_j \mid t$ & $\frac{t}{X_j} \notin \mathscr{S}(G)\}$. Two cases may arise:

Case 1. $A_t = \emptyset$. Since $t \notin \mathscr{S}(G)$, there exists $g \in G$ such that $\mathrm{LM}(g) \mid t$. Set $u := \frac{t}{\mathrm{LM}(g)}$. If there exists X_j dividing u, then, setting $v := \frac{u}{X_j}$, we would have $\frac{t}{X_j} = \mathrm{LM}(g)v \notin \mathscr{S}(G)$ and, thus, $j \in A_t$, a contradiction. It follows that $u = 1$ and $\mathrm{LM}(g) = t$.

Case 2. $A_t \ne \emptyset$, that is, there exists $1 \le j \le n$ such that $X_j \mid t$ and $t' := \frac{t}{X_j} \notin \mathscr{S}(G)$. As $t \in \mathscr{F}(G)$, there exists $e \in \mathscr{S}(G)$ such that $t = X_i e = X_j t'$. Necessarily, $i \ne j$

because $t' \notin \mathscr{S}(G)$. It follows that $X_j \mid e$, and, thus, by virtue of 1), $e' := \frac{e}{X_j} \in \mathscr{S}(G)$.

We have $t = X_j t'$ with $t' = X_i e' \in \mathscr{F}(G)$.

5) Let $G = \{g_1, \ldots, g_s\}$ be a reduced Gröbner basis for I. Since G is reduced, the LM(g_j)'s are pairwise distinct and all belong to $\mathscr{F}(G)$. Hence

$$\sharp(G) = s \leq \sharp \left(\bigcup_{1 \leq i \leq n} X_i \mathscr{S}(G) \right) \leq \sum_{1 \leq i \leq n} \sharp(\mathscr{S}(G)) \leq n\,\delta.$$

Exercise 194:

1) x is the abscissa of a point in $\mathscr{C}$ if and only if $x(x^2 - 1) \geq 0$. Thus, the projection of $\mathscr{C}$ on the x-axis is $H := [-1, 0] \cup [1, \infty[$. As H is disconnected so is $\mathscr{C}$.

2) We can prove that the polynomial $f(X, Y) = Y^2 - X(X^2 - 1)$ is irreducible in $\mathbb{R}[X, Y]$. For this, if $Y^2 - X(X^2 - 1) = (a(X)Y + b(X))(c(X)Y + d(X))$ for some polynomials $a(X)$, $b(X)$, $c(X)$, $d(X) \in \mathbb{R}[X]$, then, by identification, we would have $X(X^2 - 1) = (\frac{b(X)}{a(X)})^2$ which is impossible (take $X \geq -2$).

3) The singular points of $\mathscr{C}$ correspond to the solutions of the following system of polynomial equations:

$$S : \begin{cases} f(x, y) = y^2 - x^3 + x = 0 \\ \frac{\partial f}{\partial x}(x, y) = -3x^2 + 1 = 0 \\ \frac{\partial f}{\partial y}(x, y) = 2y = 0. \end{cases} \tag{2.9}$$

As S is impossible, $\mathscr{C}$ is smooth and its genus is 1 (equal to its arithmetical genus). We conclude that $\mathscr{C}$ is not rational (by virtue of Corollary 304).

Exercise 195:

It is easy to check that $(0, 0)$ is the unique singular point of $\mathscr{E}$. Thus, the genus of $\mathscr{E}$ is 0 and it is rational (by virtue of Corollary 304).

We have: $(x, y) \in \mathscr{E} \Leftrightarrow x + 1 = \frac{y^2}{x^2}$. Thus, putting $x + 1 = t^2$, we obtain the following parametrization of $\mathscr{E}$:

$$\begin{cases} x(t) = t^2 - 1 \\ y(t) = t(t^2 - 1). \end{cases} \tag{2.10}$$

We infer that $\mathscr{E}$ is irreducible.

Exercise 196:

1) P is a common multiple of f and g $\Leftrightarrow$ P is a multiple of h.

2) Compute the reduced Gröbner basis G of $\langle t f, (1 - t)g \rangle \subseteq \mathbf{K}[X_1, \ldots, X_n, t]$ with respect to the lexicographic order with $t > X_n > \cdots > X_1$. Then $G \cap \mathbf{K}[X_1, \ldots, X_n] = \{\mathrm{lcm}(f, g)\}$. Finally, $\gcd(f, g) = \frac{fg}{\mathrm{lcm}(f,g)}$.

Exercise 197:

1) The coordinate ring of $\mathscr{A}$ is $\mathbf{A} := \mathbb{R}[X,Y,Z]/\langle X,Y \rangle \cong \mathbb{R}[Z]$. Since $\mathbf{A}$ is integral, the variety $\mathscr{A}$ is an irreducible.

2)

  a) $\langle X^2 - Y^2 Z \rangle : \langle X, Y \rangle = (\langle X^2 - Y^2 Z \rangle : \langle X \rangle) \cap (\langle X^2 - Y^2 Z \rangle : \langle Y \rangle) = \langle X^2 - Y^2 Z \rangle$
  $\cap \langle X^2 - Y^2 Z \rangle = \langle X^2 - Y^2 Z \rangle$ (by virtue of Exercise 196).

  b) This is an immediate consequence of a).

  c) This amounts to proving that the polynomial $X^2 - Y^2 Z$ is irreducible. For this aim, it suffices to notice that each product of two polynomials which is equal to $X^2 - YZ$ contains a factor of degree 0 at Z. This latter factor must divide both X^2 and Y^2 and hence it is a nonzero constant.

Exercise 198:

Let J be the ideal $\langle x - st,\ y - t,\ z - s^2 \rangle$ of $\mathbb{R}[s,t,x,y,z]$. We calculate the reduced Gröbner basis G of J with respect to the lexicographic order with $s > t > x > y > z$ using the Computer Algebra software $\mathtt{SINGULAR}$. We obtain $G = \{x^2 - y^2 z,\ t - y,\ sy - x,\ sx - yz,\ s^2 - z\}$. The Zariski closure V is then the algebraic curve defined by the equation $x^2 - y^2 z$ (Whitney umbrella, Fig. 2.12).

Exercise 199:

For a ring $\mathbf{R}$, we will denote its subset of zero-divisors by $Z(\mathbf{R})$ (including zero).

(1)

  (a) As $\mathbf{R}$ is local, we have:
  $$x \in Z(\mathbf{R}) \;\Rightarrow\; 1 + x \in \mathbf{R}^\times, \text{ and thus,}$$
  $$\sharp Z(\mathbf{R}) \le \sharp U(\mathbf{R}), \text{ and}$$
  $$P_{(\mathbf{R})} \le \frac{1}{2}.$$

  (b) Note that If $\mathbf{R}$ and $\mathbf{T}$ are two finite rings, then:
  $$\sharp (\mathbf{R} \times \mathbf{T})^\times = \sharp \mathbf{R}^\times \cdot \sharp \mathbf{T}^\times,$$
  $$\sharp Z(\mathbf{R} \times \mathbf{T}) = \sharp Z(\mathbf{R}) \cdot \sharp(\mathbf{T}) + \sharp(\mathbf{R}) \cdot \sharp Z(\mathbf{T}) - \sharp Z(\mathbf{R}) \cdot \sharp Z(\mathbf{T}).$$

  In particular, if $\mathbf{R}$ is a finite local ring with n elements, denoting $\sharp Z(\mathbf{R}) = k$ (necessarily, $2k \le n$), we have:

$$\sharp(\mathbf{R} \times \mathbf{R})^{\times} = (n-k)^2, \ \sharp Z(\mathbf{R} \times \mathbf{R}) = k(2n-k), \text{ and thus,}$$

$$\frac{\sharp Z(\mathbf{R} \times \mathbf{R})}{\sharp(\mathbf{R} \times \mathbf{R})} = \frac{k(2n-k)}{n^2} = P_{(\mathbf{R} \times \mathbf{R})} = 2\frac{k}{n} - \frac{k^2}{n^2} = 2P_{(\mathbf{R})} - P_{(\mathbf{R})}^2.$$

(2) From the Euler function formula

$$\Phi(p^{\alpha}) = \sharp(\mathbb{Z}/p^{\alpha}\mathbb{Z})^{\times} = p^{\alpha} - p^{\alpha-1},$$

we infer that $\sharp Z(\mathbb{Z}/p^{\alpha}\mathbb{Z}) = p^{\alpha-1}$, and hence, $P_{(\mathbb{Z}/p^{\alpha}\mathbb{Z})} = \frac{1}{p}$. In addition,

$$P_{((\mathbb{Z}/p^{\alpha}\mathbb{Z}) \times (\mathbb{Z}/q^{\beta}\mathbb{Z}))} = \frac{p^{\alpha-1}q^{\beta} + p^{\alpha}q^{\beta-1} - p^{\alpha-1}q^{\beta} - 1}{p^{\alpha}q^{\beta}} = \frac{1}{p} + \frac{1}{q} - \frac{1}{pq}.$$

(3) It is clear that $P_{(\mathbb{Z}/p^{\alpha}\mathbb{Z})}$ is maximal when $p = 2$. Taking $p = q = 2$, we have $P_{((\mathbb{Z}/2^{\alpha}\mathbb{Z}) \times (\mathbb{Z}/2^{\beta}\mathbb{Z}))} = \frac{3}{4}$. For $p, q \geq 3$, we have $P_{((\mathbb{Z}/p^{\alpha}\mathbb{Z}) \times (\mathbb{Z}/q^{\beta}\mathbb{Z}))} \leq \frac{1}{p} + \frac{1}{q} \leq \frac{2}{3} < \frac{3}{4}$. For $p = 2$ and $q \geq 3$, we have $P_{((\mathbb{Z}/p^{\alpha}\mathbb{Z}) \times (\mathbb{Z}/q^{\beta}\mathbb{Z}))} = \frac{1}{2} + \frac{1}{q} - \frac{1}{2q} = \frac{1}{2} + \frac{1}{2q} \leq \frac{2}{3} < \frac{3}{4}$. We conclude that

$$\sup_{\{p,q \text{ prime numbers; } \alpha, \beta \geq 1\}} P_{((\mathbb{Z}/p^{\alpha}\mathbb{Z}) \times (\mathbb{Z}/q^{\beta}\mathbb{Z}))} = \frac{3}{4}.$$

It is in fact a maximum reached only when $p = q = 2$.

Exercise 200:

Let $f \in \mathbf{K}[X,Y]$. Note first that, for $(c,d), (c',d') \in \mathbf{K}^2 \setminus \{(0,0)\}$, two parametrizations

$$\begin{cases} x = a + ct & x = a + c't \\ y = b + dt & y = b + d't \end{cases} \tag{2.11}$$

correspond to the same line L if and only if there is a nonzero number $\lambda \in \mathbf{K}$ such that $(c,d) = \lambda(c',d')$. In that case, the corresponding polynomials

$$g(t) = f(a+ct, b+dt) \quad \text{and} \quad h(t) = f(a+c't, b+d't).$$

are related by $g(t) = h(\lambda t)$.

Exercise 201:

1) By a coordinates change, we can suppose that $P = (0,0)$ and $L : x - y = 0$. Now write $C : F(x,y) = F_0(x,y) + F_1(x,y) + F_2(x,y) + \cdots = 0$, where $F_i \in \mathbf{K}[X,Y]$ is homogeneous of degree i. Then, L is tangent to C at P if and only if $F_0 = 0$ & $F_1(X,Y) = \lambda(X - Y)$ for some $\lambda \in \mathbf{K}^{\times}$, or, in other terms, if and only if X^2 divides $F(X,X)$. This latter condition is nothing but $\mathrm{Ord}_{P,L}(F(x,y)) \geq 2$.

2) $F_0(x,y) = F_1(x,y) = 0$ since the multiplicity of P is 2, and $F_2(x,y)$ need to be equal to L^2 sine L is tangent to C at P with multiplicity 2. Hence, we can put $x = y$ and eliminate F_0, F_1, and F_2.

Exercise 202:

Let $f(X,Y) = X^4 + 2XY^2 + Y^3$, and consider a line $x = ct, y = dt$, where $(c,d) \neq (0,0)$, through the origin. Calculate $g(t) = f(ct,dt)$ so as to obtain

$$g(t) = t^3(c^4 t + d^2(2c+d)).$$

Now do a case distinction in order to see that $t = 0$ is a root of multiplicity ≥ 3 of g. The two lines that meet $V(f)$ with multiplicity 4 are the x-axis ($y = 0$), and the line $y = -2x$.

Exercise 203:

1) We find $G = \{Y + Y^3, XY, X^2 - Y^2 - 1\}$ as the reduced Gröbner basis for I accordingly to the lexicographic order with $X > Y$. Thus,

$$\mathrm{LT}(I) = \langle Y^3, XY, X^2 \rangle.$$

It follows that $\mathscr{B} := (1, Y, Y^2, X)$ is a basis for $\mathbf{A} := \mathbb{Q}[X,Y]/I$ as $\mathbb{Q}$-vector space and for $\bar{\mathbf{A}} := \mathbb{C}[X,Y]/I_\mathbb{C}$ as $\mathbb{C}$-vector space, and that $\deg(I) = 4$. Of course, the dimension of the variety $V(I)$ is zero since $\mathbf{A}$ has a finite dimension as $\mathbb{Q}$-vector space.

2) We denote by $q_I = \mathrm{QuadHerm}(I)$ the Hermite quadratic form associated to I, i.e., the map $A \to \mathbb{Q}$ such that $q_I(p) = \mathrm{Tr}(M_{p^2})$: the trace of the multiplication matrix in A by p^2 written in the basis $\mathscr{B}$.

This quadratic form is associated to the bilinear from $\varphi_I : A \times A \to \mathbb{Q}$; $(p,q) \mapsto \mathrm{Tr}(M_{pq})$ where, for $f \in A$, M_f is the matrix of the multiplication in A by f written in the basis $\mathscr{B}$. We have:

$$M := M_X = \begin{pmatrix} 0 & 0 & 0 & 1 \\ 0 & 0 & 0 & 0 \\ 0 & 0 & 0 & 1 \\ 1 & 0 & 0 & 0 \end{pmatrix}, N := M_Y = \begin{pmatrix} 0 & 0 & 0 & 0 \\ 1 & 0 & -1 & 0 \\ 0 & 1 & 0 & 0 \\ 0 & 0 & 0 & 0 \end{pmatrix}.$$

$$M_1 = \mathbf{I}_4, \mathrm{Tr}(M_1) = 4, \mathrm{Tr}(M_Y) = 0, M_{Y^2} = N^2 = \begin{pmatrix} 0 & 0 & 0 & 0 \\ 0 & -1 & 0 & 0 \\ 1 & 0 & -1 & 0 \\ 0 & 0 & 0 & 0 \end{pmatrix},$$

$\mathrm{Tr}(M_{Y^2}) = -2, \mathrm{Tr}(M_X) = 0, \mathrm{Tr}(M_{Y^3}) = \mathrm{Tr}(M_{-Y}) = 0, M_{XY} = 0,$
$\mathrm{Tr}(M_{Y^4}) = \mathrm{Tr}(M_{-Y^2}) = 2, \mathrm{Tr}(M_{XY^2}) = 0, \mathrm{Tr}(M_{X^2}) = \mathrm{Tr}(M_{Y^2+1}) = -2 + 4 = 2.$
Thus,

$$\text{Mat}(\varphi,\mathscr{B}) = \begin{pmatrix} 4 & 0 & -2 & 0 \\ 0 & -2 & 0 & 0 \\ -2 & 0 & 2 & 0 \\ 0 & 0 & 0 & 2 \end{pmatrix}.$$

We have $\sharp(V(I)) = \text{rk}(q_I) = \text{rk}(\varphi_I) = \text{rank}(\text{Mat}(\varphi,\mathscr{B})) = 4 = \deg(I)$, and, thus, I is radical.

3) We begin by testing whether X is separating for $V(I)$. We have to test whether $1, X, X^2, X^3$ are linearly independent in A. This is not the case since $X^3 = X$ modulo I. We pass to the test for $u = X + Y$. We have to test whether $1, (X + Y), (X + Y)^2, (X + Y)^3$ are linearly independent in A. The matrix

$$H := \begin{pmatrix} 1 & 0 & 1 & 0 \\ 0 & 1 & 0 & -1 \\ 0 & 0 & 2 & 0 \\ 0 & 1 & 0 & 1 \end{pmatrix}$$

being of rank 4, we infer that $u = X + Y$ is separating for $V(I)$.

4) $M_u = M_X + M_Y = M + N = \begin{pmatrix} 0 & 0 & 0 & 1 \\ 1 & 0 & -1 & 0 \\ 0 & 1 & 0 & 1 \\ 1 & 0 & 0 & 0 \end{pmatrix}.$

Its characteristic polynomial is $q(X) = X^4 - 1$. We have

$$H^{-1} = \begin{pmatrix} 0 & 0 & 0 & 1 \\ -1 & 0 & 1 & 0 \\ 0 & -1 & 0 & 1 \\ 1 & 0 & 0 & 0 \end{pmatrix},$$

and, thus, $X = 0.1 + \frac{1}{2}u + 0.u^2 + \frac{1}{2}u^3$ et $Y = 0.1 + \frac{1}{2}u + 0.u^2 - \frac{1}{2}u^3$.

5) We then obtain the following parametrization of the variety $V(I)$:

$$\begin{cases} X = \frac{1}{2}(T + T^3) \\ Y = \frac{1}{2}(T - T^3) \\ T^4 - 1 = 0. \end{cases}$$

Exercise 204:

I.1) By virtue of Exercise 238.1, the principal ideal $I \subseteq \mathbf{K}[X_1]$ contains a square-free polynomial. Therefore it is generated by a square-free polynomial (and, thus, radical).

I.2) Obviously, $I \subseteq \cap_{j=1}^{s}(I + \langle h_j \rangle)$. Conversely, let $f \in \cap_{j=1}^{s}(I + \langle h_j \rangle)$, and write, for $1 \leq j \leq s$, $f = u_j + h_j v_j$ with $u_j \in I$ and $v_j \in \mathbf{K}[X_1, \ldots, X_n]$. We have $f \cdot \prod_{\ell \neq j} h_\ell \in I$ for $1 \leq j \leq s$. As $\gcd(\prod_{\ell \neq j} h_\ell, 1 \leq j \leq s) = 1$ in $\mathbf{K}[X_1]$, we can find

$w_1, \ldots, w_s \in \mathbf{K}[X_1]$ such that $\sum_{j=1}^{s} w_j \prod_{\ell \neq j} h_\ell = 1$. It follows that $f = \sum_{j=1}^{s} w_j \cdot f \cdot \prod_{\ell \neq j} h_\ell \in I$, as desired.

I.3) We proceed by induction on n. For $n = 1$, this is 1). Now let $n > 1$. By virtue of 2), we can suppose that g_1 is irreducible. Consider the field $Ł := \mathbf{K}[X_1]/\langle g_1 \rangle$ (it is a finite $\mathbf{K}$-vector space), denote by φ the canonical surjective homomorphism $\varphi : \mathbf{K}[X_1, \ldots, X_n] \to Ł[X_2, \ldots, X_n]$, and consider the ideal $J = \varphi(I)$ of $Ł[X_2, \ldots, X_n]$. This latter is zero-dimensional since $Ł[X_2, \ldots, X_n]/J \cong \mathbf{K}[X_1, \ldots, X_n]/I$. For every $2 \leq i \leq n$, the polynomial $\tilde{g}_i := \varphi(g_i)$ satisfies the hypothesis $\tilde{g}_i \wedge \tilde{g}'_i = 1$. So, by the induction hypothesis, J is a radical ideal of $Ł[X_2, \ldots, X_n]$, or, in other words, we have no nonzero nilpotent elements in $Ł[X_2, \ldots, X_n]/J$. By the above isomorphism, also in $\mathbf{K}[X_1, \ldots, X_n]/I$ there are no nonzero nilpotent elements, that is, I is a radical ideal.

II) For $1 \leq i \leq n$, compute a generator $g_i \in \mathbf{K}[X_i]$ of the elimination ideal $I \cap \mathbf{K}[X_i]$. Then

$$\sqrt{I} = I + \langle \text{sqfree}(g_1), \ldots, \text{sqfree}(g_n) \rangle.$$

Exercise 205:

$$
\begin{aligned}
H_I(t) &= \frac{1-t^4}{(1-t)^4} + H_{\langle x_1^2 x_3^2, x_2^2 x_3^2 \rangle}(t) - H_{\langle x_1^2 x_3^2 x_3^2 \rangle}(t) \\
&= \frac{1-t^4}{(1-t)^4} + \frac{1-t^4}{(1-t)^4} + \frac{1-t^4}{(1-t)^4} - H_{\langle x_1^2 x_2^2 x_3^2 \rangle}(t) - H_{\langle x_1^2 x_2^2 x_3 \rangle}(t) \\
&= \frac{3(1-t^4)}{(1-t)^4} - \frac{2(1-t^6)}{(1-t)^4} = \frac{2t^6 - 3t^4 + 1}{(1-t)^4} \\
&= (2t^6 - 3t^4 + 1) \sum_{d=0}^{\infty} \binom{d+3}{d} t^d \\
&= 1 + 4t + 10t^2 + 20t^3 + 32t^4 + 44t^5 + \sum_{d \geq 6} \left(\binom{d+3}{d} - 3\binom{d-1}{d-4} + 2\binom{d-3}{d-6} \right) t^d.
\end{aligned}
$$

The Hilbert polynomial of I is

$$p_I(X) = \binom{X+3}{X} - 3\binom{X-1}{X-4} + 2\binom{X-3}{X-6} = 12X - 16.$$

The first d for which the Hilbert function and the Hilbert polynomial coincide is $d = 3$:

$$H_I(t) = 1 + 4t + 10t^2 + \sum_{d \geq 3} (12t - 16)t^d.$$

$$\text{Kdim}(\mathbf{K}[x_1, x_2, x_3]/I) = \deg(p_I) = 1.$$

Exercise 206:

We have $H_I(t) = H_{\mathrm{LT}(I)}(t) = \sum_{d=0}^{\infty} h(d) t^d$ and $H_J(t) = H_{\mathrm{LT}(J)}(t) = \sum_{d=0}^{\infty} \tilde{h}(d) t^d$ with $h(d), \tilde{h}(d) \in \mathbb{N}$. We have $H_{\mathrm{LT}(I)} = H_{\mathrm{LT}(J)}$ and $\mathrm{LT}(I) \subseteq \mathrm{LT}(J)$.

For $d \geq 0$, denote by V_d (resp. $\tilde{V}_d$) the $\mathbf{K}$-vector subspace of $\mathbf{K}[X_1, \ldots, X_k]$ generated by the monomials of total degree $\leq d$ which don't belong to $\mathrm{LT}(I)$ (resp. $\mathrm{LT}(J)$). We also denote by $V = \cup_{d \geq 0} V_d$ (resp. $\tilde{V} = \cup_{d \geq 0} \tilde{V}_d$) the $\mathbf{K}$-vector subspace of $\mathbf{K}[X_1, \ldots, X_k]$ generated by the monomials which don't belong to $\mathrm{LT}(I)$ (resp. $\mathrm{LT}(J)$).

We have $h(d) = \dim_{\mathbf{K}} V_d$ and $\tilde{h}(d) = \dim_{\mathbf{K}} \tilde{V}_d$. For $d \geq 0$, as $V_d \subseteq \tilde{V}_d$ and $h(d) = \tilde{h}(d)$, the two $\mathbf{K}$-vector spaces V_d and $\tilde{V}_d$ coincide. It follows that $V = \tilde{V}$ and, thus, $\mathrm{LT}(I) = \mathrm{LT}(J)$.

Now, since $\mathrm{LT}(I) = \mathrm{LT}(J)$ and $I \subseteq J$, considering a Gröbner basis G of J, it is also a Gröbner basis of I, and, thus, $I = J = \langle G \rangle$.

Chapter 3

Finite fields and field extensions

Finite field are algebraic structure which are both rich and easily manipulated. For these reasons they have numerous applications most notably in cryptography and coding theory.

3.1 Construction of finite fields

Recall the a field is a unitary ring in which every nonzero element is invertible (we don't suppose that it is commutative). Recall also that every finite field is commutative. As for now, the only finite fields we know are the $\mathbb{F}_p := \mathbb{Z}/p\mathbb{Z}$ where p is a prime number. We will see hereafter how to construct other examples of finite fields.

Let $\mathbf{K}$ be a finite field and denote by 1 its unit element. The set $\{n.1 = \underbrace{1 + \cdots + 1}_{n \text{ times}} \mid n \in \mathbb{Z}\}$ is included in $\mathbf{K}$ and we have the following ring-homomorphism:

$$\varphi : \begin{array}{ccc} \mathbb{Z} & \longrightarrow & \mathbf{K} \\ n & \longmapsto & n.1 \end{array}$$

Its kernel $\mathrm{Ker}(\varphi)$ is an ideal of $\mathbb{Z}$, and, thus, it is of the form $p\mathbb{Z}$ for some $p \in \mathbb{N}$. Necessarily, $p > 0$ (since $\mathbf{K}$ is finite) and p is prime (by minimality of p). The prime number p is called the *characteristic* of $\mathbf{K}$, and denote by $\mathrm{char}(\mathbf{K})$. It is the smallest positive integer n such that $n.1 = 0$ in $\mathbf{K}$. If for a field $\mathbf{F}$ such p does not exist, we say that $\mathbf{F}$ has characteristic zero. For example, $\mathbb{F}_p$ (with p prime) has characteristic p. The infinite field $\mathbb{F}_p(t)$ of algebraic fractions with coefficients in $\mathbb{F}_p$ has characteristic p, while the fields $\mathbb{Q}, \mathbb{R}, \mathbb{C}$ have characteristic 0.

Let $\mathbf{K}$ be a field of characteristic p. Keeping the above notation, by the first isomorphism theorem, we have $\mathbb{Z}/\mathrm{Ker}(\varphi) = \mathbb{F}_p \cong \varphi(\mathbb{Z})$. Thus, every field $\mathbf{K}$ of characteristic $p > 0$ contains a subfield which is isomorphic to $\mathbb{F}_p$. It is called the *prime subfield* of **K**, and denote by $\mathbf{K}_0$. This allows us to specify the cardinality of a finite field.

Proposition 207. *The cardinality of a finite field is a power of a prime number (equal to its characteristic).*

Proof. Let **K** be a finite field with characteristic p. As it is a finite-dimensional $\mathbf{K}_0$-vector space, it is isomorphic to $\mathbf{K}_0^d$ for some $d > 0$, and, thus, its cardinality is p^d.

Example 208. If f is an irreducible polynomial in $\mathbb{F}_p[X]$ (p a prime number) of degree d, then it is immediate that $\mathbb{F}_p[X]/\langle f \rangle$ is a field of cardinality p^d.

The following identity is central when computing in a field with characteristic p.

Proposition 209. *Let **K** be a field with characteristic p. Then, for all $a, b \in$ **K**, we have:*
$$(a+b)^p = a^p + b^p.$$

Proof. We have $(a+b)^p = \sum_{k=0}^{p} \binom{p}{k} a^k b^{p-k}$. But it is well-known that for $1 \le k \le p-1$, p divides $\binom{p}{k}$.

3.2   Field extensions and Galois groups

Definition 210. (Minimal polynomial)

Let **K** be a field, Ł an extension field of **K**, and $\alpha \in$ Ł. The set $I_\alpha := \{f \in \mathbf{K}[X] \mid f(\alpha) = 0\}$ is an ideal of $\mathbf{K}[X]$. If $I_\alpha = (0)$ we say that α is *transcendental* over **K**, otherwise we say that α is *algebraic* over **K**. In that case I_α is generated by a unique monic polynomial called *the minimal polynomial* of α over **K**. For example, the minimal polynomial of $\sqrt{2}$ over $\mathbb{Q}$ is $X^2 - 2$.

Definition 211. (Rupture field, splitting field)

Let **K** be a field and $P \in \mathbf{K}[X]$. A *rupture field* of P over **K** is an extension field of **K** generated by a root a of P.

For instance, if $\mathbf{K} = \mathbb{Q}$ and $P = X^3 - 2$, then $\mathbb{Q}(\sqrt[3]{2})$ is a rupture field for P. If P is irreducible over **K**, all rupture fields of P over **K** are isomorphic to $\mathbf{K}[X]/\langle P \rangle$.

A rupture field of a polynomial does not necessarily contain all the roots of that polynomial: in the above example, the field $\mathbb{Q}(\sqrt[3]{2})$ does not contain the other two (complex) roots of P (namely, $w\sqrt[3]{2}$ and $w^2\sqrt[3]{2}$, where $w = e^{\frac{2i\pi}{3}}$).

A *splitting field* of a polynomial with coefficients in a field is a minimal (for inclusion) extension field of that field over which the polynomial splits or decomposes into linear factors. For example, in the above example, $\mathbb{Q}(\sqrt[3]{2})$ is a rupture field for P but not a splitting field ($\mathbb{Q}(\sqrt[3]{2}, w)$ is a splitting field for P). A rupture field of $X^2 + 1$ over $\mathbb{R}$ is $\mathbb{C}$. It is also a splitting field.

The following proposition explains how finite fields are built.

Proposition 212. *Let* **K** *be a finite field with* q *elements* ($q = p^d$, *where* p *is the characteristic of* **K**)*. Then,* **K** *is a splitting field for* $X^q - X$ *over its prime subfield* $\mathbf{K}_0$*. Conversely, if* $q = p^d$*, where* p *is a prime number, then a splitting field for* $X^q - X$ *over* $\mathbb{F}_p$ *is a finite field with* q *elements.*

Proof. As $\mathbf{K}^\times := \mathbf{K} \setminus \{0\}$ is a finite multiplicative group of order $q - 1$, then, by Lagrange's theorem, for all $x \in \mathbf{K}^\times$, we have $x^{q-1} - 1 = 0$. It follows that for all $x \in \mathbf{K}$, we have $x^q - x = 0$. Thus, $X^q - X = \prod_{x \in \mathbf{K}} (X - x)$ and **K** is a splitting field for $X^q - X$ over $\mathbf{K}_0$ (due to the cardinality of **K**).

Conversely, let **F** be a splitting field for $X^q - X$ over $\mathbb{F}_p$ with $q = p^d$ and p is a prime number. As $(X^q - X)' = qX^{q-1} - 1 = -1$, all the roots of $X^q - X$ are simple in any extension field of $\mathbb{F}_p$, and, thus, **F** has at least q elements. To prove the desired result, it suffices to prove that the set $Ł$ of roots of $X^q - X$ in **F** is a subfield of **F**. For this, let $a, b \in Ł$, that is, $a, b \in \mathbf{F}$ with $a^q = a$ and $b^q = b$. It is clear that $ab \in Ł$ since $(ab)^q = ab$. On the other hand, by virtue of Proposition 209, we have

$$(a+b)^q = (a+b)^{p^d} = ((a+b)^p)^{p^{d-1}} = (a^p + b^p)^{p^{d-1}}$$
$$= ((a^p + b^p)^p)^{p^{d-2}} = \cdots = a^{p^d} + b^{p^d} = a^q + b^q = a + b.$$

Moreover, if $a \neq 0$ then $a^{-1} \in Ł$, and it is clear that 0 and 1 are in L. We conclude that $Ł$ is a subfield of **F**. Necessarily, $\mathbf{F} = Ł$ by minimality of **F**. ∎

Remark 213.

- From Proposition 212, we infer that if **K** is a finite field with q elements and $Ł$ is an extension field of **K**, then an element $x \in Ł$ belongs to **K** if and only if $x^q = x$. We also deduce that two finite fields with the same cardinality q are isomorphic as they are splitting fields for $X^q - X$ over fields isomorphic to $\mathbb{F}_p$.

- A finite field **K** with $q = p^d$ elements has a subfield of cardinality p^ℓ if and only if ℓ divides d (see Exercise 233). Moreover such a field is unique as it is equal to $\{x \in \mathbf{K} \mid x^{p^\ell} = x\}$. For example, denoting by $\mathbb{F}_q$ a field with q elements, as $64 = 2^6$ and the divisors of 6 are $1, 2, 3, 6$, then the subfields of $\mathbb{F}_{64}$ are $\mathbb{F}_{64}, \mathbb{F}_8, \mathbb{F}_4$, and $\mathbb{F}_2$. Moreover, $\mathbb{F}_8 \cap \mathbb{F}_4 = \mathbb{F}_2$ since $2 \nmid 3$.

The following proposition shows the multiplicative group of a finite field is cyclic. In other words, any finite field $\mathbf{K}$ of cardinality q contains an element α (called *primitive*) of order $q - 1$, i.e., such that $\mathbf{K}^\times = \{1, \alpha, \ldots, \alpha^{q-2}\}$. For example, 2 is a primitive element of the fields $\mathbb{F}_3$ and $\mathbb{F}_5$, but not of $\mathbb{F}_7$ since it generates the cyclic subgroup $\{2, 4, 1\}$ of order 3. In the field $\mathbb{F}_9 = \mathbb{F}_3[X]/\langle X^2 + 1 \rangle$ with 9 elements, $\bar{X}$ is not primitive as it is of order 4. However, it is a *generator* of $\mathbb{F}_9$, that is, $\mathbb{F}_9 = \mathbb{F}_3[\bar{X}] = \mathbb{F}_3(\bar{X})$. The minimal polynomial of a primitive element is called a *primitive polynomial*.

Proposition 214. *If $\mathbf{K}$ is a finite field with q elements then its multiplicative group $\mathbf{K}^\times$ is cyclic ($\cong \mathbb{Z}/(q-1)\mathbb{Z}$).*

Proof. Let Ω be the exponent of the group $\mathbf{K}^\times$, that is, the LCM of the orders of its elements. We know that there exists $g \in \mathbf{K}^\times$ whose order is Ω (this is a classical result in finite groups theory). It follows that $x^\Omega = 1$ for all $x \in \mathbf{K}^\times$. As the polynomial $X^\Omega - 1$ has at least $q - 1$ roots, we deduce that $q - 1 \leq \Omega$, and, thus, $\Omega = q - 1$.

We will see in Exercise 243 how to find a primitive element in a finite field.

3.3   Automorphisms group of a finite field

Definition 215. Let $\mathbf{K}$ be a field. An *automorphism* of $\mathbf{K}$ is a bijection $\sigma : \mathbf{K} \to \mathbf{K}$ such that for all $x, y \in \mathbf{K}$, we have:

$$\begin{cases} \sigma(x+y) = \sigma(x) + \sigma(y) \\ \sigma(xy) = \sigma(x)\,\sigma(y) \\ \sigma(1) = 1. \end{cases} \tag{3.1}$$

The set $\mathrm{Aut}(\mathbf{K})$ of automorphisms of $\mathbf{K}$ equipped with composition is a group.

Definition 216. Let $\mathbf{K}$ be a field and $Ł$ an extension field of $\mathbf{K}$. A $\mathbf{K}$-*automorphism* of $Ł$ is an automorphism σ of $Ł$ such that $\sigma(x) = x$ for all $x \in \mathbf{K}$.

We denote by $\mathrm{Gal}(Ł/\mathbf{K})$ the set of $\mathbf{K}$-automorphisms of $Ł$. It is a subgroup of $\mathrm{Aut}(Ł)$ called the *Galois group* of $Ł$ over $\mathbf{K}$. For example, $\mathrm{Gal}(\mathbb{C}/\mathbb{R}) = \{\mathrm{id}_\mathbb{C}, \mathbf{c}\}$ where $\mathbf{c}(z) = \bar{z}$.

The following proposition is immediate.

Proposition 217. *Let $\mathbf{K}$ be a field and $f \in \mathrm{Aut}(\mathbf{K})$. Then $\mathrm{Fix}(f) := \{x \in \mathbf{K} \mid f(x) = x\}$ is a subfield of $\mathbf{K}$. In particular, it contains the prime subfield $\mathbf{K}_0$ of $\mathbf{K}$ (the subfield of $\mathbf{K}$ generated by 1) and we have $\mathrm{Aut}(\mathbf{K}) = \mathrm{Gal}(\mathbf{K}/\mathbf{K}_0)$.*

For example, we have $\mathrm{Aut}(\mathbb{R}) = \mathrm{Gal}(\mathbb{R}/\mathbb{Q})$ ($= \{\mathrm{id}_\mathbb{R}\}$, see Exercise 247). The following proposition (that we give without proof) gives a bound on the order of Galois group in case of an extension field with finite degree.

Proposition 218. *Let* **K** *be a field and* Ł *an extension field of* **K** *of finite degree* [Ł : **K**] *(that is, the dimension* [Ł : **K**] *of* Ł *as* **K***-vector space is finite). Then*

$$|\mathrm{Gal}(Ł/\mathbf{K})| \leq [Ł : \mathbf{K}].$$

Definition 219. Let **K** be a field and Ł an extension field of **K** of finite degree [Ł : **K**]. We say that the field extension Ł/**K** is *Galoisian* if $|\mathrm{Gal}(Ł/\mathbf{K})| = [Ł : \mathbf{K}]$.

Example 220.

(1) Every field **K** is a Galoisian extension of itself since $|\mathrm{Gal}(\mathbf{K}/\mathbf{K})| = [\mathbf{K} : \mathbf{K}] = 1$.

(2) $\mathbb{R}$ is not a Galoisian extension of $\mathbb{Q}$ since $\mathrm{Gal}(\mathbb{R}/\mathbb{Q}) = \{\mathbf{id}_{\mathbb{R}}\}$ and $[\mathbb{R} : \mathbb{Q}] = \infty$.

(3) $\mathbb{C}$ is a Galoisian extension of $\mathbb{R}$ since $\mathrm{Gal}(\mathbb{C}/\mathbb{R}) = \{\mathbf{id}_{\mathbb{C}}, \mathbf{c}\}$ where $\mathbf{c}(z) = \bar{z}$, and $[\mathbb{C} : \mathbb{R}] = 2$.

Definition 221. Let **K** be a field and Ł an algebraic extension of **K**. We say that the field extension Ł/**K** is *normal* if at each time an irreducible polynomial $f(X) \in \mathbf{K}[X]$ has a root in Ł, all its roots are in Ł (i.e., splits in Ł[X]).

Example 222.

(1) Every field is a normal extension of itself.

(2) Every quadratic field extension Ł/**K** (i.e., [Ł : **K**] = 2) is normal.

(3) Let $\mathbb{Q}(\sqrt{2})$ be the splitting field of $X^2 - 2$ over $\mathbb{Q}$. The extension $\mathbb{Q}(\sqrt{2})/\mathbb{Q}$ is normal since it is quadratic. Let $\mathbb{Q}(\sqrt[4]{2})$ be the splitting field of $X^2 - \sqrt{2}$ over $\mathbb{Q}(\sqrt{2})$. The extension $\mathbb{Q}(\sqrt[4]{2})/\mathbb{Q}(\sqrt{2})$ is normal since it is quadratic. But the extension $\mathbb{Q}(\sqrt[4]{2})/\mathbb{Q}$ is not normal because the polynomial $X^4 - 2$, which is irreducible over $\mathbb{Q}$, has only two roots $\pm\sqrt[4]{2}$ in $\mathbb{Q}(\sqrt[4]{2})$; the other two roots $\pm i\sqrt[4]{2}$ are not in $\mathbb{Q}(\sqrt[4]{2})$. We deduce that being normal is not transitive.

(4) Let **K** be a field and Ł an extension field of **K** of finite degree [Ł : **K**]. Then Ł/**K** is normal if and only if Ł is the splitting field over **K** of a polynomial in **K**[X].

For example, let α be a real root of the irreducible polynomial $P = X^3 - 2$ over $\mathbb{Q}$ and set $j = e^{\frac{2i\pi}{3}}$. Then the extension $\mathbb{Q}(\alpha, j)/\mathbb{Q}$ is normal since $\mathbb{Q}(\alpha, j)$ is a splitting field of P over $\mathbb{Q}$. But the extension $\mathbb{Q}(\alpha)/\mathbb{Q}$ since P has a root α in $\mathbb{Q}(\alpha)$ but does not split over $\mathbb{Q}(\alpha)$. If ω is a primitive n^{th} root of unity ($n \geq 1$) then the extension $\mathbb{Q}(\omega)/\mathbb{Q}$ is normal since $\mathbb{Q}(\omega)$ is a splitting field of $X^n - 1$ over $\mathbb{Q}$.

(5) If **K** is a finite field and Ł an algebraic extension of **K** then the extension Ł/**K** is normal (see Exercise 240).

Definition 223.

(1) Let **K** be a field and $P \in \mathbf{K}[X]$ of degree ≥ 1. We say that P is *separable* if P has only simple roots in any extension field Ł of **K**, or, equivalently, if $P \wedge P' = 1$. Note that an irreducible polynomial is separable if and only if its formal derivative is not zero $(P \wedge P' = P \Rightarrow P \mid P' \Rightarrow P' = 0)$.

(2) Let Ł be an extension field of a field **K** and $\alpha \in$ Ł. We say that α is *separable* if there exists a separable polynomial $P \in \mathbf{K}[X]$ of degree ≥ 1 such that $P(\alpha) = 0$.

(3) Let Ł be an algebraic extension of a field **K**. We say that the extension Ł/**K** is *separable* if every element in Ł is separable over **K**. If Ł be an algebraic extension of a field **K** and **F** an algebraic extension Ł, then **F**/**K** is separable if and only if both **F**/Ł and Ł/**K** are separable.

(4) Let **K** be a field. We say that **K** is *perfect* if every irreducible polynomial in **K**[X] is separable, or equivalently, if every algebraic extension of **K** is separable.

Examples of perfect fields are finite fields, fields of characteristic 0 (see Theorem 224 below), and algebraically closed fields (this is obvious). The field $\mathbb{F}_p(T)$ is an example of an infinite field of characteristic $p > 0$ which is not perfect (see Exercise 237). Note also that every algebraic extension field of a perfect field is perfect.

(5) Given a field extension Ł/**K**, we consider the field

$$\mathbf{S} = \{\alpha \in \text{Ł}: \ \alpha \text{ is separable over } \mathbf{K}\}.$$

It is clearly an algebraic separable extension of **K**. The separable degree of Ł/**K** is defined as $[\mathbf{S} : \mathbf{K}]$.

Theorem 224. *A field* **K** *is perfect if and only it is of characteristic* 0, *or of characteristic* $p > 0$ *and the Frobenius endomorphism* $x \mapsto x^p$ *of* **K** *is surjective. In particular, by virtue of Proposition 227, every finite field is perfect.*

Proof. If **K** has characteristic 0 then it is perfect. As a matter of fact, if $P(X)$ is an irreducible monic polynomial in **K**[X] of degree $n > 0$, then the leading term of its formal derivative is $nX^{n-1} \neq 0$. Now, let **K** be a field of characteristic $p > 0$. Suppose that the Frobenius endomorphism $\mathscr{F} : \mathbf{K} \to \mathbf{K}; \ x \mapsto x^p$ is surjective and consider an irreducible polynomial $P(X)$ in **K**[X]. If $P' = 0$, then all the monomials appearing in P would have degrees multiples of p, and thus, since

$\mathscr{F}$ is surjective and by virtue of the identity $(\sum_{i=0}^{n} a_i X^i)^p = \sum_{i=0}^{n} a_i^p X^{pi}$ for $a_i \in \mathbf{K}$, there would exist $Q \in \mathbf{K}[X]$ such that $P = Q^p$, in contradiction with the fact that P is irreducible. Conversely, suppose that $\mathbf{K}$ is perfect and let us prove that $\mathscr{F}$ is surjective, that is, every $a \in \mathbf{K}$ has a pth root in $\mathbf{K}$. Consider the polynomial $P(X) = X^p - a$ and let r be a root of $P(X)$ in a rupture field. Then $P(X) = X^p - a = X^p - r^p = (X - r)^p$ is not separable and, thus, is not irreducible (since $\mathbf{K}$ is perfect). It follow that there exists $k \in [\![1, p-1]\!]$ such that $(X - r)^k \in \mathbf{K}[X]$, and, hence $r^k \in \mathbf{K}$. Using a Bézout identity $up + vk = 1$ between p and k in $\mathbb{Z}$, we obtain $r = (r^p)^u (r^k)^v \in \mathbf{K}$ as desired.

It is worth pointing out that the above Frobenius endomorphism is always injective since for $a, b \in \mathbf{K}$ with $\mathbf{K}$ a field of characteristic p:

$$a^p = b^p \Rightarrow (a - b)^p = 0 \Rightarrow a = b.$$

Theorem 225. *Let $\mathbf{K}$ be a field and $Ł$ an extension field of $\mathbf{K}$ of finite degree $[Ł : \mathbf{K}]$. Then $Ł/\mathbf{K}$ is Galoisian if and only if it is both normal and separable.*

In particular, if $\mathbf{K}$ is perfect then $Ł/\mathbf{K}$ is Galoisian if and only if $Ł$ is the splitting field over $\mathbf{K}$ of a polynomial in $\mathbf{K}[X]$.

Example 226.

1) As seen above, if $\mathbf{K}$ is a finite field and $Ł$ an extension field of $\mathbf{K}$ of finite degree $[Ł : \mathbf{K}]$, then $Ł/\mathbf{K}$ is Galoisian.

2) (*The Klein group as a Galois group*)

The field extension $\mathbb{Q}(i, \sqrt{2})/\mathbb{Q}$ is Galoisian of degree 4 with Galois group $\mathrm{Gal}(\mathbb{Q}(i, \sqrt{2})/\mathbb{Q}) \cong \mathbb{F}_2 \times \mathbb{F}_2$ (the Klein group).

To see this, first note that $(1, \sqrt{2}, i, i\sqrt{2})$ is a basis for $\mathbb{Q}(i, \sqrt{2})$ as a $\mathbb{Q}$-vector space, and, thus, $[\mathbb{Q}(i, \sqrt{2}) : \mathbb{Q}] = 4$. On the other hand, $\mathbb{Q}(i, \sqrt{2})$ is a splitting field of $(X^2 + 1)(X^2 - 2)$ over $\mathbb{Q}$, and, thus, the extension $\mathbb{Q}(i, \sqrt{2})/\mathbb{Q}$ is normal. As $\mathbb{Q}$ is perfect (because of characteristic 0), we infer that the extension $\mathbb{Q}(i, \sqrt{2})/\mathbb{Q}$ is Galoisian.

Let $\sigma \in \mathrm{Gal}(\mathbb{Q}(i, \sqrt{2})/\mathbb{Q})$. Identifying σ amounts to identifying $\sigma(\sqrt{2})$ and $\sigma(i)$. But, as $(\sqrt{2})^2 = 2$ and $i^2 = -1$, then $\sigma(\sqrt{2}) = \pm\sqrt{2}$ and $\sigma(i) = \pm i$. Therefore, $\mathrm{Gal}(\mathbb{Q}(i, \sqrt{2})/\mathbb{Q}) = \{\mathrm{id}, \sigma_1, \sigma_2, \sigma_3\}$ with

$$\begin{cases} \sigma_1(\sqrt{2}) = -\sqrt{2} \\ \sigma_1(i) = i \end{cases} \quad ; \quad \begin{cases} \sigma_2(\sqrt{2}) = \sqrt{2} \\ \sigma_2(i) = -i \end{cases} \quad ; \quad \begin{cases} \sigma_3(\sqrt{2}) = -\sqrt{2} \\ \sigma_3(i) = -i. \end{cases}$$

The group table of $\mathrm{Gal}(\mathbb{Q}(i, \sqrt{2})/\mathbb{Q})$ below shows that it is isomorphic to the Klein group.

$\circ$	id	σ_1	σ_2	σ_3
id	id	σ_1	σ_2	σ_3
σ_1	σ_1	id	σ_3	σ_2
σ_2	σ_2	σ_3	id	σ_1
σ_3	σ_3	σ_2	σ_1	id

We turn now to automorphisms of a finite field.

Proposition 227. *Let* **K** *be a field of characteristic p (p a prime number). Then the map*

$$\mathscr{F} : \begin{array}{ccc} \mathbf{K} & \longrightarrow & \mathbf{K} \\ x & \longmapsto & x^p \end{array}$$

is an $\mathbb{F}_p$-*endomorphism of* **K**. *It is called the Frobenius endomorphism of* **K**. *Moreover, if* **K** *is finite then* $\mathscr{F}$ *is an automorphism, and if* $\mathbf{K} = \mathbb{F}_p$ *then* $\mathscr{F}$ *is the identity.*

Proof. The fact that $\mathscr{F}$ is an $\mathbb{F}_p$-endomorphism of **K** is already proved. Suppose now that **K** is finite and consider two distinct elements $a, b \in \mathbf{K}$. We have:

$$a \neq b \Rightarrow \exists \alpha \in \mathbf{K} \mid \alpha(a-b) = 1 \Rightarrow \mathscr{F}(\alpha)(\mathscr{F}(a) - \mathscr{F}(b)) = 1 \Rightarrow \mathscr{F}(a) \neq \mathscr{F}(b).$$

It follows that $\mathscr{F}$ is injective, and thus, bijective.

Propositions 218 and 227 allow us to completely describe the automorphisms group of a finite field.

Theorem 228. *Let* **K** *be a finite field of characteristic p and cardinality* $q = p^n$. *Then* $\mathrm{Aut}(\mathbf{K}) = \mathrm{Gal}(\mathbf{K}/\mathbf{K}_0)$ *is cyclic of order* $n = [\mathbf{K} : \mathbf{K}_0]$ *and generated by the Frobenius automorphism of* **K**.

Proof. We know, by Proposition 218, that $|\mathrm{Aut}(\mathbf{K})| \leq n$. Let ω be the order of the Frobenius automorphism $\mathscr{F}$ of **K**. The desired result follows from the following implications:

$$\mathscr{F}^\omega = \mathbf{id_K} \Rightarrow \forall x \in \mathbf{K}, x^{p^\omega} = x \Rightarrow \deg(X^{p^\omega} - X) \geq p^n \Rightarrow p^\omega \geq p^n \Rightarrow \omega \geq n \Rightarrow \omega = n.$$

Remark 229. Similarly to Theorem 228, if $\mathbf{K} \subseteq \mathbf{L}$ are two finite fields of characteristic p and cardinality p^ℓ and q^d, respectively, with $d = n\ell$. Then $\mathrm{Gal}(\mathbf{L}/\mathbf{K})$ is cyclic of order $n = [\mathbf{L} : \mathbf{K}]$ and generated by the $\mathscr{F}^\ell$.

3.4 Exercises

Exercise 230. Show that a finite field is not algebraically closed.

Exercise 231. Let $\mathbb{F}_3$ be the field with three elements and let α be a 7th root of unity (such a root exists in a rupture field of the polynomial $X^7 - 1 \in \mathbb{F}_3[X]$). We consider the field $\mathbf{K} = \mathbb{F}_3(\alpha)$.

1) Show that $\mathbf{K}$ is a finite field.

2) Find the minimal polynomial of α over $\mathbb{F}_3$ and determine the field $\mathbf{K}$.

Exercise 232. Let $\alpha \in \mathbb{F}_{27} \setminus \{0, 1, -1\}$. Show that $\{\alpha, -\alpha\}$ contains a primitive element of $\mathbb{F}_{27}$.

Exercise 233. Let p, m, n be positive integers with $p > 1$. Show that:

$$(p^m - 1) \mid (p^n - 1) \iff m \mid n.$$

Exercise 234. Show that if b is a primitive element in $\mathbb{F}_{p^n}$ (p a prime number and $n \geq 2$) and if $m \mid n$ then $b^{\frac{p^n-1}{p^m-1}}$ is a primitive element in $\mathbb{F}_{p^m}$.

Exercise 235. 1) Let Q be a polynomial of degree 5 in $\mathbb{F}_2[X]$.

a) Show that Q is not irreducible over $\mathbb{F}_2$ if and only if it is divisible by a polynomial of degree 1 or 2 in $\mathbb{F}_2[X]$.

b) Deduce that if Q has no root in $\mathbb{F}_2$, then Q is not irreducible over $\mathbb{F}_2$ if and only if it is divisible by $X^2 + X + 1$.

2) Consider the polynomial $P = X^5 + X^2 + 1$ in $\mathbb{F}_2[X]$. Denote by $\mathbf{K}$ the ring $\mathbb{F}_2[X]/\langle P \rangle$. Show that $\mathbf{K}$ is a field. Give its characteristic and cardinality. What are its primitive elements?

3) Let α be the class of X in $\mathbf{K}$.

a) Show that $P(\alpha^{2^i}) = 0$ for all $i \in \mathbb{N}$.

b) Show that the elements α, α^2, α^4, α^8 and α^{16} are all distinct in $\mathbf{K}$.

c) Deduce that the decomposition of P as the product of irreducible factors in $\mathbf{K}[X]$ is
$$P = (X - \alpha)(X - \alpha^2)(X - \alpha^4)(X - \alpha^8)(X - \alpha^{16}).$$

Exercise 236. Let $\mathbf{F}$ be a finite field with q elements, $\mathbf{K}$ an extension field of $\mathbf{F}$, and $f \in \mathbf{K}[X]$. Show that
$$f \in \mathbf{F}[X] \iff f(X)^q = f(X^q).$$

Exercise 237. Let p be a prime number and q a power of p.

1) Show that the Frobenius map

$$\mathscr{F}: \quad \begin{array}{ccc} \mathbb{F}_p(T) & \longrightarrow & \mathbb{F}_p(T) \\ f(T) & \longmapsto & f(T)^p \end{array}$$

is not surjective. Deduce that the field $\mathbb{F}_p(T)$ is not perfect.

2) Show that the Frobenius map

$$\mathscr{F}: \quad \begin{array}{ccc} \bar{\mathbb{F}}_q & \longrightarrow & \bar{\mathbb{F}}_q \\ x & \longmapsto & x^p \end{array}$$

is bijective.

Exercise 238. (Univariate square-free polynomials)

Let $P \in \mathbf{K}[X]$, where $\mathbf{K}$ is a field. Recall that P is said to be *square-free* if there is no $Q \in \mathbf{K}[X]$ such that Q^2 divides P.

Write $P = cP_1^{\alpha_1} \cdots P_s^{\alpha_s}$, where the P_i's are pairwise distinct irreducible polynomials in $\mathbf{K}[X]$, and $c \in \mathbf{K}^\times$. *The square-free part* is defined as $\mathrm{sqfree}(P) := cP_1 \cdots P_s$.

1) Show that if $P \wedge P' = 1$ then P is square-free.

2) Show that if all the α_i's are nonzero in $\mathbf{K}$ (in other words, $\mathrm{char}(\mathbf{K}) = 0$ or $\mathrm{char}(\mathbf{K}) = p > 0$ and $p \nmid \alpha_i$ in $\mathbb{N}$ for $1 \le i \le s$), then $\mathrm{sqfree}(P) = \frac{P}{P \wedge P'}$.

3) Let $P \in \mathbf{K}[X]$, where $\mathbf{K}$ is a perfect field of characteristic $p > 0$.

   a) Show that: $P' = 0 \Leftrightarrow \exists Q \in \mathbb{F}_p[X] \mid P = Q^p$.

   b) Deduce that if P is an irreducible polynomial in $\mathbf{K}[X]$ then $P' \ne 0$.

   c) Show that: P is square-free $\Leftrightarrow P \wedge P' = 1$.

4) Let $P \in \mathbf{K}[X]$, where $\mathbf{K}$ is a perfect field. Show that: P is square-free $\Leftrightarrow P \wedge P' = 1$.

Exercise 239. Show that for every $r \ge 1$, there exists an irreducible polynomial in $\mathbb{F}_q[X]$ of degree r.

Exercise 240. Let $f(X)$ be a monic irreducible polynomial over $\mathbb{F}_q$ of degree r. Show that:

1) All the roots of $f(X)$ lie in $\mathbb{F}_{q^r}$ and in every field containing $\mathbb{F}_q$ along with one root of $f(X)$.

2) $f(X) = \prod_{i=1}^{r} (X - \alpha_i)$, where $\alpha_i \in \mathbb{F}_{q^r}$.

3) $f(X) \mid X^{q^r} - X$.

Exercise 241. *(Cyclotomic cosets and minimal polynomials)*

Let $\mathbb{F}_{q^t}$ be an extension field of $\mathbb{F}_q$ and let $\alpha \in \mathbb{F}_{q^t}$ with minimal polynomial $M_\alpha(X) \in \mathbb{F}_q[X]$ over $\mathbb{F}_q$.

1) Show that $M_\alpha(X) \mid X^{q^t} - X$.

2) Show that $M_\alpha(X)$ has distinct roots all lying in $\mathbb{F}_{q^t}$.

3) Show that the degree of $M_\alpha(X)$ divides t.

4) Show that $X^{q^t} - X = \prod_\alpha M_\alpha(X)$, where α runs through some subsets of $\mathbb{F}_{q^t}$ which enumerate the minimal polynomials of all elements in $\mathbb{F}_{q^t}$ exactly once.

5) Show that $X^{q^t} - X = \prod_f f(X)$, where f runs through all monic irreducible polynomials in $\mathbb{F}_q[X]$ whose degree divide t.

6) Recall that two elements of $\mathbb{F}_{q^t}$ which have the same minimal polynomial over $\mathbb{F}_q$ are called *conjugate* over $\mathbb{F}_q$. Show that if $\alpha \in \mathbb{F}_{q^t}$ then $\alpha, \alpha^q, \alpha^{q^2}, \ldots$ are conjugates of α over $\mathbb{F}_q$.

7) Let $\alpha = \gamma^s \in \mathbb{F}_{q^t}$ with minimal polynomial $M_\alpha(X)$ over $\mathbb{F}_q$, where γ is a primitive element in $\mathbb{F}_{q^t}$ and $0 \le s \le q^t - 2$. By 7), we know that $\gamma^s, \gamma^{sq}, \gamma^{sq^2}, \ldots$ are all roots of $M_\alpha(X)$. The sequence stops exactly after r terms where r is the smallest positive integer such that $sq^r \equiv s \bmod q^t - 1$. Based on this, we define the *q-cyclotomic coset of s modulo $q^t - 1$* to be the set

$$C_s := \{s, sq, \ldots, sq^{r-1}\} \bmod q^t - 1.$$

The sets C_s partition the set $\{0, 1, 2, \ldots, q^t - 2\}$ into disjoint sets. When listing the cyclotomic cosets, it is usual to list C_s only once, where s is the smallest element of the coset.

Compute the 2-cyclotomic cosets modulo $15 = 2^4 - 1$.

8) Show that if γ is a primitive element in $\mathbb{F}_{q^t}$ and $0 \le s \le q^t - 2$, then the minimal polynomial of γ^s over $\mathbb{F}_q$ is $M_{\gamma^s}(X) = \prod_{i \in C_s} (X - \gamma^i)$.

9) Deduce that the size r of a cyclotomic coset modulo $q^t -$ divides t.

10) Check that $\mathbf{K} := \mathbb{F}_2[t]/\langle t^2 + t + 1 \rangle$ is a field with 8 elements in which $\alpha = \bar{t}$ is primitive. Give a factorization of $X^7 - 1$ as a product of irreducible polynomials in $\mathbb{F}_2[X]$.

11) Without factoring $X^{15} - 1$, how many irreducible factors does it have over $\mathbb{F}_2$ and what are their degrees? How many binary cyclic codes of length 15 can we construct?

Exercise 242. Let $\mathbf{K}$ be a finite field with cardinality $q = p^d$ and characteristic p.

1) An element $x \in \mathbf{K}^\times$ is called a *perfect square* if there exists $y \in \mathbf{K}^\times$ such that $x = y^2$. The set of prefect squares in $\mathbf{K}$ is denoted by H (it is a subgroup of $\mathbf{K}^\times$). Show that if $p = 2$ then $|H| = q - 1$ and if $p > 2$ then $|H| = \frac{q-1}{2}$.

2) Show that there exist $a, b \in \mathbf{K}$ such that $a^2 + b^2 = -1$.

Exercise 243. (How to find a primitive element)

Let $\mathbf{K}$ be a finite field with cardinality $q = p^d$ and characteristic p.

1) Denote by $U = \gcd(X^{q-1} - 1, \displaystyle\prod_{r|q-1; \, r \neq q-1} (X^r - 1))$. Show that the polynomial

$\phi_{p,q-1} = \frac{X^{q-1}-1}{U}$ has exactly as roots the primitive elements of $\mathbf{K}$. This polynomial is called the *cyclotomic polynomial of order* $q - 1$ *modulo* p.

2) Show that $\deg(\phi_{p,q-1}) = \varphi(q-1)$ (the Euler function of $q - 1$, see Definition and Proposition 250).

3) Denote by $\mathbf{K} = \mathbb{F}_9 = \mathbb{F}_3[t]/\langle t^2 + 1 \rangle$. Find primitive elements in $\mathbf{K}$.

4) Show that $\phi_{p,q-1}$ splits as the product of primitive polynomials all of degree d.

5) Build a field $\mathbb{F}_{16}$ of cardinality 16 and give a primitive element in it.

Exercise 244. (Constructing sliced orthogonal arrays using finite fields [3])

Latin hypercube designs have been widely used in computer experiments with quantitative factors. When there are both qualitative and quantitative factors in computer experiments, sliced space-filling designs have been proposed by Ai, Jiang and Li (among others). Their construction is based on a class of sliced orthogonal arrays (partially presented in this exercise) in which each slice is balanced and becomes an orthogonal array after some level-collapsing.

An *orthogonal array*, denoted by $OA(n, s^m, t)$, with n runs, m factors and strength t ($m \geq t \geq 1$) is an $n \times m$ matrix in which each column has s levels from a set of s elements, such that all possible level combinations occur equally often as rows in every $n \times t$ submatrix. An array is called *balanced* if it is an orthogonal array of strength one.

A *Latin hypercube* with n runs and m factors is an $OA(n, n^m, 1)$ in which each column is a permutation of n levels from a set S of n elements. Usually the set S is taken to be $\{1, \ldots, n\}$.

An $r \times c$ *difference matrix*, denoted by $D(r, c, g)$, is an array with entries from an Abelian group G of g elements such that every element of G appears equally often in the vector difference between any two columns of the array.

For a matrix B, ${}^t B$ denotes its transpose, $B(:, j)$ denotes its jth column, $B(i, :)$ denotes its ith row, $B(i, j)$ denotes its $(i, j)^{\text{th}}$ entry, and $a + B$ denotes the element-wise sum of B and a scalar a.

Let $\mathbf{K} \subsetneq \mathbf{L}$ be two finite fields of characteristic p and cardinality $s_2 = p^{u_2}$ and $s_1 = p^{u_1}$, respectively, with $u_1 = \lambda u_2$, $\lambda \in \{2, 3, \ldots\}$, and set $q = \frac{s_1}{s_2} = \frac{s_2^{\lambda}}{s_2} = s_2^{\lambda-1}$. Let α be a primitive element in $\mathbf{L}$. We saw in Exercise 234, that $\beta := \alpha^{\frac{s_1-1}{s_2-1}}$ is a primitive element in $\mathbf{K}$. Let $\alpha_0, \ldots, \alpha_{s_1-1}$ denote the elements of $\mathbf{L}$ with $\alpha_0 = 0$ and $\alpha_i = \alpha^i$ for $1 \le i \le s_1 - 1$. Also, let $\beta_0, \ldots, \beta_{s_2-1}$ denote the elements of $\mathbf{K}$ with $\beta_0 = 0$ and $\beta_i = \beta^i$ for $1 \le i \le s_2 - 1$.

1) Let A_0 be the multiplication table of $\mathbf{L}$, where the rows and columns are labeled with the s_1 elements of $\mathbf{L}$. Show that A_0 is a $D(s_1, s_1, s_1)$.

2) Show that any element $a \in \mathbf{L}$ can be uniquely represented by

$$a = b_0 + b_1 \alpha + \cdots + b_{\lambda-1} \alpha^{\lambda-1}, \quad b_i \in \mathbf{K}, \ 0 \le i \le \lambda - 1. \tag{3.2}$$

3) Keeping the above notation, we consider the projection

$$\phi : \ \mathbf{L} \ \longrightarrow \ \mathbf{K}$$
$$a \ \longmapsto \ b_0 + b_1 \beta + \cdots + b_{\lambda-1} \beta^{\lambda-1}.$$

Check that

a) $\phi(b) = b$ for any $b \in \mathbf{K}$;

b) $\phi(a_1 + a_2) = \phi(a_1) + \phi(a_2)$ for any $a_1, a_2 \in \mathbf{L}$;

c) $\phi(ba) = b\phi(a)$ for any $b \in \mathbf{K}$ and $a \in \mathbf{L}$;

d) $\sharp(\phi^{-1}(b)) = q$ for any $b \in \mathbf{K}$.

4) Define the kernel matrix Γ of ϕ to be the $s_2 \times q$ matrix whose i^{th} row is $\Gamma(i, :) = \phi^{-1}(\beta_{i-1})$, where the elements in each row are arranged in lexicographical order when written as in (3.2). Check that

a) each element of $\mathbf{L}$ appears precisely once in Γ;

b) ϕ collapses the entries in the same row of Γ into a common element in $\mathbf{K}$;

c) for $1 \le j \le s_2$, $\phi(\Gamma(:, j)) = \mathbf{K}$ (there is a small abuse of language here).

5) Take $p = 3$, $\mathbf{K} = \mathbb{F}_3$, and $\mathbf{L} = \mathbb{F}_3[X]/\langle X^2 + X + 2 \rangle$ with primitive element $\alpha = \bar{X}$. Give Γ.

6) Let $u = {}^t(1, \alpha, \ldots, \alpha^{\lambda-1})$. Obtain an $s_1 \times \lambda$ matrix A by taking the columns of A_0 labeled with the elements of u. Let $Q = \{1, \ldots, q\}$ and Q^{λ} be the set of all possible λ-tuples from Q. For any $(\ell_1, \ldots, \ell_{\lambda}) \in Q^{\lambda}$, let

$$C_{(\ell_1, \ldots, \ell_{\lambda})} = A + \mathbf{1}_{s_1} v_{(\ell_1, \ldots, \ell_{\lambda})},$$

where $\mathbf{1}_{s_1}$ is the s_1-vector of ones and $v_{(\ell_1, \ldots, \ell_{\lambda})} = (\Gamma(1, \ell_1), \ldots, \Gamma(1, \ell_{\lambda}))$. Finally, obtain an array C by the row juxtaposition of all the $C_{(\ell_1, \ldots, \ell_{\lambda})}$'s.

a) From Formula (3.2), the label of the ith row of A can be uniquely represented as $(b_{i,0}, b_{i,1}, \ldots, b_{i,\lambda-1}) u$ for $1 \le i \le s_1$. Let B be the $s_1 \times \lambda$ matrix with $(b_{i,0}, b_{i,1}, \ldots, b_{i,\lambda-1})$ as the ith row. Check that $A = Bu^t u$ and show that $\phi(u^t u)$ has full rank over **K**. Deduce that $\phi(A)$ is an $OA(s_2^\lambda, s_2^\lambda, \lambda)$.

b) Show that each slice $C_{(\ell_1,\ldots,\ell_\lambda)}$ is a balanced $D(s_1, \lambda, s_1)$ and $\phi(C_{(\ell_1,\ldots,\ell_\lambda)})$ is an $OA(s_2^\lambda, s_2^\lambda, \lambda)$ for any $(\ell_1, \ldots, \ell_\lambda) \in Q^\lambda$.

c) Show that C is an $OA(s_1^\lambda, s_1^\lambda, \lambda)$.

7) Give the matrices $C_{(2,3)}$, $C_{(2,2)}$, $\phi(C_{(2,3)})$, and $\phi(C_{(2,2)})$ corresponding to the example in 5).

Exercise 245. We consider a finite field $\mathbb{F}_q$ with q elements. We define on $\mathbb{F}_q^3 \setminus \{(0,0,0)\}$ the following equivalence relation:

$$(x,y,z) \sim (x',y',z') \Leftrightarrow \exists \lambda \in \mathbb{F}_q \setminus \{0\} \mid (x',y',z') = \lambda(x,y,z).$$

The set $\mathbb{P}^2(\mathbb{F}_q) := (\mathbb{F}_q^3 \setminus \{(0,0,0)\})/\sim$ is called the projective plane of order q.

1) Give $\sharp(\mathbb{P}^2(\mathbb{F}_q))$.

2) Give all the points of the projective plane of order 3.

Exercise 246. Let p be a prime number and $a \in \mathbb{F}_p^\times$. We consider the polynomial

$$f(X) = X^p - X + a.$$

1) Show that $f(X)$ is separable.

2) Show that $f(X)$ does not have a root in $\mathbb{F}_p$.

3) Show that if α is a root of $f(X)$ then $\alpha + j$ is a root of $f(X)$ for any $j \in \mathbb{F}_p$.

4) Show that $f(X)$ is irreducible over $\mathbb{F}_p$.

Exercise 247. Show that $\text{Aut}(\mathbb{R}) = \text{Gal}(\mathbb{R}/\mathbb{Q}) = \{\text{id}_\mathbb{R}\}$.

Exercise 248. Let $n > 2$. A *cyclotomic field* is a number field obtained by adjoining a complex primitive root of unity to $\mathbb{Q}$. The nth cyclotomic field $\mathbb{Q}(\xi_n)$ is obtained by adjoining a primitive n^{th} root of unity ξ_n (i.e., $\xi_n^n = 1$ and $\xi_n^k \neq 1$ for $1 \le k \le n-1$) to the rational numbers.

1) Let ξ_n be a primitive n^{th} root of unity. Show that the field extension $\mathbb{Q}(\xi_n)/\mathbb{Q}$ is Galoisian, and the groups $\text{Gal}(\mathbb{Q}(\xi_n)/\mathbb{Q})$ and $(\mathbb{Z}/n\mathbb{Z})^\times$ are isomorphic.

2) Give $[\mathbb{Q}(e^{\frac{2i\pi}{n}}) : \mathbb{Q}]$ and $[\mathbb{Q}(\cos(\frac{2\pi}{n})) : \mathbb{Q}]$.

3.5   Solutions to the exercises

Exercise 230:

Denoting the elements of the finite field $\mathbf{F}$ by $x_1,\ldots,x_q$, the polynomial $(X - x_1)\cdots(X - x_q) + 1 \in \mathbf{F}[X]$ has no root in $\mathbf{F}$.

Exercise 231:

1) The element $\alpha \in \mathbf{K}$ is algebraic over $\mathbb{F}_3$ as it satisfies $\alpha^7 = 1$. Therefore, the field $\mathbb{F}_3(\alpha)$ is a finite extension of $\mathbb{F}_3$. It is a finite-dimensional $\mathbb{F}_3$-vector space, and, thus, has finitely many elements.

2) Denote by P the minimal polynomial of α over $\mathbb{F}_3$. We have $\alpha^7 - 1 = (\alpha - 1)(\alpha^6 + \alpha^5 + \cdots + 1) = 0$. If $\alpha = 1$, then $P = X - 1$ and $\mathbf{K} = \mathbb{F}_3$. Suppose that $\alpha \neq 1$. Denote by Q the cyclotomic polynomial $X^6 + X^5 + \cdots + 1 \in \mathbb{F}_3[X]$. As $Q(\alpha) = 0$, Q is a multiple of P. To prove that $Q = P$, it suffices to prove that Q is irreducible in $\mathbb{F}_3[X]$. It is immediate that Q is irreducible in $\mathbb{F}_3[X]$ if and only if it has no roots in any extension field of $\mathbb{F}_3$ of degree at most $\frac{\deg Q}{2} = 3$. There are only three possibilities for these extensions, namely, $Ł_1 = \mathbb{F}_3$, $Ł_2 = \mathbb{F}_{3^2}$, and $Ł_3 = \mathbb{F}_{3^3}$. By way of contradiction, suppose that Q has a root x in one of the $Ł_d$'s, $1 \leq d \leq 3$. As $x \neq 0$, x is an element of the multiplicative group $Ł_d^\times$. Since $x \neq 1$ (because $Q(1) = 1 \neq 0$) and $x^7 = 1$, the order of x is 7. By Lagrange's theorem, this would imply that 7 divides $|Ł_d^\times| = 3^d - 1$, which impossible for $1 \leq d \leq 3$. We infer that Q is irreducible in $\mathbb{F}_3[X]$, and $Q = P$. As a conclusion, $\mathbf{K}$ is the finite field $\mathbb{F}_{3^6}$ with 3^6 elements.

Exercise 232:

$\mathbb{F}_{27}^\times$ is a multiplicative group of order 26. The only element of $\mathbb{F}_{27}^\times$ of order 2 is -1. As $\alpha \in \mathbb{F}_{27}^\times \setminus \{1, -1\}$, the order ω of α is either 13 or 26. If $\omega = 26$ then α is primitive. If $\omega = 13$ then the order of $-\alpha$ is $\mathrm{lcm}(2, 13) = 26$, that is, $-\alpha$ is primitive.

Exercise 233:

"$\Leftarrow$" Straightforward.

"$\Rightarrow$" Write $n = qm + r$ with $q \in \mathbb{N}$ and $0 \leq r \leq m - 1$. We have $p^r \equiv 1 \bmod p^m - 1$ with $p^r - 1 < p^m - 1$. Thus, $p^r - 1 = 0$, that is, $r = 0$.

Exercise 234:

Denoting by $b' = b^{\frac{p^n-1}{p^m-1}}$, as $b'^{p^m} = b'$, we have $b' \in \mathbb{F}_{p^m}$. Moreover, since the b'^j, $j = 0, 1, \ldots, p^m - 2$, are all distinct, we have $\mathbb{F}_{p^m}^\times = \langle b' \rangle$.

Exercise 235:

1.a) Q is not irreducible over $\mathbb{F}_2$ if and only if there exist $A, B \in \mathbb{F}_2[X]$ such that $Q = AB$ and $1 \leq \deg(A), \deg(B) \leq 4$. The equality $\deg(A) + \deg(B) = 5$ shows that we cannot have $\deg(A) \geq 3$ and $\deg(B) \geq 3$.

1.b) $X^2 + X + 1$ is the only irreducible polynomial of degree 2 in $\mathbb{F}_2[X]$.

2) As the polynomial $P = X^5 + X^2 + 1$ has no root in $\mathbb{F}_2$ and is not divisible by $X^2 + X + 1$, it is irreducible in $\mathbb{F}_2[X]$. Thus, $\mathbf{K}$ is a field. Its characteristic is 2 and cardinality is $2^5 = 32$. As the group $\mathbf{K}^\times$ has order 31 (a prime number), all its elements other than 1 are primitive in $\mathbf{K}$.

3.a) $P(\alpha) = 0$ by definition. We have $P(X)^2 = P(X^2)$ and, thus, by induction, $P(\alpha^{2^i}) = 0$ for all $i \in \mathbb{N}$.

3.b) The elements $\alpha, \alpha^2, \alpha^4, \alpha^8$ and α^{16} are all distinct in $\mathbf{K}$ since α has order 31 in the group $\mathbf{K}^\times$.

3.c) The polynomial P is monic of degree 5 and is divisible by $A = (X - \alpha)(X - \alpha^2)(X - \alpha^4)(X - \alpha^8)(X - \alpha^{16})$. Thus, $P = A$.

Exercise 236:

Denoting $f(X) = a_0 + a_1 X + \cdots + a_n X^n \in \mathbf{K}[X]$, we have $f(X)^q = a_0^q + a_1^q X^q + \cdots + a_n^q X^{nq}$. So,

$$f(X)^q = f(X^q) \Leftrightarrow a_i^q = a_i \,\forall\, 0 \le i \le n \Leftrightarrow a_i \in \mathbf{F} \,\forall\, 0 \le i \le n \Leftrightarrow f \in \mathbf{F}[X].$$

Exercise 237:

1) $\mathscr{F}\left(\frac{f(T)}{g(T)}\right) = T \Rightarrow p(\deg(f) - \deg(g)) = 1$. So, $T \notin \mathscr{F}(\mathbb{F}_p(T))$. The conclusion that $\mathbb{F}_p(T)$ is not perfect follows from Theorem 224.

2) Let $x, x' \in \bar{\mathbb{F}}_q$ such that $\mathscr{F}(x) = \mathscr{F}(x')$. By Proposition 227 with $\mathbf{K} = \mathbb{F}_q(x, x')$, we obtain that $x = x'$. Now, let $y \in \bar{\mathbb{F}}_q$. As $\bar{\mathbb{F}}$ is algebraically closed, the polynomial $X^q - y$ has a root $x \in \bar{\mathbb{F}}$, that is, $\exists x \in \bar{\mathbb{F}} \mid \mathscr{F}(x) = y$.

Exercise 238:

1) P has a square factor $\Rightarrow \exists \alpha_i \ge 2 \Rightarrow P_i \mid \frac{P}{P_i} \overset{\text{since } P_i \mid \frac{P}{P_j} \text{ for } j \ne i}{\Rightarrow} P_i \mid P \wedge P' \Rightarrow P \wedge P' \ne 1$.

2) $P \wedge P' = P_1^{\alpha_1 - 1} \cdots P_s^{\alpha_s - 1}$, and $\frac{P}{P \wedge P'} = c P_1 \cdots P_s = \text{sqfree}(P)$.

3.a) If $P' = 0$ then $P = \sum b_i X^{pi}$, with $b_i \in \mathbf{K}$. As $\mathbf{K}$ is perfect, we can write $b_i = a_i^p$ with $a_i \in \mathbf{K}$ (see Theorem 224). It follows that $P = (\sum a_i X^i)^p$.

3.b) P irreducible $\Rightarrow \nexists Q \in \mathbb{F}_p[X] \mid P = Q^p \Rightarrow P' \ne 0$.

3.c) We have $P' = c \sum_i \alpha_i P_i' \frac{P}{P_i}$. Note that $P_i' \ne 0$ since P_i is irreducible.

"$\Leftarrow$" This is 1).

"$\Rightarrow$" Suppose that $P \wedge P' \ne 1$, that is, $\exists P_i$ dividing P', and, thus, $P_i \mid \alpha_i P_i' \frac{P}{P_i}$. Since $P_i' \ne 0$ and $\deg(P_i') < \deg(P_i)$, we deduce that $P_i \mid \alpha_i \frac{P}{P_i}$. If $\alpha_i \ne 0$ in $\mathbf{K}$, then $P_i \mid \frac{P}{P_i}$, and, thus, $P_i^2 \mid P$. If $\alpha_i = 0$ in $\mathbf{K}$, then $p \mid \alpha_i$ in $\mathbb{N}$, and, thus, $\alpha_i \ge 2$ and $P_i^2 \mid P$.

4) If char$(\mathbf{K}) = p > 0$, this is 3.c). If char$(\mathbf{K}) = 0$, this is clear from the proof of 3.c).

Exercise 239:

The minimal polynomial over $\mathbb{F}_q$ of a primitive element in $\mathbb{F}_{q^r}$ is irreducible in $\mathbb{F}_q[X]$ of degree r.

Exercise 240:

1) Let α, β be two roots of $f(X)$ in any extension field of $\mathbb{F}_q$. As the fields $\mathbb{F}_q(\alpha)$ and $\mathbb{F}_q(\beta)$ are subfields of the finite field $\mathbb{F}_q(\alpha, \beta)$ with the same cardinality q^r, they are equal.

2) $\mathbb{F}_q$ is perfect.

3) $X^{q^r} - X = \prod_{\alpha \in \mathbb{F}_{q^r}} (X - \alpha)$.

Exercise 241:

1) Denoting $P(X) = X^{q^t} - X \in \mathbb{F}_q[X]$, since $P(\alpha) = 0$, we have that $M_\alpha(X) \mid P(X)$.

2) $M_\alpha(X)$ has distinct roots because $\mathbb{F}_q$ is perfect and they are all lying in $\mathbb{F}_{q^t}$ by virtue of Exercise 240.

3) This is because $\mathbb{F}_q(\alpha)$ is subfield of $\mathbb{F}_{q^t}$ of cardinality $q^{\deg M_\alpha}$.

4) Write $X^{q^t} - X = \prod_{i=1}^{n} P_i(X)$, where the P_i's are irreducible monic polynomials in $\mathbb{F}_q[X]$. Since $X^{q^t} - X$ has distinct roots, the P_i's are pairwise distinct. So, $P_i(X) = M_{\alpha_i}(X)$ for any $\alpha_i \in \mathbb{F}_{q^t}$ with $P_i(\alpha_i) = 0$.

5) It suffices to prove that every monic irreducible polynomials $f(X)$ in $\mathbb{F}_q[X]$ of degree r dividing t is a factor of $X^{q^t} - X$. But $f(X) \mid X^{q^r} - X$ by Exercise 240 and $X^{q^r} - X \mid X^{q^t} - X$ since $r \mid t$.

6) This follows from Exercise 236.1.

7) $C_0 = \{0\}, C_1 = \{1, 2, 4, 8\}, C_3 = \{3, 6, 12, 9\}, C_5 = \{5, 10\}$, and $C_7 = \{7, 14, 13, 11\}$.

8) Let us first prove that $g(X) := \prod_{i \in C_s} (X - \gamma^i)$ is in $\mathbb{F}_q[X]$ (not merely in $\mathbb{F}_{q^t}[X]$). We have $g(X)^q = \prod_{i \in C_s} (X^q - \gamma^{qi}) = \prod_{i \in C_s} (X^q - \gamma^i) = g(X^q)$, and, thus, by Exercise 236.1, $g(X) \in \mathbb{F}_q[X]$. Now, since $g(\gamma^s) = 0$, $M_{\gamma^s}(X) \mid g(X)$. The reverse inclusion holds since the roots of $g(X)$ are all simple and they are also roots of $M_{\gamma^s}(X)$.

9) $\sharp(C_s) = \deg M_{\gamma^s}$ and $\deg M_{\gamma^s}$ divides t by 3).

10) $\mathbf{K} = \mathbb{F}_8 = \mathbb{F}_2[t]/\langle t^2 + t + 1 \rangle$ is a field since $t^2 + t + 1$ has no roots in $\mathbb{F}_2$. Its cardinality is $2^3 = 8$. The element $\alpha = \bar{t}$ is primitive since $\alpha \neq 1$ and $\mathbb{F}_8 \setminus \{0\}$ is a group of order 7 (a prime number). We have $\mathbb{F}_8 = \{0, 1, \alpha, \alpha^2, \alpha^3 = 1 + \alpha, \alpha^4 = \alpha + \alpha^2, \alpha^5 = 1 + \alpha + \alpha^2, \alpha^6 = 1 + \alpha^2\}$.

Root	Minimal Polynomial	2-cyclotomic Coset
0	X	
$1 = \alpha^0$	$X - 1 = X + 1$	$C_0 = \{0\}$
$\alpha, \alpha^2, \alpha^4$	$(X - \alpha)(X - \alpha^2)(X - \alpha^4) = X^3 + X + 1$	$C_1 = \{1, 2, 4\}$
$\alpha^3, \alpha^6, \alpha^5$	$(X - \alpha^3)(X - \alpha^6)(X - \alpha^5) = X^3 + X^2 + 1$	$C_3 = \{3, 6, 5\}$

It follows that $X^8 - X = X(X + 1)(X^3 + X + 1)(X^3 + X^2 + 1)$ is the factorization of $X^8 - X$ as the product of irreducible monic polynomials in $\mathbb{F}_2[X]$, and $X^7 - 1 = (X + 1)(X^3 + X + 1)(X^3 + X^2 + 1)$ is that of $X^7 - 1$.

11) By virtue of 7) and 8), over $\mathbb{F}_2$, the polynomial $X^{15} - 1$ has one irreducible factor of degree 1, one irreducible factor of degree 2, and three irreducible factors of degree 4.

Binary cyclic codes of length 15 correspond to ideals of the ring $\mathbb{F}_2[X]/\langle X^{15} - 1 \rangle$ which, in turn, correspond to divisors of $X^{15} - 1$ in $\mathbb{F}_2[X]$. Since $X^{15} - 1$ has 5 irreducible factors, the number of binary cyclic codes of length 15 is $2^5 = 32$.

Exercise 242:

1) If $p = 2$ then $H = \mathbf{K}^\times$ since, for all $x \in \mathbf{K}$, we have $x = x^{2^d} = (x^{2^{d-1}})^2$. If $p > 2$ then the surjective group homomorphism $\varphi : \mathbf{K}^\times \to H$ sending y onto y^2 has $\{1, -1\}$ as kernel. The result follows from the first isomorphism theorem.

2) If $p = 2$ take $a = 1$ and $b = 0$. If $p > 2$ then the two subsets $E = \{a^2 \mid a \in \mathbf{K}\}$ and $F = \{-1 - b^2 \mid a \in \mathbf{K}\}$ of $\mathbf{K}$ have the same cardinality $1 + |H| = \frac{q+1}{2}$. As a consequence, they cannot be disjoint.

Exercise 243:

1) A non primitive element has an order $r \neq q - 1$ which divides $q - 1$, thus, a root of $X^r - 1$.

2) All the roots of $\phi_{p,q-1}$ are simple because $\phi_{p,q-1} \mid (X^{q-1} - 1)$ and all the roots of $X^{q-1} - 1$ are simple. We have $(\mathbf{K}^\times, \cdot) \cong (\mathbb{Z}/(q-1)\mathbb{Z}, +)$ and the primitive elements in $\mathbf{K}$ correspond to the generators of $\mathbb{Z}/(q-1)\mathbb{Z}$ as a group.

3) $U = \gcd(X^8 - 1, \prod_{r|8;\, r\neq 8} (X^r - 1)) = \gcd(X^8 - 1, (X - 1)(X^2 - 1)(X^4 - 1)) = X^4 - 1$, and, thus, $\phi_{3,8} = X^4 + 1$. Its irreducible factorization is

$$\phi_{3,8} = (X^2 - X - 1)(X^2 + X - 1).$$

It follows that any root ξ of $X^2 - X - 1$ (or of $X^2 + X - 1$) is a primitive element of $\mathbb{F}_9$ with $\xi^0 = 1$, $\xi^0 = \xi$, $\xi^2 = \xi + 1$, $\xi^3 = -\xi + 1$, $\xi^4 = -1$, $\xi^5 = -\xi$, $\xi^6 = -\xi - 1$, $\xi^7 = \xi - 1$.

4) Let $P \in \mathbb{F}_p[X]$ be an irreducible monic factor of $\phi_{p,q-1}$ and denote by $\delta = \deg P$. Let ξ be a root of P (it is also a root of $\phi_{p,q-1}$) contained in **K**. The polynomial P is then the minimal polynomial of ξ over $\mathbb{F}_p$. The field $\mathbb{F}_p(\xi) \cong \mathbb{F}_p[X]/\langle P \rangle$ is a subfield of **K** of cardinality p^δ. Since ξ is primitive, $\mathbf{K} = \mathbb{F}_p(\xi)$, and, thus, $d = \delta$.

5) Let $p = 2$, $q = 2^4 = 16$, $\varphi(q-1) = \varphi(15) = (3-1)(5-1) = 8$. We compute

$$U = \gcd(X^{15} - 1, \prod_{r|15;\, r\neq 15} (X^r - 1)) = \gcd(X^{15} - 1, (X-1)(X^3-1)(X^5-1))$$

$$= (X^5 - 1) \cdot \gcd(X^{10} + X^{10} + 1, (X-1)^2(X^2+X+1))$$

$$= (X^5 - 1) \cdot \gcd(X^{10} + X^{10} + 1, X^2 + X + 1) = (X^5 - 1)(X^2 + X + 1),$$

and thus,

$$\phi_{2,15} = X^8 + X^7 + X^5 + X^4 + X^3 + X + 1.$$

By 4), $\phi_{2,15}$ has two irreducible factors of degree 4. Writing $X^8 + X^7 + X^5 + X^4 + X^3 + X + 1 = (X^4 + X^3 + aX^2 + bX + 1)(X^4 + cX^2 + dX + 1)$, with $a,b,c,d \in \mathbb{F}_2$, we obtain by identification $a = b = c = 0$ and $d = 1$. Thus, we obtain the irreducible factorization

$$\phi_{2,15} = (X^4 + X + 1)(X^4 + X^3 + 1).$$

We conclude that $\mathbb{F}_{16} = \mathbb{F}_2[X]/\langle X^4 + X + 1 \rangle$ or $\mathbb{F}_2[X]/\langle X^4 + X^3 + 1 \rangle$ in which $\bar{X}$ is primitive.

Exercise 244:

1) This is due to the fact that for $a \in \mathbf{Ł}^\times$, the map $\mathbf{Ł} \to \mathbf{Ł}$; $x \mapsto ax$, is a bijection.

2) This is because $\mathbf{Ł}$ is a **K**-vector space of dimension λ with basis $(1, \alpha, \ldots, \alpha^{\lambda-1})$.

3.a), 3.b) and 3.c) are straightforward.

3.d) Let $b \in \mathbf{K}$. As ϕ is **K**-linear (by 3.b) and 3.c)), using 3.a), we have $\phi^{-1}(b) = b + \mathrm{Ker}(\phi)$, and, thus, $\sharp(\phi^{-1}(b)) = \sharp(\mathrm{Ker}(\phi))$. Since $\dim_{\mathbf{K}} \mathrm{Ker}(\phi) = \lambda - 1$, we have $\sharp(\mathrm{Ker}(\phi)) = s_2^{\lambda-1} = q$.

4.a), 4.b) and 4.c) are straightforward.

5) We have $\mathbf{K} = \mathbb{F}_3 = \{0, 1, 2\}$ with primitive element $\beta = 2$. Here ϕ is: $\{0, \alpha + 1, 2\alpha + 2\} \mapsto 0$, $\{1, \alpha + 2, 2\alpha\} \mapsto 1$, $\{2, \alpha, 2\alpha + 1\} \mapsto 2$. The kernel matrix of ϕ is

$$\Gamma = \begin{pmatrix} 0 & \alpha+1 & 2\alpha+2 \\ 1 & \alpha+2 & 2\alpha \\ 2 & \alpha & 2\alpha+1 \end{pmatrix}.$$

6.a) It is immediate that $A = Bu^t u$. Let us prove that $\phi(u^t u)$ has full rank over **K**. First note that $\phi(\alpha^i) = \beta^i$ for $0 \leq i \leq \lambda - 1$. Performing on $\phi(u^t u)$ the elementary operations $R_i \leftarrow R_i - \beta^{i-1}R_1$ for $2 \le i \le \lambda$ (R_i stands for the ith row), and then $C_j \leftarrow C_j - \beta^{j-1}C_1$ for $2 \le j \le \lambda$ (C_j stands for the jth column), we obtain the matrix

$$\begin{pmatrix} 1 & 0 & \cdots & & 0 \\ 0 & 0 & \cdots & 0 & \gamma \\ \vdots & \vdots & \ddots & & \reflectbox{$\ddots$} \\ & & 0 & & \\ 0 & \gamma & & & \end{pmatrix}$$

with $\gamma = \phi(\alpha^\lambda) - \beta^\lambda$. To obtain the desired result, it suffices to show that $\gamma \ne 0$. For this, by way of contradiction, suppose that $\phi(\alpha^\lambda) = \beta^\lambda$, and write $\alpha^\lambda = b_0 + b_1\alpha + \cdots + b_{\lambda-1}\alpha^{\lambda-1}$, with $b_i \in \mathbf{K}$, $0 \le i \le \lambda - 1$. Then $\phi(\alpha^{\lambda+1}) = \phi(b_0\alpha + b_1\alpha^2 + \cdots + b_{\lambda-1}\alpha^\lambda) = b_0\beta + b_1\beta^2 + \cdots + b_{\lambda-1}\beta^\lambda = \beta\phi(\alpha^\lambda) = \beta^{\lambda+1}$, and so on we obtain that $\phi(\alpha^k) = \beta^k$ for any k. This would imply that $\mathrm{Ker}(\phi) = \{0\}$, in contradiction with the fact that $\dim_{\mathbf{K}} \mathrm{Ker}(\phi) = \lambda - 1 > 0$.

As $A = Bu^tu$, then, by virtue of 3.c), $\phi(A) = B\phi(u^tu)$. Because B has no repeated rows and $\phi(u^tu)$ has full rank over $\mathbf{K}$, we infer that $\phi(A)$ also has no repeated rows and consists of all the s_2^λ possible λ-tuples from $\mathbf{K}$, i.e., $\phi(A)$ is an $OA(s_2^\lambda, s_2^\lambda, \lambda)$.

6.b) Since A is a balanced $D(s_1, \lambda, s_1)$ and $C_{(\ell_1,\dots,\ell_\lambda)}(:,j) = A(:j) + \Gamma(1,\ell_j)$ for $1 \le j \le \lambda$, we deduce that $C_{(\ell_1,\dots,\ell_\lambda)}$ is a balanced $D(s_1, \lambda, s_1)$.

For $(\ell_1,\dots,\ell_\lambda) \in Q^\lambda$, as $\phi(v_{(\ell_1,\dots,\ell_\lambda)}) = (0,\dots,0)$, $\phi(C_{(\ell_1,\dots,\ell_\lambda)}) = \phi(A)$ is an $OA(s_2^\lambda, s_2^\lambda, \lambda)$.

6.c) Pick any two distinct λ-tuples $(\ell_1,\dots,\ell_\lambda)$, $(\ell_1',\dots,\ell_\lambda') \in Q^\lambda$. Obviously, the ith rows of $C_{(\ell_1,\dots,\ell_\lambda)}$ and $C_{(\ell_1',\dots,\ell_\lambda')}$ are distinct for $1 \le i \le s_1$. Since $\phi(C_{(\ell_1,\dots,\ell_\lambda)}) = \phi(C_{(\ell_1',\dots,\ell_\lambda')}) = \phi(A)$ and $\phi(A)$ has no repeated rows, we infer that $C_{(\ell_1,\dots,\ell_\lambda)}$ and $C_{(\ell_1',\dots,\ell_\lambda')}$ have no same rows. Thus, C has no repeated rows and consists of all the s_1^λ possible λ-tuples from $Ł$, i.e., C is an $OA(s_1^\lambda, s_1^\lambda, \lambda)$.

$$7)\ A = \begin{pmatrix} 0 & 0 \\ 1 & \alpha \\ \alpha & 2\alpha+1 \\ 2\alpha+1 & 2\alpha+2 \\ 2\alpha+2 & 2 \\ 2 & 2\alpha \\ 2\alpha & \alpha+2 \\ \alpha+2 & \alpha+1 \\ \alpha+1 & 1 \end{pmatrix},\ C_{(2,3)} = \begin{pmatrix} \alpha+1 & 2\alpha+2 \\ \alpha+2 & 2 \\ 2\alpha+1 & \alpha \\ 2 & \alpha+1 \\ 0 & 2\alpha+1 \\ \alpha & \alpha+2 \\ 1 & 1 \\ 2\alpha & 0 \\ 2\alpha+2 & 2\alpha \end{pmatrix},$$

$$C_{(2,2)} = \begin{pmatrix} \alpha+1 & \alpha+1 \\ \alpha+2 & 2\alpha+1 \\ 2\alpha+1 & 2 \\ 2 & 0 \\ 0 & \alpha \\ \alpha & 1 \\ 1 & 2\alpha \\ 2\alpha & 2\alpha+2 \\ 2\alpha+2 & \alpha+2 \end{pmatrix}, \phi(C_{(2,3)}) = \begin{pmatrix} 0 & 0 \\ 1 & 2 \\ 2 & 2 \\ 2 & 0 \\ 0 & 2 \\ 2 & 1 \\ 1 & 1 \\ 1 & 0 \\ 0 & 1 \end{pmatrix}, \phi(C_{(2,2)}) = \begin{pmatrix} 0 & 0 \\ 1 & 2 \\ 2 & 2 \\ 2 & 0 \\ 0 & 2 \\ 2 & 1 \\ 1 & 1 \\ 1 & 0 \\ 0 & 1 \end{pmatrix}.$$

Exercise 245:

1) $\sharp(\mathbb{P}^2(\mathbb{F}_q)) = \frac{q^3-1}{q-1} = q^2+q+1$ since $\sharp(\mathbb{F}_q^3 \setminus \{(0,0,0)\}) = q^3 - 1$ and all the equivalence classes have cardinality $\sharp(\mathbb{F}_q \setminus \{0\}) = q-1$.

2) For $q = 3$, $\sharp(\mathbb{P}^2(\mathbb{F}_q)) = 13$. The points of the projective plane of order 3 are (we denote $(a:b:c)$ instead of $\overline{(a,b,c)}$):

$$(0:1:0)\ (1:1:0)\ (2:1:0)\ (1:0:0)\ \text{(points at } \infty)$$
$$(0:0:1)\ (0:1:1)\ (0:2:1)$$
$$(1:0:1)\ (1:1:1)\ (1:2:1)$$
$$(2:0:1)\ (2:1:1)\ (2:2:1)$$

Exercise 246:

1) $f'(X) = -1$ is coprime with $f(X)$, and, thus, $f(X)$ is separable.

2) If α is a root of $f(X)$ in $\mathbb{F}_p$, then $0 = f(\alpha) = \alpha^p - \alpha + a = a$, in contradiction with $a \neq 0$.

3) $f(\alpha + j) = (\alpha + j)^p - (\alpha + j) + a = \alpha^p + j^p - \alpha - j + a = f(\alpha) = 0$.

4) Let α be a root of $f(X)$. Then, by virtue of 3), we have $f(X) = \prod_{j \in \mathbb{F}_p} (X - \alpha - j)$. Denote by $m(X)$ the minimal polynomial of α over $\mathbb{F}_p$. We have $f(X) = m(X)f_1(X)$ for some monic polynomial $f_1(X) \in \mathbb{F}_p[X]$. If $f_1(X) = 1$, then $f(X) = m(X)$ is irreducible over $\mathbb{F}_p$. Else, $f_1(X)$ has some root of the form $\alpha + j$ for some $j \in \mathbb{F}_p$. It is obvious that $m(X - j)$ is the minimal polynomial of $\alpha + j$ over $\mathbb{F}_p$. Thus, $f_1(X) = m(X - j)f_2(X)$ for some monic polynomial $f_2(X) \in \mathbb{F}_p[X]$. And so on, after a finite number of iterations, we find $f(X) = \prod_{j \in J} m(X - j)$ for some subset J of $\mathbb{F}_p$. Let n be the degree of $m(X)$. Then, by comparing the degree of both sides, we obtain $p = n \cdot |J|$. The case $n = 1$ is impossible since $\alpha \notin \mathbb{F}_p$. So, $n = p$, $|J| = 1$, and $f(X) = m(X)$ is irreducible over $\mathbb{F}_p$. ·

Exercise 247:

As $\mathbb{Q}$ is the prime subfield of $\mathbb{R}$, we have $\text{Aut}(\mathbb{R}) = \text{Gal}(\mathbb{R}/\mathbb{Q})$. Let $f \in \text{Gal}(\mathbb{R}/\mathbb{Q})$. We will first prove that f is increasing. For $x > 0$, we have $f(x) = f(\sqrt{x})^2 >$. Thus, for $a > b$, we have $f(a) - f(b) = f(a-b) > 0$.

Now, let $x \in \mathbb{R} \setminus \mathbb{Q}$, and suppose that $f(x) \neq x$. If $f(x) < x$, then there exists $r \in \mathbb{Q}$ with $f(x) < r < x$. Thus, $f(f(x)) < f(r) = r < f(x)$ in contradiction with $f(x) < r$. If $f(x) > x$, then there exists $r \in \mathbb{Q}$ with $x < r < f(x)$. Thus, $f(x) < f(r) = r < f(f(x))$ in contradiction with $r < f(x)$. We conclude that $\text{Gal}(\mathbb{R}/\mathbb{Q}) = \{\text{id}_{\mathbb{R}}\}$.

Exercise 248:

1) The cyclotomic field $\mathbb{Q}(\xi_n)$ is the splitting field over $\mathbb{Q}$ of the polynomial $X^n - 1$. As $\mathbb{Q}$ is perfect, the field extension $\mathbb{Q}(\xi_n)/\mathbb{Q}$ is then Galoisian by Theorem 225. The minimal polynomial of ξ_n is the cyclotomic polynomial ϕ_n whose roots (they are all simple) are the ξ_n^k, $k \in [\![1, n-1]\!]$ with $k \wedge n = 1$. Denoting by $\tau_k \in \text{Gal}(\mathbb{Q}(\xi_n)/\mathbb{Q})$ the $\mathbb{Q}$-automorphism of $\mathbb{Q}(\xi_n)$ corresponding to ξ_n^k, we have $\tau_k \circ \tau_{k'} = \tau_{kk'}$. The groups $\text{Gal}(\mathbb{Q}(\xi_n)/\mathbb{Q})$ and $(\mathbb{Z}/n\mathbb{Z})^\times$ are then isomorphic.

2) Let us denote by $\xi = e^{\frac{2i\pi}{n}}$ and $\alpha = \cos(\frac{2\pi}{n})$. As the field extension $\mathbb{Q}(\xi)/\mathbb{Q}$ is Galoisian and $\text{Gal}(\mathbb{Q}(\xi)/\mathbb{Q}) \cong (\mathbb{Z}/n\mathbb{Z})^\times$, we have

$$[\mathbb{Q}(\xi) : \mathbb{Q}] = |\text{Gal}(\mathbb{Q}(\xi)/\mathbb{Q})| = |(\mathbb{Z}/n\mathbb{Z})^\times| = \varphi(n),$$

the Euler function of n.

Since $\alpha = \frac{\xi + \xi^{-1}}{2}$, we have the field extensions $\mathbb{Q} \subseteq \mathbb{Q}(\alpha) \subseteq \mathbb{Q}(\xi)$. The monic polynomial $(X - \xi)(X - \xi^{-1}) = X^2 - 2\alpha X + 1 \in \mathbb{Q}(\alpha)[X]$ vanishes at ξ and is irreducible over $\mathbb{Q}(\alpha)$ (since $\xi \notin \mathbb{Q}(\alpha) \subseteq \mathbb{R}$). Hence it is the minimal polynomial of ξ over $\mathbb{Q}(\alpha)$, and $[\mathbb{Q}(\xi) : \mathbb{Q}(\alpha)] = 2$. By the multiplicativity of the degree, we conclude that

$$[\mathbb{Q}(\alpha) : \mathbb{Q}] = \frac{[\mathbb{Q}(\xi) : \mathbb{Q}]}{[\mathbb{Q}(\xi) : \mathbb{Q}(\alpha)]} = \frac{\varphi(n)}{2}.$$

Chapter 4

Algorithms for cryptography

Cryptography is the science of encryption and decryption of messages. The characters of the original text are often transformed, for transmission, into numbers. Cryptography has been used well before our era. For example, during the reign of Julius Caesar, the Romans used the following method that one will seize by the deciphering of the following short message:

YHQL YLGL YLFL.

We try to replace each letter by its predecessor in the alphabet (by circular permutation) until we find a meaningful text. We find successively:

XGPK XKFK XKEK,
WFOJ WJEJ WJDJ,
VENI VIDI VICI,

(I came, I saw, I won).

A difficult algorithmic problem can be used for creating secret codes for data transmission via public communication channels. The view would be that those who are sending messages are going to have additional information allowing them to solve the problem rapidly, while an adversary would face an exponential time consuming problem (or, at least, a problem for which no polynomial time algorithm is known).

4.1 Public-key cryptosystems — RSA method

This scheme was first introduced by Diffie and Hellman. We consider a set of individuals $i, j, k, \ldots$ wishing to communicate between them. But when j sends a message to i, this latter should be the only one capable to decipher the received message. To each individual i correspond two procedures, the first is E_i and is public, the second is D_i and is secret known only by i. The list of procedures E_i is placed on the equivalent of a directory.

Suppose that Alice wants to send a message m to Bob. She proceeds as follows: she consults the directory to find the procedure, say E_B, indicating the coding for Bob. She computes $m' = E_B(m)$ and sends it to Bob. In order to decipher m', Bob computes $D_B(m')$. In addition, we consider that Alice and Bob communicate in the presence of a spy, Eve, who wants to know the exchanged messages and possibly to modify them by the occasion.

The modern problems of cryptography are the following:

1. **Confidentiality:** A message sent by Alice to Bob should not be read by another person.

2. **Authentification:** Bob should be able to check that it is really Alice who sent the message.

3. **Non-repudiation:** It should be impossible for Alice to pretend that she has never sent the message.

The RSA method:

RSA (Rivest-Shamir-Adleman) is one of the first practical public-key cryptosystems and is widely used for secure data transmission. It is based on the practical difficulty of the factorization of the product of two large prime numbers. This method relies on the following mathematical result:

Proposition 249.

(1) *Let p be a prime number, and consider an integer $k \equiv 1 \bmod (p-1)$. Then, $\forall x \in \mathbb{Z}$, we have:*
$$x^k \equiv x \bmod p.$$

(2) *Let $n = p_1.....p_r$ where the p_i are pairwise distinct prime numbers, and consider an integer $k \equiv 1 \bmod (p_1-1)\cdots(p_r-1)$. Then, $\forall x \in \mathbb{Z}$, we have:*
$$x^k \equiv x \bmod n.$$

Proof. (1) If p/x then $x^k \equiv x \equiv 0 \bmod p$. Now suppose that $p \nmid x$ and write $k = 1 + (p-1)\ell$ for some $\ell \in \mathbb{Z}$. By Fermat's little theorem (or by Lagrange's theorem), we know that $x^{p-1} \equiv 1 \bmod p$, and, thus, $x^k = x(x^{p-1})^\ell \equiv x \bmod p$.

(2) By the Chinese remainder theorem, we know that $\mathbb{Z}/n\mathbb{Z} \cong \prod_{i=1}^{r}(\mathbb{Z}/p_i\mathbb{Z})$. So, to obtain the desired result, it suffices to check that $x^k \equiv x \bmod p_i$ for all $x \in \mathbb{Z}$ and $1 \leq i \leq r$. This follows from (1) since $k \equiv 1 \bmod (p_i-1)$.

Note that we can obtain an alternative proof of the previous proposition using the expression of the Euler function as follows.

Definition and Proposition 250. Let n be an integer ≥ 2, and denote by $\varphi(n)$ the order of the multiplicative group $(\mathbb{Z}/n\mathbb{Z})^{\times}$ of invertible elements in the ring $\mathbb{Z}/n\mathbb{Z}$. It is called *the Euler function* of n. We obviously have $(\mathbb{Z}/n\mathbb{Z})^{\times} = \{\bar{k} \in [\![1,n]\!] \mid k \wedge n = 1\}$, $\varphi(n) = \sharp\{k \in [\![1,n]\!] \mid k \wedge n = 1\}$, and for every $\bar{k} \in (\mathbb{Z}/n\mathbb{Z})^{\times}$, $\bar{k}^{\varphi(n)} = \bar{1}$. Moreover, if we know the prime factorization $n = p_1^{\alpha_1} \cdots p_r^{\alpha_r}$ of n, then

$$\varphi(n) = n\left(1 - \frac{1}{p_1}\right) \cdots \left(1 - \frac{1}{p_r}\right).$$

In particular, if $n = p_1 \cdots p_r$ is the product of r distinct prime number, then $\varphi(n) = (p_1 - 1) \cdots (p_r - 1)$.

Proof. Let $n = p_1^{\alpha_1} \cdots p_r^{\alpha_r}$ be the prime factorization of n. We want to count the positive integers $k \in [\![1,n]\!]$ such that $k \wedge n = 1$. There are n possibilities for such an integer k. Of these we throw away $\frac{n}{p_1}$ of them because they are divisible by p_1. Then we discard $\frac{n}{p_2}$ multiples of p_2, etc. This leaves us with $\varphi(n) = n - \frac{n}{p_1} - \cdots - \frac{n}{p_r}$ possible k's. But this formula is erroneous because we have thrown away too much. An integer k that is a multiple of both p_1 and p_2 has been discarded at least twice. So let's correct these errors as follows $\varphi(n) = n - \frac{n}{p_1} - \cdots - \frac{n}{p_r} + \frac{n}{p_1 p_2} + \frac{n}{p_1 p_3} + \cdots + \frac{n}{p_{r-1} p_r}$. But this formula is still erroneous as we added back too much, because an integer that is divisible by $p_1 p_2 p_3$, for instance, would have been re-entered at least twice. And so on, we end up with the correct formula $\varphi(n) = n - \frac{n}{p_1} - \cdots - \frac{n}{p_r} + \frac{n}{p_1 p_2} + \frac{n}{p_1 p_3} + \cdots + \frac{n}{p_{r-1} p_r} - \cdots + (-1)^r \frac{n}{p_1 \cdots p_r} = n(1 - \frac{1}{p_1}) \cdots (1 - \frac{1}{p_r})$.

We are now in position to present the RSA method. To each individual i, we associate integers e_i, d_i and n_i, where e_i and n_i are public while d_i is private (secret, known only by i). The messages m sent to i are integers modulo n_i. The coding and decoding procedures are defined, respectively, as:

$$E_i(m) = m^{e_i}, \quad D_i(c) = c^{d_i}.$$

Each integer n_i is the product of two distinct prime numbers p_i and q_i. In order for the decoding to be correct, we should have $D_i \circ E_i = \mathbf{id}$, that is, $(m^{e_i})^{d_i} \equiv m \bmod n_i$, and, thus, we need to have $e_i d_i \equiv 1 \bmod (p_i - 1)(q_i - 1)$. The integer e_i is random put in public domain by the recipient who made sure beforehand that e_i is coprime with $(p_i - 1)(q_i - 1)$ by computing their gcd and choosing randomly a new e_i until the gcd becomes 1. This is quite fast as the recipient knows p_i and q_i. He should also compute d_i the inverse of e_i modulo $(p_i - 1)(q_i - 1)$ (by the extended Euclidean algorithm).

Concerning complexity, we know that the calculation of x^k requires at most $\emptyset(\log k)$ multiplications. This proves that the calculation time of coding and decoding functions are polynomial at $\log(n_i)$, and, thus, possible for very large values of n_i. Concerning security, it is exactly related to the difficulty to factorize the integer n_i. Despite the spectacular progress in the factorization methods (the most rapid algorithms are probabilistic of sub-exponential complexity), when choosing the p_i's and q_i's greater than 10^{170}, we should have an excellent security for at least the next few years.

Let us build the following small scenario: $n = 143 = 11 \times 13$, $(p-1)(q-1) = 120$, $e = 7$, $7 \times (-17) \equiv 1 \bmod 120$, $d = 103$,

$$m = 3 \xrightarrow{\text{coding}} m^7 = 3^7 = 2187 \equiv 42 \bmod 143,$$

$$\xrightarrow{\text{decoding}} (42)^{103} = \underbrace{15655 \cdots 088}_{\text{more than 100 digits}} \xrightarrow{\text{remainder on division by } 143} 3.$$

4.2   Groups based cryptography

In this section, we consider a finite abelian group G of order N. We often suppose that G is cyclic ($\cong \mathbb{Z}/N\mathbb{Z}$) and we consider one generator g of G. The *discrete logarithm problem* is the following: we consider $h \in G$ and we look for the smallest x (if it does exist) such that $h = g^x$.

It is important in cryptography that this problem is difficult.

4.2.1   Diffie-Hellman keys exchange

Alice and Bob want to share a common key which could be used for example in an encrypted communication via a symmetric algorithm such as DES (Data-Encryption-Standard) without circulating the key on the network. For this, we can fix a group G of order N and an element $g \in G$ of high order. They proceed as follows:

* Alice chooses randomly an integer $a \in [\![1, N-1]\!]$ (that she keeps secret) and transmits g^a to Bob.

* Bob chooses randomly an integer $b \in [\![1, N-1]\!]$ (that he keeps secret) and transmits g^b to Alice.

* On her side, Alice computes $(g^b)^a$, while Bob computes $(g^a)^b$. Thus, they have shared the same key g^{ab} without circulating it on the network.

For most groups, recovering the key from the knowledge of G, g, g^a, and g^b, is as difficult as solving the discrete logarithm problem.

Discrete Logarithm Problem 251. Let a, b be two elements in a group G denoted multiplicatively. An integer k that solves the equation $b^k = a$ is called a *discrete logarithm* of a to the base b, and we write $k = \log_b a$. Finding k knowing a and b is called the *discrete logarithm problem*.

The discrete logarithm problem is considered to be computationally intractable. That is, no efficient classical algorithm is known for computing discrete logarithms in general. A general naïve exponential-time algorithm for computing $\log_b a$ in a finite group G is to compute $b^2, b^3, \ldots$ until finding the desired a. Its complexity is linear in the size of the group G and thus exponential in the number of digits in the size of the group.

Faster algorithms (some of them are linear in the square root of the size of the group, and thus exponential in half the number of digits in the size of the group) exist (e.g., Baby-step giant-step, Function field sieve, Pollard's rho algorithm, ...) and are usually inspired by similar algorithms for integer factorization. However none of them run in polynomial time.

4.2.2 El-Gamal's encryption

Alice wishes to send a message $m \in G$ to Bob whose public key is (g, h) with $h = g^x$ and x is secret.

Alice chooses an integer $k \in [\![1, N-1]\!]$ ($N = |G|$) then computes and sends to Bob the pair $(a, b) = (g^k, h^k m)$. Then, Bob performs the following calculation:

$$ba^{-x} = h^k m g^{-kx} = g^{kx-kx} m = m,$$

and retrieves the initial message m.

4.2.3 El-Gamal's signature

The El-Gamal signature scheme allows a third-party to confirm the authenticity of a message sent over an insecure channel. We consider this time a message $m \in \mathbb{Z}/N\mathbb{Z}$ that Bob wants to send signed. As above, we suppose that he has a public key $h = g^x$ with x secret. Moreover, we suppose that f is a given bijection between G and $\mathbb{Z}/N\mathbb{Z}$. The procedure is the following:

* Bob chooses a random $k \in [\![1, N-1]\!]$ with $\gcd(k, N) = 1$, and computes $a = g^k$.

* Bob computes a solution $b \in \mathbb{Z}/N\mathbb{Z}$ of the congruence

$$m \equiv xf(a) + bk \bmod N,$$

($b = (m - xf(a))k^{-1} \bmod N$; if $b = 0$ start over again). Then he sends the pair (a, b) together with the message m to Alice.

* Alice has only to check the relation

$$h^{f(a)} a^b = g^{xf(a)+kb} = g^m,$$

ensuring that Bob is the expeditor.

The El-Gamal signature algorithm is rarely used in practice. A variant developed at NSA and known as the Digital Signature Algorithm (DSA) is much more widely used. The effective procedures of DSA use, in addition, a hash function compressing the initial message to make it illegible.

4.2.4 Massey-Omura coding

Here Alice wants to send a message $m \in G$ to Bob. The procedure is as follows:

* Alice chooses an integer $x \in [\![1, N-1]\!]$ coprime with $N = |G|$ and sends $a = m^x$ to Bob.

* Bob chooses an integer $y \in [\![1, N-1]\!]$ coprime with N and sends $b = a^y = m^{xy}$ to Alice.

* Alice computes $x' \in [\![1, N-1]\!]$ with $xx' \equiv 1 \bmod N$ and sends $a' = b^{x'} = m^{xyx'} = m^y$ to Bob.

* Finally, Bob computes $y' \in [\![1, N-1]\!]$ with $yy' \equiv 1 \bmod N$, computes $a'^{y'} = m^{yy'} = m$, and, thus, retrieves the initial message m.

Note that this procedure does not use any public key neither an exchange of a secret key. However, each of the actors uses a private secret key (a double-key algorithm).

Let us build the following small scenario: suppose that Alice wants to send secretely her examination's mark $13 \in [\![0, 20]\!]$ to Bob using Massey-Omura coding. The considered group is then $G = \mathbb{Z}/21\mathbb{Z}$. Imagine that Alice chooses $x = 2$ (coprime with 21 with inverse $x' = 11$ modulo 21), and that Bob chooses $y = 10$ (coprime with 21 with inverse $y' = 19$ modulo 21). The exchange will be as follows:

$$\text{Alice} \xrightarrow{5} \text{Bob} \xrightarrow{8} \text{Alice} \xrightarrow{4} \text{Bob}.$$

Finally, Bob computes $19 \times 4 \bmod 21 = 13$ and retrieves Alice's mark.

4.2.5 The choice of the group

To make the coding and decoding operations feasible rapidly, it is necessary for the calculations in the group G to be relatively simple. However, the discrete logarithm problem should be difficult in G, and this excludes all groups of the form $\mathbb{Z}/N\mathbb{Z}$. For these reasons, the first groups used were the groups $\mathbb{F}_q^\times$ of nonzero elements of finite fields $\mathbb{F}_q$ of cardinality q. But for these groups, there are algorithms of sub-exponential complexity calculating the discrete logarithm, imposing to take very large values of q. That is why, since the mid-1980s, Miller and Koblitz proposed to use the group of points on an elliptic curve over a finite field $\mathbb{F}_q$. In that case, there is no known sub-exponential method for calculating the discrete logarithm and, thus, we can work with substantially smaller values of q.

4.3 Exercises

Exercise 252. The number of available keys in a coding system gives a maximal bound on its security (measure of the complexity of an exhaustive search). Note that it is only an upper bound and rarely a good measure of the security.

1) How many possible keys are in a Caesar's code?

2) What about an affine coding? ($C(x) = ax + b$ mod 26 for each character $x \in \mathbb{Z}/26\mathbb{Z}$).

3) What about a coding by substitution (arbitrary substitution, character by character).

Exercise 253. Alice changes her RSA key every 25 days while Bob changes his RSA key every 31 days. Knowing that Alice has changed her key today and that Bob has changed it three days ago, determine when will be the next time Alice and Bob change their keys the same day.

Exercise 254.

1) Show for any prime number $p > 3$, $p^2 - 1$ is divisible by 24.

2) Deduce that if $n = 35$ is used as an RSA module then the encryption exponent e always equals the decryption exponent d.

Exercise 255. We consider an RSA module $n = pq$, where p and q are unknown. Show how the knowledge of $\varphi(n)$ (Euler function of n, see Definition and Proposition 250) allows to retrieve the factorization of n.

Exercise 256. Consider an RSA system with $p = 19$ and $q = 23$.

1) Calculate n and $\varphi(n)$.

2) Calculate the exponent d associated to $e = 9$ and to $e = 17$.

Exercise 257. 1) Show that for $n \geq 4$, a natural number with at most $\log_4(n)$ digits in basis 4 is $\leq n - 1$, and the number of digits of a natural number $\leq n - 1$ is at most $1 + \log_4(n - 1)$.

2) We consider an RSA module $n = pq$ with $p = 23$ and $q = 11$.

(i) Compute the exponent d associated to $e = 3$.

(ii) We propose to encrypt a message containing one word made up of the letters a, b, c. We transform it into a number written in basis 4 by matching a to 1, b to 2, and c to 3.

What is the maximal size of a plaintext (its maximal number of digits) and the maximal size of the corresponding ciphertext?

(iii) Code with the above RSA module ($e = 3$) the message abb (the output is a word made up of the alphabet $\{a, b, c\}$).

(iv) Explain how the recipient can decode the received ciphertext with the key d computed in (i).

Exercise 258. (RSA with two close factors)

Suppose that n is the product of two close prime numbers p and q (we can suppose that $p > q > 2$). Set $t = \frac{p+q}{2}$ and $s = \frac{p-q}{2}$. Show that:

1) $n = t^2 - s^2$,

2) s is small,

3) t is slightly greater than the square root of n,

4) we can use this information to factorize n (the algorithm is called "Fermat's algorithm").

5) Apply this algorithm to factorize 24960007.

6) Give the complexity of the algorithm as a function of s or p, and n.

7) Determine the number of iterations of the algorithm when p differs from $\sqrt{n}$ by less than $\sqrt[4]{4n}$.

Exercise 259. Consider a coding RSA system where all the users choose the same exponent $e = 3$ and suppose that the same message m is sent to three different recipients (using different RSA modules n_1, n_2, and n_3).

Show how a hacker intercepting these three coded messages c_1, c_2, and c_3 can retrieve the message m without knowing the secret keys of the recipients (we suppose that he possesses a numerical method (such as Newton's method) allowing him to extract a cubic root).

Exercise 260. Alice and Bob chose the same RSA module $n_A = n_B = n$ but they selected two coprime public exponents e_A and e_B ($e_A \wedge e_b = 1$).

1) Can you send a message to Alice or Bob without the other being aware of it using his public RSA key?

2) We suppose now that Alice and Bob send to each other the same message m. Can you retrieve this message?

Exercise 261. (The number of fixed points in an RSA module)

Consider an RSA module $n = pq$ (p, q are distinct prime numbers) with public exponent e.

1) Show that $\text{Card}(\{m \in \mathbb{F}_p \mid m^e = m\}) = 1 + \gcd(e - 1, p - 1)$.

2) Deduce the number of fixed-points of the RSA module, that is,

$$\text{Card}(\{m \in \mathbb{Z}/n\mathbb{Z} \mid m^e = m\}).$$

Exercise 262. (Baby-step giant-step)

Let p be a prime number and g a primitive element in $\mathbb{F}_p$. Recall that the discrete logarithm problem consists in expressing an element $\beta \in \mathbb{F}_p^\times$ as $\beta = g^x$ with $0 \leq x \leq p-2$. The integer x is denoted by $\log_g(\beta)$. Let $m = \lceil \sqrt{p-1} \rceil$ and write $x = mi + j$ with $i \in \mathbb{N}$ and $0 \leq j < m$ (by Euclidean division). Note that: $g^x = \beta \Leftrightarrow \beta(g^{-m})^i = g^j$.

The algorithm baby-step giant-step proceeds as follows: in order to find x, one computes the list (j, g^j) (the baby-steps) then the $(i, \beta(g^{-m})^i)$'s (the giant-steps) until finding a second member which is already in the baby-steps list. The corresponding pair (i, j) gives x. Of course the group $\mathbb{F}_p^\times$ can be replaced by any cyclic group.

1) Explain why $i < m$.

2) Shanks's algorithm for computing $\log_g(\beta)$:

Input: A prime number p, a primitive element g in $\mathbb{F}_p$, and $\beta \in \mathbb{F}_p^\times$.

Output: A value x satisfying $g^x = \beta$.

```
1  m ← ⌈√(p-1)⌉
2  Compute g^j for 0 ≤ j < m
3  Sort the pairs (j, g^j) accordingly to their second coordinates;
4  let L₁ be the list (g^j) of their second coordinates
5  Compute g^{-m}
6  γ ← β
7  for i from 0 to m-1 do
8      (a) Check whether γ ∈ L₁
9      (b) If so, return x = mi + j
10     (c) If not, γ ← γ·g^{-m}
```

2.a) What is the complexity of the algorithm (in the number of bits of p)? Is it polynomial, exponential, sub-exponential?

2.b) Use Shanks's algorithm to compute $\log_3(111)$ in $\mathbb{F}_{113}$.

4.4 Solutions to the exercises

Exercise 252:

1) 26.

2) $\varphi(26) \times 26 = (13-1)(2-1) \times 26 = 312$.

3) 26!.

Exercise 253:

Denote by d the number of days until Alice and Bob change their keys the same day. Since Alice changes her key every 25 days and she has changed it today, d has to be divisible by 25. Since Bob changes his key every 31 days and he has changed it three days ago, $d + 3$ has to be divisible by 31. Thus, d satisfies the congruence system:

$$\begin{cases} d & \equiv & 0 & \mod & 25 \\ d & \equiv & -3 & \mod & 31. \end{cases} \tag{4.1}$$

By the Chinese remainder theorem, this system is equivalent to the congruence

$$d \equiv 400 \mod 775,$$

and hence Alice and Bob will change their keys the same day in 400 days.

Exercise 254:

1) This follows simply from the fact that $p^2 - 1 = (p-1)(p+1)$. Since p is not divisible by either 2 or 3,
   - the factors $p - 1$ and $p + 1$ must both be even,
   - one of them must be divisible by 4, and
   - one of them must be divisible by 3.
Thus, their product must be divisible by $2 \cdot 4 \cdot 3 = 24$.

2) $n = 35 = pq$ with $p = 5$, $q = 7$; its Euler function is $\varphi(35) = (p-1)(q-1) = 24$. Then to find e we need the fact that $\gcd(e, 24) = 1$. All the possible values of e are prime: $5, 7, 11, 13, 17, 19, 23$. The desired result follows from 1).

Exercise 255:

We have: $n = pq$ and $\varphi(n) = (p-1)(q-1) = n - p - q + 1$. Thus, p and q are the roots of the polynomial $X^2 - (n - \varphi(n) + 1)X + n$.

Exercise 256:

1) $n = 19 \times 23 = 437$, $\varphi(n) = 18 \times 22 = 396$.

2) 9 is not coprime with 396. By the Extended Euclidean algorithm, we can find a Bézout identity $1 = 7 \times 396 - 163 \times 17$ between 17 and 396. Thus, $d = 233$.

Exercise 257:

1) Let N be the number of digits in basis 4 of a natural number $n \geq 4$. We have $4^{N-1} \leq n \leq 3(1 + 4 + \cdots + 4^{N-1}) = 4^N - 1$, or also, $4^{N-1} \leq n < 4^N$, and thus $\log_4(n) < N \leq \log_4(n) + 1$. The desired result easily follows.

2) We consider an RSA module $n = pq$ with $p = 23$ and $q = 11$.

(i) $n = 253, d = 147$.

(ii) By virtue of 1), the maximal size of a plaintext is $k = \lfloor \log_4(253) \rfloor = 3$ while the maximal size of the corresponding ciphertext is $k + 1 = 4$.

(iii) abb $\rightarrow (122)_4 = 26 = m \xrightarrow{\text{encryption}} 26^3 \bmod 253 = 119 = (1313)_4 \rightarrow acac$.

(iv) acac $\rightarrow (1313)_4 = 119 = c \xrightarrow{\text{decryption}} 119^{147} \bmod 253 \overset{\text{exponentiation by squaring}}{=}$
$26 = (122)_4 \rightarrow abb$.

Exercise 258:

1) $\frac{1}{4}((p+q)^2 - (p-q)^2) = pq$.

2) s is small because p and q are close.

3) $t^2 - n = s^2$ is small.

4) Fermat's algorithm:
   (i) $t \leftarrow \lceil \sqrt{n} \rceil$
   (ii) If $t^2 - n$ is a perfect square then return $(t, \sqrt{t^2 - n})$
   (iii) Do $t \leftarrow t + 1$ and go to (ii)

5) $\lceil \sqrt{24960007} \rceil = 4996$; $t^2 - n = 9 = 3^2$. So we obtain the factorization of n in one shot: $n = 24960007 = (t - s)(t + s) = 4993 \times 4999$.

6) The number N of iterations is $\leq t - \sqrt{n} = \sqrt{n + s^2} - \sqrt{n} = \sqrt{n}(\sqrt{1 + \frac{s^2}{n}} - 1) \sim$
$\frac{s^2}{2n}$. Or also, $N \leq t - \sqrt{n} = \frac{p+q}{2} - \sqrt{pq} = \frac{1}{2}(\sqrt{p} - \sqrt{q})^2 = \frac{1}{2}\frac{(p - \sqrt{n})^2}{p}$.

7) $N \leq \frac{1}{2}\frac{\sqrt[4]{4n}^2}{p} = \frac{\sqrt{n}}{p}$.

Exercise 259:

We can suppose that n_1, n_2, and n_3 are pairwise coprime (otherwise one could factorize two of the n_i's). We have $c_i \equiv m^3 \bmod n_i$ for $i = 1, 2, 3$. Using Bézout identities between the n_i's, one can find $x \in \mathbb{Z}$ such that $x \equiv c_i \bmod n_i$ for $i = 1, 2, 3$. Let c be the remainder on division of x by $n_1 n_2 n_3$. As $c \equiv m^3 \bmod n_1 n_2 n_3$ (by the Chinese theorem) and $0 \leq c, m^3 < n_1 n_2 n_3$, we deduce that $c = m^3$. The hacker has only to extract an integer cubic root of c to retrieve m.

Exercise 260:

1) No because Bob can compute $d_A = e_A^{-1} \bmod \varphi(n)$.

2) Let $u_A, u_B \in \mathbb{Z}$ such that $u_A e_A + u_B e_B = 1$. Let us denote by c_A (resp. c_B) the cipher-text sent to Alice (resp. to Bob). Then

$$(c_A)^{u_A}(c_B)^{u_B} = m^{u_A e_A + u_B e_B} = m.$$

Exercise 261:

1) We know that $\mathbb{F}_p^\times$ is cyclic of order $p - 1$. Let ξ be a primitive element in $\mathbb{F}_p$, that is, whose powers generate $\mathbb{F}_p^\times$. For $x = \xi^r \in \mathbb{F}_p^\times$ with $r \in [\![0, p - 2]\!]$, we denote $r = \log x$. Let $m \in \mathbb{F}_p$ such that $m^e = m$. Then, either $m = 0$ or $e \log m = \log m \bmod (p - 1)$ (or, equivalently, $p - 1$ divides $(e - 1) \log m$). The values of $\log m$ satisfying the second condition is equal to the number of multiples of $\frac{p-1}{\gcd(e-1, p-1)}$ in $[\![0, p - 2]\!]$. This number is $\gcd(e - 1, p - 1)$. The conclusion follows.

2) As $\mathbb{Z}/n\mathbb{Z} \cong \mathbb{F}_p \times \mathbb{F}_q$ (by the Chinese remainder theorem), then, by virtue of 1), we have $\mathrm{Card}(\{m \in \mathbb{Z}/n\mathbb{Z} \mid m^e = m\}) = (1 + \gcd(e - 1, p - 1))(1 + \gcd(e - 1, q - 1))$.

Exercise 262:

1) $i \geq m \Rightarrow x \geq m^2 \Rightarrow x \geq p - 1$.

2.a) The complexity is of the order of $2m$, thus, $\emptyset(\sqrt{p})$. It is exponential in the size of p.

2.b) We initialize $m = \lceil \sqrt{112} \rceil = 11$. Then we obtain: $g^0 = 1$, $g^1 = 3$, $g^2 = 9$, $g^3 = 27$, $g^4 = 81$, $g^5 = 17$, $g^6 = 51$, $g^7 = 40$, $g^8 = 7$, $g^9 = 21$, and $g^{10} = 63$. The list obtained is

$$(0,1),(1,3),(8,7),(2,9),(5,17),(9,21),(3,27),(7,40),(6,51),(10,63),(4,81).$$

We then have $g^m = g^{11} = 76$. By the extended Euclidean algorithm, we get $g^{-m} = -55 = 58$. We initialize γ at 111.

$i = 0$: $\gamma \notin L_1$, $\gamma \leftarrow 111 \times 58 = 110$
$i = 1$: $110 \notin L_1$, $\gamma \leftarrow 110 \times 58 = 52$
$i = 2$: $52 \notin L_1$, $\gamma \leftarrow 52 \times 58 = 78$
$i = 3$: $78 \notin L_1$, $\gamma \leftarrow 78 \times 58 = 4$
$i = 4$: $4 \notin L_1$, $\gamma \leftarrow 4 \times 58 = 6$
$i = 5$: $6 \notin L_1$, $\gamma \leftarrow 6 \times 58 = 9$
$i = 6$: $9 \in L_1$ for $j = 2$, so $x = \log_3(111) = 11 \times 6 + 2 = 68$.

Chapter 5

Algebraic plane curves

In this chapter, **K** denotes a (commutative) field and $\bar{\mathbf{K}}$ an algebraic closure of **K**. This chapter relies essentially on [9, 61].

5.1 Projective space and projective varieties

Let us first define the projective plane $\mathbb{P}^2(\mathbb{R})$ over the real numbers $\mathbb{R}$. Two distinct lines in $\mathbb{A}^2(\mathbb{R})$ meet in a point except when they are parallel. We can eliminate this exception if we see that two parallel lines meet at a certain point at ∞ (see Fig. 5.1). This can be visualized by a perspective drawing. The horizon represents points at ∞ except the point at ∞ corresponding to lines parallel to the horizon[1] (see Fig. 5.2).

Definition 263. The *projective plane* over the real numbers is $\mathbb{P}^2(\mathbb{R}) := (\mathbb{R}^3 \setminus \{(0,0,0)\})/\sim$, where $\sim$ is the equivalence relation:

$$(x,y,z) \sim (x',y',z') \Leftrightarrow \exists \lambda \in \mathbb{R} \setminus \{0\} \mid (x',y',z') = \lambda(x,y,z).$$

We denote by $(x:y:z)$ the class of the triple $(x,y,z) \in \mathbb{R}^3 \setminus \{(0,0,0)\}$ in $\mathbb{P}^2(\mathbb{R})$. A system of homogeneous coordinates of a point $p \in \mathbb{P}^2(\mathbb{R})$ is an element in $\mathbb{R}^3 \setminus \{(0,0,0)\}$ whose class in $\mathbb{P}^2(\mathbb{R})$ is p.

Definition 264. Let $(a,b,c) \in \mathbb{R}^3 \setminus \{(0,0,0)\}$. The set

$$\{(x:y:z) \in \mathbb{P}^2(\mathbb{R}) \mid ax+by+cz = 0\},$$

is called a *projective line* of $\mathbb{P}^2(\mathbb{R})$.

The following formalizes the observation given at the beginning of this section.

---

[1]Figures 5.1 and 5.2 are reproduced from [14].

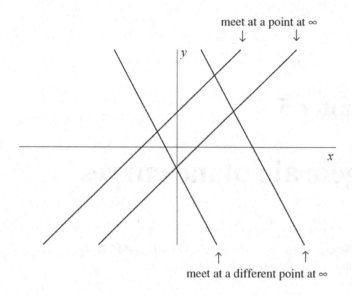

Figure 5.1: Parallel lines in $\mathbb{P}^2(\mathbb{R})$

Proposition 265. *The map* $\varphi : \mathbb{A}^2(\mathbb{R}) \to \mathbb{P}^2(\mathbb{R})$ *sending* (x,y) *onto* $(x : y : 1)$ *is one-to-one.*

$H_\infty := \mathbb{P}^2(\mathbb{R})) \setminus \varphi(\mathbb{A}^2(\mathbb{R})) = \{(x : y : z) \in \mathbb{P}^2(\mathbb{R}) \mid z = 0\}$ is called *the line at* ∞. By identifying $\mathbb{A}^2(\mathbb{R})$ with its image by φ, we can see the projective plane as the disjoint union

$$\mathbb{P}^2(\mathbb{R}) = \mathbb{A}^2(\mathbb{R}) \cup H_\infty.$$

Hereafter, we explain how affine lines (in $\mathbb{A}^2(\mathbb{R})$) are related to projective ones (in $\mathbb{P}^2(\mathbb{R})$):

affine line	projective line	point at ∞
$y = mx + b$	$y = mx + bz$	$(1 : m : 0)$
$x = c$	$x = cz$	$(0 : 1 : 0)$

The above construction of the projective plane over $\mathbb{R}$ can be generalized to any dimension n and any base field $\mathbf{K}$.

Definition 266. The *projective space* of dimension n over a field $\mathbf{K}$ is $\mathbb{P}^n(\mathbf{K}) := (\mathbf{K}^{n+1} \setminus \{0\})/\sim$, where $0 = (0, \ldots, 0)$ and $\sim$ is the equivalence relation:

$$(x_0, \ldots, x_n) \sim (x_0', \ldots, x_n') \Leftrightarrow \exists \lambda \in \mathbf{K}^\times \mid (x_0', \ldots, x_n') = \lambda(x_0, \ldots, x_n).$$

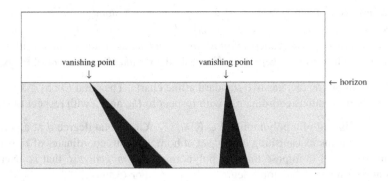

Figure 5.2: Points at ∞ in $\mathbb{P}^2(\mathbb{R})$

We denote by $(x_0 : \cdots : x_n)$ the class of the $(n+1)$-tuple $(x_0, \ldots, x_n) \in \mathbf{K}^{n+1} \setminus \{0\}$ in $\mathbb{P}^n(\mathbf{K})$. A system of homogeneous coordinates of a point $p \in \mathbb{P}^n(\mathbf{K})$ is an element in $\mathbf{K}^{n+1} \setminus \{0\}$ whose class in $\mathbb{P}^n(\mathbf{K})$ is p.

Analogously to the fact that $\mathbb{P}^2(\mathbb{R})$ contains the affine plane $\mathbb{A}^2(\mathbb{R})$ as a subset, by the following, we see that $\mathbb{P}^n(\mathbf{K})$ contains the affine space $\mathbb{A}^n(\mathbf{K})$.

Proposition 267. *Let $U_0 := \{(x_0 : \cdots : x_n) \in \mathbb{P}^n(\mathbf{K}) \mid x_0 \neq 0\}$, then the map ϕ : $\mathbb{A}^n(\mathbf{K}) \to \mathbb{P}^n(\mathbf{K})$ sending $(a_1, \ldots, a_n)$ onto $(1 : a_1 : \cdots : a_n)$ is one-to-one.*

In particular, by identifying U_0 with $\phi(\mathbb{A}^n(\mathbf{K}))$, we have

$$\mathbb{P}^n(\mathbf{K}) = U_0 \cup H, \text{ where } H = \{p \in \mathbb{P}^n(\mathbf{K}) \mid p = (0 : x_1 : \cdots : x_n)\}.$$

Identifying U_0 with $\mathbb{A}^n(\mathbf{K})$ and H with $\mathbb{P}^{n-1}(\mathbf{K})$, we have:

$$\mathbb{P}^n(\mathbf{K}) = \mathbb{A}^n(\mathbf{K}) \cup \mathbb{P}^{n-1}(\mathbf{K}) \quad (*).$$

Note that a point $p \in \mathbb{P}^{n-1}(\mathbf{K})$ corresponds to a line $L \subseteq \mathbb{A}^n(\mathbf{K})$ through the origin. Thus, in (*), we can think that p represents the asymptotical direction of all lines in $\mathbb{A}^n(\mathbf{K})$ parallel to L. We then see p as a point at ∞ (as in the projective plane case). As special cases, we have:

$$\mathbb{P}^0(\mathbf{K}) = \{\infty\} \quad \& \quad \mathbb{P}^1(\mathbf{K}) = \mathbb{A}^1(\mathbf{K}) \cup \mathbb{P}^0(\mathbf{K}) = \mathbb{A}^1(\mathbf{K}) \cup \{\infty\}.$$

Since every point in $\mathbb{P}^n(\mathbf{K})$ has at least one nonzero homogeneous coordinate, it lies in at least one of the U_i's, where for $0 \leq i \leq n$,

$$U_i := \{(x_0 : \cdots : x_n) \in \mathbb{P}^n(\mathbf{K}) \mid x_i \neq 0\}.$$

This implies

$$\mathbb{P}^n(\mathbf{K}) = \cup_{i=0}^n U_i.$$

So, $\mathbb{P}^n(\mathbf{K})$ is covered by $n+1$ subsets, each of which looks just like $\mathbb{A}^n(\mathbf{K})$.

Definition 268. Each subset U_i of $\mathbb{P}^n(\mathbf{K})$ is called a *standard affine chart* of $\mathbb{P}^n(\mathbf{K})$. For every point $p = (a_0 : \cdots : a_n) \in U_i$, the n-tuple $(a_0 a_i^{-1}, \ldots, a_{i-1} a_i^{-1}, a_{i+1} a_i^{-1}, \ldots, a_n a_i^{-1})$ are called the *non-homogeneous coordinates* of p with respect to U_i. The cover $\mathbb{P}^n(\mathbf{K}) = \cup_{i=0}^n U_i$ is called a *standard affine cover* of $\mathbb{P}^n(\mathbf{K})$.

For example, $\mathbb{P}^1(\mathbb{R})$ has two standard affine charts. The point $(3 : 5) \in \mathbb{P}^1(\mathbb{R})$ has non-homogeneous coordinate $\frac{5}{3}$ with respect to U_0, and $\frac{3}{5}$ with respect to U_1.

The vanishing of a polynomial $f \in \mathbf{K}[X_0, \ldots, X_n]$ of total degree d at a point $p \in \mathbb{P}^n(\mathbf{K})$ means its vanishing on any set of homogeneous coordinates of p. But, for this, we need to impose that our polynomial is *homogeneous*, that is, every term appearing in f has total degree exactly d. For example, $X_1^3 + X_1 X_2 X_3$ is homogeneous while $X_1^3 - X_2^2$ is not. This latter can be homogenized by adding a new (homogenizing) variable Z (it becomes $X_1^3 - X_2^2 Z$). We can consider subsets of $\mathbb{P}^n(\mathbf{K})$ defined by the vanishing of a system of homogeneous polynomials (possibly of different total degrees). The generalization of affine varieties defined in Chapter 2 is the following:

Definition 269. Let $f_1, \ldots, f_s \in \mathbf{K}[X_0, \ldots, X_n]$ be homogeneous polynomials. We set

$$V(f_1, \ldots, f_s) := \{(a_0 : \cdots : a_n) \in \mathbb{P}^n(\mathbf{K}) \mid f_j(a_0, \ldots, a_n) = 0 \ \forall \ 1 \leq j \leq s\}.$$

It is called the *projective variety* defined by $f_1, \ldots, f_s$.

Note that $V(f_1, \ldots, f_s) \cap U_0$ can be identified with the affine variety $V(g_1, \ldots, g_s) \subseteq \mathbb{A}^n(\mathbf{K})$, where $g_j(X_1, \ldots, X_n) = f_j(1, X_1, \ldots, X_n)$ for $1 \leq j \leq s$.

Example 270. How do we draw the projective plane curve $C = V(f)$ with $f(X, Y, Z) = X^2 + Y^2 - Z^2$? The easiest way is to draw the piece of $V(f)$ that sits in affine space. For example, look at the set of $(1 : y : z)$ such that $f(1, y, z) = 1 + y^2 - z^2 = 0$. If $\mathbf{K} = \mathbb{R}$ this is a hyperbola (Fig. 5.3). However, it is missing the points when $x = 0$, i.e. solutions to $f(0, y, z) = 0$, i.e. the two points $(0 : 1 : 1)$ and $(0 : 1 : -1)$ at ∞. Alternatively, we could look at a different chart $\{z \neq 0\}$. Then we get $x^2 + y^2 = 1$ which is a circle (Fig. 5.4). Is this the whole curve? Yes, as one can see by noting that $C(\mathbb{R}) \cap \{z = 0\} = \emptyset$.

5.2 Algebraic plane curves

Definition 271. We say that $C \subseteq \mathbb{A}^2(\bar{\mathbf{K}})$ is an *affine plane curve* if there exists an nonconstant polynomial $F \in \bar{\mathbf{K}}[X, Y]$ such that $C = V(F)$.

We say that $C \subseteq \mathbb{P}^2(\bar{\mathbf{K}})$ is a *projective plane curve* if there exists an nonconstant homogeneous polynomial $H \in \bar{\mathbf{K}}[X, Y, Z]$ such that $C = V(H)$.

As $\bar{\mathbf{K}}[X, Y]$ is factorial, every affine plane curve can be written $C = V(F)$ where $F \in \bar{\mathbf{K}}[X, Y]$ is nonconstant and square-free (i.e., not divisible by the square of an irreducible polynomial; see Exercise 238). The *degree of C* is defined as

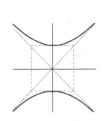

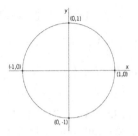

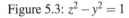

Figure 5.3: $z^2 - y^2 = 1$ Figure 5.4: $x^2 + y^2 = 1$

$\deg(F)$. Moreover, since $\bar{\mathbf{K}}[X,Y,Z]$ is factorial and factors of nonzero homogeneous polynomials are homogeneous, then every projective plane curve D can be written $D = V(H)$ where $H \in \bar{\mathbf{K}}[X,Y,Z]$ is nonconstant homogeneous and square-free. The *degree of D* is defined as $\deg(H)$.

Definition 272. Let $C = V(H)$ be a projective plane curve, with $H \in \mathbf{K}[X,Y,Z]$. We say that C has a *rational point* (over **K**) if $C(\mathbf{K}) := \{(\alpha : \beta : \gamma) \in \mathbb{P}^2(\mathbf{K}) \mid H(\alpha,\beta,\gamma) = 0\} \neq \emptyset$. An analogous definition applies for an affine curve. Hereafter some examples:

- A curve has always rational points (infinitely many) over $\bar{\mathbf{K}}$.

- The affine curve $x^2 + y^2 + 1 = 0$ over the reals has no rational points.

- The projective curve $y^2z + yz^2 = x^3 + xz^2 + z^3$ defined over $\mathbb{F}_2$ has a unique rational point $(0 : 1 : 0)$.

- The set of rational points of the projective curve $E : y^2z = x^3 + 2z^3$ defined over $\mathbb{F}_7$ is $E(\mathbb{F}_7) = \{(0 : 1 : 0), (0 : 3 : 1), (0 : 4 : 1), (3 : 1 : 1), (3 : 6 : 1), (5 : 1 : 1), (5 : 6 : 1), (6 : 1 : 1), (6 : 6 : 1)\}$.

- *Fermat's last theorem* (proved by Taylor and Wiles) is equivalent to the statement that for an integer n at least 3, the only rational points of the curve $x^n + y^n = z^n$ in $\mathbb{P}^2(\mathbb{Q})$ are the obvious ones: $(0 : 1 : 1)$ and $(1 : 0 : 1)$; $(0 : 1 : -1)$ and $(1 : 0 : -1)$ for n even; and $(1 : -1 : 0)$ for n odd.

- *Mordell's Conjecture*, proved by Faltings, says that a projective smooth curve of genus at least two, defined on a *number field* (i.e., a finite degree extension field of $\mathbb{Q}$), has only a finite number of rational points.

Lemma 273. *Let $F, G \in \bar{\mathbf{K}}[X,Y]$ without a common irreducible factor. Then $V(F,G)$ is finite.*

Proof. Since $\gcd(F,G) = 1$ in $\bar{\mathbf{K}}(X)[Y]$ and $\bar{\mathbf{K}}(Y)[X]$, there exist $U_1, V_1, U_2, V_2 \in \bar{\mathbf{K}}[X,Y]$, $\varphi_1 \in \bar{\mathbf{K}}[X] \setminus \{0\}$, $\varphi_2 \in \bar{\mathbf{K}}[Y] \setminus \{0\}$ such that

$$U_1 F + V_1 G = \varphi_1(X) \quad \& \quad U_2 F + V_2 G = \varphi_2(Y).$$

It follows that $V(F,G)$ is finite of cardinality at most $\deg(\varphi_1)\deg(\varphi_2)$.

Lemma 274. *Let $F, G \in \bar{\mathbf{K}}[X,Y]$ with F irreducible and $V(F) \subseteq V(G)$. Then $F \mid G$.*

Proof. We have $V(F,G) = V(F)$ infinite since $\bar{\mathbf{K}}$ is infinite (see Exercise 230). The desired result follows from Lemma 273.

As a consequence, if $C = V(F)$ is an algebraic plane curve where $F \in \bar{\mathbf{K}}[X,Y]$ is square-free, then $\mathscr{I}(C) = \langle F \rangle$. In particular, C is irreducible if and only if F is irreducible.

Definition 275. Let C be an affine plane curve. A map $f : C \to \bar{\mathbf{K}}$ is said to be a *regular function* over C if there exists $F \in \bar{\mathbf{K}}[X,Y]$ such that $f = F_{|C}$. We denote by $\bar{\mathbf{K}}[C]$ the ring of regular functions over C. If C is defined over $\mathbf{K}$, we denote by $\mathbf{K}[C]$ the image of $\mathbf{K}[X,Y]$ in $\bar{\mathbf{K}}[C]$.

As the restriction morphism $\bar{\mathbf{K}}[X,Y] \to \bar{\mathbf{K}}[C]$ is surjective with kernel $\mathscr{I}(C)$, we have the rings isomorphism

$$\bar{\mathbf{K}}[C] \cong \bar{\mathbf{K}}[X,Y]/\mathscr{I}(C).$$

The images of X, Y in $\bar{\mathbf{K}}[C]$ will be denoted by x, y.

It is useful to know the link between the definitions of projective and affine plane curves. As already explained, we have a one-to-one map $\mathbb{A}^2(\bar{\mathbf{K}}) \to \mathbb{P}^2(\bar{\mathbf{K}})$, $(x,y) \mapsto (x:y:1)$. Its image $U_{x,y}$ is an affine chart of $\mathbb{P}^2(\bar{\mathbf{K}})$. Recall that $U_{x,y} = \mathbb{P}^2(\bar{\mathbf{K}}) \setminus V(Z)$ and that $V(Z) \cong \mathbb{P}^1(\bar{\mathbf{K}})$. In the following, we identify $\mathbb{A}^2(\bar{\mathbf{K}})$ with $U_{x,y}$ and $\mathbb{P}^1(\bar{\mathbf{K}})$ with $V(Z)$.

Proposition 276. *Let $C = V(H)$ be a projective plane curve with $H \in \bar{\mathbf{K}}[X,Y,Z]$ homogeneous nonconstant. If C meets $\mathbb{A}^2(\bar{\mathbf{K}})$ then $C \cap \mathbb{A}^2(\bar{\mathbf{K}})$ is an affine plane curve $V(F)$ with $F(X,Y) = H(X,Y,1) \in \bar{\mathbf{K}}[X,Y]$. Conversely, if $C = V(F)$ is an affine plane curve with $d = \deg F \geq 1$, then the smallest projective plane curve containing C is $\bar{C} = V(H)$ where $H(X,Y,Z) = Z^d F(\frac{X}{Z}, \frac{Y}{Z}) \in \bar{\mathbf{K}}[X,Y,Z]$ (the homogenized of F).*

The curve $\bar{C}$ is *the projective completion* of C. The points in $\bar{C} \setminus C$ are the points at ∞ of C.

The maps $C \mapsto \bar{C}$ and $\bar{C} \mapsto C \cap \mathbb{A}^2(\bar{\mathbf{K}})$ are bijections between affine plane curves and projective plane curves which do not contain $\mathbb{P}^1(\bar{\mathbf{K}}) = V(Z)$ (that is, $Z \nmid H$).

Recall that if C is an irreducible plane curve, the *field of rational functions* over C, denoted by $\bar{\mathbf{K}}(C)$, is the quotient field of the integral ring $\bar{\mathbf{K}}[C]$. If C is defined over $\mathbf{K}$, then $\mathbf{K}(C)$ is the quotient field of $\mathbf{K}[C]$.

Definition 277. Let C be an irreducible affine plane curve. For a rational function $f \in \bar{\mathbf{K}}(C)$ and $P \in C$, we say that f is *regular at P* if there exist $g, h \in \bar{\mathbf{K}}[C]$ with $h(P) \neq 0$ such that $f = \frac{g}{h}$; we then set $f(P) = \frac{g(P)}{h(P)}$.

We say that f is regular over a subset $S \subseteq C$ if it is regular at every point of S.

Here, it is worth pointing out that it may happen that $f = \frac{g}{h}$ with $h(P) = 0$ but f is regular at P. For example, $C = V(X^2 + Y^2 - 1)$ is irreducible and the rational function $f = \frac{x-1}{y}$ is regular at $P = (1,0)$. In fact, we also have $f = \frac{-y}{x+1}$. In characteristic different from 2, $x + 1$ does not vanish at P. In characteristic 2, $C = V(X + Y + 1)$ and $f = 1$.

Lemma 278. *Let C be an irreducible affine plane curve.*

(1) For every $f \in \bar{\mathbf{K}}(C)$ there exists $S \subseteq C$ finite such that f is regular over $C \setminus S$.
(2) If $f_1, f_2 \in \bar{\mathbf{K}}(C)$ agree out of a finite set then $f_1 = f_2$.

Proof.

(1) If $f = \frac{g}{h}$ with $h \neq 0$ then the set of zeroes of h in C is finite by virtue of Lemma 273.
(2) Suppose that $f_1 = \frac{g_1}{h_1}$ and $f_2 = \frac{g_2}{h_2}$ agree over $C \setminus S$ with S finite. Then $g_1 h_2 - g_2 h_1$ is zero over the infinite set $C \setminus S$. Thus, $g_1 h_2 = g_2 h_1$ by Lemma 273.

The following proposition justifies the terminology of regular functions.

Proposition 279. *Let C be an irreducible affine plane curve. If $f \in \bar{\mathbf{K}}(C)$ is regular over C then $f \in \bar{\mathbf{K}}[C]$.*

Proof. Consider the ideal $I = \{h \in \bar{\mathbf{K}}[C] \mid hf \in \bar{\mathbf{K}}[C]\}$ and suppose that $I \neq \bar{\mathbf{K}}[C]$. Then I is contained in a maximal ideal M of $\bar{\mathbf{K}}[C]$. We know, by Ideal Variety Correspondence 150, that $M = \langle x - a, y - b \rangle$ for some $a, b \in \bar{\mathbf{K}}$ with $(a,b) \in C$. As f is regular at (a,b), f can be written $f = \frac{g_1}{h_1}$ with $g_1, h_1 \in \bar{\mathbf{K}}[C]$ and $h_1(a,b) \neq 0$. But, $h_1 \in I \Rightarrow h_1(a,b) = 0$, a contradiction.

Let C be an irreducible affine plane curve. If U is an affine chart of $\mathbb{P}^2(\bar{\mathbf{K}})$ meeting C, we can see $\bar{C}$ as the projective completion of the affine curve $C' = \bar{C} \cap U$ (Proposition 276). Moreover, the fields of rational functions over C and C' coincide.

Let C be an irreducible projective plane curve. The field $\bar{\mathbf{K}}(C)$ of rational functions over C is the field of rational functions on an affine chart of C. This definition does not depend on the chosen affine chart. A function $f \in \bar{\mathbf{K}}(C)$ is said to be *regular* at $P \in C$ if it is regular in an affine chart containing P.

A projective plane curve C is *nonsingular* (or *regular*) at $P \in C$ if it is nonsingular at P in an affine chart containing P. This definition does not depend on the chosen affine chart. If $C = V(H)$ is a projective plane curve with H homogeneous nonconstant square-free, then C is nonsingular at $(x : y : z) \in C$ if and only if $(\frac{\partial H}{\partial X}, \frac{\partial H}{\partial Y}, \frac{\partial H}{\partial Z})(x,y,z) \neq 0$. If $\mathrm{char}(\mathbf{K}) = 0$, the set of singular points of C is $V(\frac{\partial H}{\partial X}, \frac{\partial H}{\partial Y}, \frac{\partial H}{\partial Z})$ (see Exercise 360).

Definition 280. Let C be an irreducible affine plane curve and $P \in C$. We denote by $\mathfrak{M}_P$ the ideal of $\bar{\mathbf{K}}[C]$ formed by functions vanishing at P. The evaluation at P furnishes an isomorphism $\bar{\mathbf{K}}[C]/\mathfrak{M}_P \cong \bar{\mathbf{K}}$, and, thus, $\mathfrak{M}_P$ is a maximal ideal of $\bar{\mathbf{K}}[C]$. Moreover, $\mathfrak{M}_P/\mathfrak{M}_P^2$ is a $\bar{\mathbf{K}}[C]/\mathfrak{M}_P$-module, that is, a $\bar{\mathbf{K}}$-vector space.

Proposition 281. *Let C be an irreducible affine plane curve and $P \in C$.*
Then $\dim_{\bar{\mathbf{K}}} \mathfrak{M}_P/\mathfrak{M}_P^2 \geq 1$ with equality of and only if C is nonsingular at P.

Proof. See Exercise 362.

Definition 282. Let C be an irreducible projective plane curve and $P \in C$. The local ring of C at P is $\mathscr{O}_{C,P} := \{f \in \bar{\mathbf{K}}(C) \mid f \text{ is regular at } P\}$. We denote by $\mathfrak{M}_{C,P}$ the ideal of $\mathscr{O}_{C,P}$ formed by functions vanishing at P. The evaluation at P furnishes an isomorphism $\mathscr{O}_{C,P}/\mathfrak{M}_{C,P} \cong \bar{\mathbf{K}}$ as in the affine case. The local ring $\mathscr{O}_{C,P}$ is nothing but the localization of the ring $\bar{\mathbf{K}}[C]$ at $\mathfrak{M}_P$.

Proposition 283. *Let $C = V(F)$ be an irreducible projective plane curve and P a nonsingular point on C. Then the ideal $\mathfrak{M}_{C,P}$ is principal generated by any $t \in \mathfrak{M}_P \setminus \mathfrak{M}_P^2$.*

Proof. By Proposition 281, we know that $\dim_{\bar{\mathbf{K}}} \mathfrak{M}_P/\mathfrak{M}_P^2 = 1$, and, hence, $\mathfrak{M}_P/\mathfrak{M}_P^2$ (and also $\mathfrak{M}_{C,P}/\mathfrak{M}_{C,P}^2$) is generated as $\bar{\mathbf{K}}$-vector space by any $\bar{t}$ with $t \in \mathfrak{M}_P \setminus \mathfrak{M}_P^2$. We conclude, using Nakayama's lemma (see Exercise 388), that the ideal $\mathfrak{M}_{C,P}$ is principal generated by any $t \in \mathfrak{M}_P \setminus \mathfrak{M}_P^2$.

Definition 284.

Keeping the notation of Proposition 283, a generator t of $\mathfrak{M}_{C,P}$ is called a *uniformizing parameter* at P (or a *local parameter* at P, or also, a *uniformizer* of the local ring of C at P). It can be any element in $\mathfrak{M}_P \setminus \mathfrak{M}_P^2$.

Every $f = \frac{g}{h} \in \bar{\mathbf{K}}(C) \setminus \{0\}$, with $g, h \in \bar{\mathbf{K}}[C]$, can be written in a unique way $f = t^n u$ with $n \in \mathbb{Z}$ and $u \in \mathscr{O}_{C,P}^\times$.

The integer n is called the *order* of f at P and denoted by $\mathrm{Ord}_P(f)$. It is equal to $\mathrm{Ord}_P(g) - \mathrm{Ord}_P(h)$. If $n > 0$, we say that P is *zero* of f of multiplicity n. If $n < 0$, we say that P is *a pole* of f of order $-n$.

Note that, for $g \in \bar{\mathbf{K}}[C] \setminus \{0\}$, denoting by $P = (a,b)$, we have

$$\mathrm{Ord}_P(g) = \ell_{\mathscr{O}_{C,P}}(\mathscr{O}_{C,P}/\langle g \rangle) = \dim_{\bar{\mathbf{K}}}(\bar{\mathbf{K}}[X,Y]_{\langle X-a, Y-b \rangle}/\langle F(X,Y), g(X,Y) \rangle),$$

where $\ell_{\mathscr{O}_{C,P}}(M)$ denotes the length of the $\mathscr{O}_{C,P}$-module M.

5.3 Riemann-Roch theorem

If C is a smooth irreducible projective plane curve, the *divisor* of $f \in \bar{\mathbf{K}}(C) \setminus \{0\}$ is the formal sum

$$(f) = \operatorname{div}(f) := \sum_{P \in C} \operatorname{Ord}_P(f).[P] \in \mathbb{Z}[C].$$

The above sum is finite by virtue of Lemma 278 applied to f and $\frac{1}{f}$. Such a divisor is called *principal*. The degree of $\operatorname{div}(f)$ is defined as

$$\deg(\operatorname{div}(f)) := \sum_{P \in C} \operatorname{Ord}_P(f) \in \mathbb{Z}.$$

Proposition and Definition 285. *Let C be an irreducible projective plane curve. If $f \in \bar{\mathbf{K}}(C)$ is nonconstant then $\bar{\mathbf{K}}(f) \subseteq \bar{\mathbf{K}}(C)$ is a finite extension, that is, $\delta :=$ $\dim_{\bar{\mathbf{K}}(f)} \bar{\mathbf{K}}(C) = [\bar{\mathbf{K}}(C) : \bar{\mathbf{K}}(f)] < \infty$.*

The positive integer δ is called the *degree of f* and denoted by $\deg(f)$. If f is a nonzero constant, we set $\deg(f) = 0$.

Proof. Since $f \in \bar{\mathbf{K}}(C) \setminus \bar{\mathbf{K}}$, $\bar{\mathbf{K}}(f)$ has transcendence degree 1 over $\bar{\mathbf{K}}$. But also $\bar{\mathbf{K}}(C)$ has transcendence degree 1 over $\bar{\mathbf{K}}$, thus, $\bar{\mathbf{K}}(C)/\bar{\mathbf{K}}(f)$ is an algebraic extension. Finally, $\bar{\mathbf{K}}(C)$ is a finite algebraic extension of $\bar{\mathbf{K}}(f)$ since $\bar{\mathbf{K}}(C)$ is finitely-generated over $\bar{\mathbf{K}}$. $\qquad\blacksquare$

Theorem 286. *Let C be an irreducible smooth projective plane curve. Then for all $f \in \bar{\mathbf{K}}(C) \setminus \{0\}$, we have:*

$$\deg(f) = \sum_{\substack{P \in C \\ \operatorname{Ord}_P(f) \geq 0}} \operatorname{Ord}_P(f).$$

Corollary 287. *Let C be an irreducible smooth projective plane curve. Then each principal divisor has degree zero.*

Proof. $\deg(\operatorname{div}(f)) = \deg(f) - \deg(\frac{1}{f}) = 0$ since $\bar{\mathbf{K}}(f) = \bar{\mathbf{K}}(\frac{1}{f})$. $\qquad\blacksquare$

Here, it worth pointing that the "irreducibility" hypothesis in Theorem 286 is superfluous since a smooth projective plane curve is necessarily irreducible (see Exercise 358).

Remark 288. Corollary 287 suggests implicitly that when we want to compute the order of a nonconstant rational function $f(x,y) = \frac{g(x,y)}{h(x,y)}$ at a point at ∞ (in $V(Z)$), we have to homogenize it in the form $\tilde{f} = z^{\deg(h)-\deg(g)} \frac{\tilde{g}(x,y,z)}{\tilde{h}(x,y,z)}$ where $\tilde{g}$ and $\tilde{h}$ are the homogenizations of g and h, respectively. To see this, remember that if

$z \neq 0$, the projective point $(x : y : z)$ corresponds to the affine point $(\frac{x}{z}, \frac{y}{z})$ in the affine chart $\{z \neq 0\}$. So, the homogenization of $f(x,y)$ is

$$f\left(\frac{x}{z}, \frac{y}{z}\right) = \frac{g(\frac{x}{z}, \frac{y}{z})}{h(\frac{x}{z}, \frac{y}{z})} = z^{\deg(h)-\deg(g)} \frac{z^{\deg(g)} g(\frac{x}{z}, \frac{y}{z})}{z^{\deg(h)} h(\frac{x}{z}, \frac{y}{z})} = z^{\deg(h)-\deg(g)} \frac{\tilde{g}(x,y,z)}{\tilde{h}(x,y,z)}.$$

For example, if $f = \frac{x+y^2}{x}$ then $\tilde{f} = \frac{xz+y^2}{xz}$; if $f = x^2$ then $\tilde{f} = \frac{x^2}{z^2}$. We see that over $C = V(Y) \subseteq \mathbb{P}^2(\bar{\mathbf{K}})$, this latter function has a double zero at $(0 : 0 : 1)$ and a double pole at $(1 : 0 : 0)$ (at ∞). This is obvious with the expression $\frac{x^2}{z^2}$ but much less obvious with the expression x^2. The principal divisor associated to f is then

$$\mathrm{div}(f) = 2[(0 : 0 : 1)] - 2[(1 : 0 : 0)].$$

Corollary 289. *Let C be an irreducible smooth projective plane curve. If $f \in \bar{\mathbf{K}}(C)$ is regular over C then it is constant.*

Proof. Let $f \in \bar{\mathbf{K}}(C) \setminus \{0\}$ and suppose that it is regular over C. As $\mathrm{Ord}_P(\frac{1}{f}) \leq 0$ for all $P \in C$, we deduce from Theorem 286 that $\frac{1}{f}$ is constant.

Example 290.

(1) $C = V(Y) \subseteq \mathbb{P}^2(\bar{\mathbf{K}})$ with $\mathrm{char}(\mathbf{K}) = 0$, $\bar{\mathbf{K}}[C] = \bar{\mathbf{K}}[x]$, $\bar{\mathbf{K}}(C) = \bar{\mathbf{K}}(x)$, $f = \frac{(x-1)^2}{x-3}$, $g = \frac{x+1}{(x+2)^3}$, $h = x - 2$, $\ell = \frac{1}{x^2}$,

$$\mathrm{div}(f) = 2[(1 : 0 : 1)] - [(3 : 0 : 1)] - [(1 : 0 : 0)],$$

$$\mathrm{div}(g) = [(-1 : 0 : 1)] - 3[(-2 : 0 : 1)] + 2[(1 : 0 : 0)],$$

$$\mathrm{div}(h) = [(2 : 0 : 1)] - [(1 : 0 : 0)], \ \mathrm{div}(\ell) = -2[(0 : 0 : 1)] + 2[(1 : 0 : 0)].$$

(2) Consider the smooth (elliptic) curve $C = V(F)$ with $F = Y^2 - X^3 - 1 \in \mathbb{F}_{13}[X,Y]$, and $f = \frac{x^2}{y} \in \mathbb{F}_{13}(C)$. We have:

$$\mathrm{div}(f) = 2[(0 : 1 : 1)] + 2[(0 : -1 : 1)] - [(-1 : 0 : 1)]$$
$$- [(4 : 0 : 1)] - [(-3 : 0 : 1)] - [(0 : 1 : 0)].$$

Let us give some details. The point $P = (0 : 1 : 1)$ has affine coordinates $(0,1)$ and at that point x is a uniformizing parameter because $\frac{\partial F}{\partial y}(0,1) \neq 0$. The order of f at P is then 2 (note that y neither vanishes nor has a pole at P so that its order is 0 at P). The same computations apply to the point $(0 : -1 : 1)$.

Note that over $\mathbb{F}_{13}$, we have $x^3 + 1 = (x+1)(x-4)(x+3)$. The point $Q = (-1 : 0 : 1)$ has affine coordinates $(-1,0)$ and at that point y is a uniformizing parameter because $\frac{\partial F}{\partial x}(-1,0) \neq 0$. The order of f at Q is then -1 (note that x neither vanishes nor has a pole at Q so that its order is 0 at Q).

Note that $\text{Ord}_P(f)$ can be computed differently:

$$\begin{aligned}
\text{Ord}_P(f) &= \text{Ord}_P(x^2) - \text{Ord}_P(y) \\
&= \dim_{\mathbb{F}_{13}}(\mathbb{F}_{13}[X,Y]_{\langle X,Y-1\rangle}/\langle Y^2 - X^3 - 1, X^2\rangle) \\
&\quad - \dim_{\mathbb{F}_{13}}(\mathbb{F}_{13}[X,Y]_{\langle X,Y-1\rangle}/\langle Y^2 - X^3 - 1, Y\rangle) \\
&= \dim_{\mathbb{F}_{13}}(\mathbb{F}_{13}[X]/\langle X^2\rangle) - \dim_{\mathbb{F}_{13}}(\mathbb{F}_{13}[X,Y]_{\langle X,Y-1\rangle}/\langle 1\rangle) = 2 - 0 = 2.
\end{aligned}$$

The same computations apply to the points $(4 : 0 : 1)$ and $(-3 : 0 : 1)$.

A question: we have $f = \frac{x^2}{y} = \frac{x^2 y}{y^2} = \frac{x^2 y}{x^3 + 1} = \frac{x^2 y}{(x+1)(x-4)(x+3)}$. Is this new form compatible with our previous computation $\text{Ord}_Q(f) = -1$?

The answer is yes of course: $\text{Ord}_Q(f) \overset{\text{since } x,(x-4),(x+3)\in \mathscr{O}_{C,Q}^{\times}}{=} \text{Ord}_Q(y) - \text{Ord}_Q(x^3 + 1) = \text{Ord}_Q(y) - \text{Ord}_Q(y^2) = 1 - 2 = -1$.

One more question: in general, how to find a uniform parameter t at a nonsingular point $A = (x_0, y_0)$ of plane curve $V(F)$?

An answer: If $\frac{\partial F}{\partial y}(A) \neq 0$ then $x - x_0$ is a uniform parameter at A; if $\frac{\partial F}{\partial x}(A) \neq 0$ then $y - y_0$ is a uniform parameter at A.

Definition 291. Let C be an irreducible projective plane curve. The space $\Omega^1(\bar{\mathbf{K}}(C))$ of rational differential forms over C is the $\bar{\mathbf{K}}(C)$-vector space generated by symbols df, with $f \in \bar{\mathbf{K}}(C)$, quotiented by the following relations:

$$\begin{aligned}
d(f+g) &= df + dg, \ d(\lambda f) = \lambda df, \\
d(fg) &= f dg + g df, \ \forall f, g \in \bar{\mathbf{K}}(C), \ \forall \lambda \in \bar{\mathbf{K}}.
\end{aligned}$$

If C is defined over $\mathbf{K}$, we denote by $\Omega^1(\mathbf{K}(C))$ the sub-$\mathbf{K}(C)$-vector space of $\Omega^1(\bar{\mathbf{K}}(C))$ generated by the df, with $f \in \mathbf{K}(C)$.

Proposition 292. *Let C be an irreducible projective plane curve. Then the vector space $\Omega^1(\bar{\mathbf{K}}(C))$ has dimension 1 over $\bar{\mathbf{K}}(C)$. If t is a uniformizing parameter at a nonsingular point of C, then dt is a basis for $\Omega^1(\bar{\mathbf{K}}(C))$.*

It is worth pointing out that C always has nonsingular points ($C = \bar{C}_0$, $C_0 = V(F)$, and apply Lemma 273 with $G = \frac{\partial F}{\partial X}$ or $\frac{\partial F}{\partial Y}$).

Definition and Proposition 293. Let C be an irreducible projective plane curve. We say that $w \in \Omega^1(\bar{\mathbf{K}}(C))$ is *regular* at a point $P \in C$ if $w = \sum_{i=1}^{n} f_i dg_i$ where all the f_i and g_i are regular at P.

If C is nonsingular at P and t is a uniformizing parameter at P, then w is regular at P if and only if $w = f dt$ where f is regular at P.

Definition 294. Let C be an irreducible projective plane curve. We denote by $\Omega^1(C)$ the $\bar{\mathbf{K}}$-vector space of rational differential forms over C which are regular at each point of C. The *genus* of C is $\dim_{\bar{\mathbf{K}}} \Omega^1(C)$. It is an invariant of the function field $\bar{\mathbf{K}}(C)$. So, the genus of C is also the genus of its nonsingular model, i.e., that of a nonsingular not necessarily plane projective curve which is birationally equivalent to C (we skip the details).

Note that the genus of a nonsingular irreducible plane curve C can be defined as the least possible value of $g \in \mathbb{N}$ such that $\ell(D) \geq \deg(D) - g - 1$ for all divisors D of C (see Riemann-Roch Theorem 301).

Note also that the genus of a curve C can also be introduced in different ways. So, for instance, we might view a complex curve as a Riemann surface, and its genus is the number of topological handles of this surface.

Theorem 295. *Every smooth projective plane curve is irreducible with genus $\frac{(d-1)(d-2)}{2}$ (the so-called arithmetical genus), where d is the degree of an irreducible polynomial defining it.*

Note that the integer $\frac{(d-1)(d-2)}{2}$ stems from putting $\Omega^1(C)$ in bijection with the $\bar{\mathbf{K}}$-vector subspace of $\bar{\mathbf{K}}[X,Y]$ formed by polynomials of total degree at most $d - 3$. Recall that the number of monomials at k variables of total degree at most d is $\binom{k+d}{d}$.

Example 296. (Example 165 continued)

The Fermat curve $x^n + y^n = z^n$, defined over a field of characteristic 0 or coprime with n, has genus $\frac{(n-1)(n-2)}{2}$.

Riemann-Roch theorem is a fundamental result in algebraic geometry for the computation of the dimension of the space of rational functions over C with prescribed zeroes and allowed poles. First, we need to introduce some notation.

The *Picard group* $\text{Pic}(C)$ is the quotient of $\mathbb{Z}[C]$ by the group $\text{div}(\bar{\mathbf{K}}(C)^\times)$ of principal divisors. If D_1 and D_2 are divisors over C, we say that $D_1 \geq D_2$ if for all $P \in C$, we have $\text{Ord}_P(D_1) \geq \text{Ord}_P(D_2)$. For $D \in \mathbb{Z}[C]$, we set:

$$\mathscr{L}(D) := \{ f \in \bar{\mathbf{K}}(C)^\times \mid \text{div}(f) \geq -D \} \cup \{0\}.$$

It is a finite-dimensional $\bar{\mathbf{K}}$-vector space whose dimension is denoted by $\ell(D)$. This dimension does not depend on the class of D in $\text{Pic}(C)$ since $\mathscr{L}(D) = h.\mathscr{L}(D + \text{div}(h))$ for $h \in \bar{\mathbf{K}}(C)^\times$.

Intuitively, we can think of $\mathscr{L}(D)$ as being all rational functions over C whose poles at every point are not worse than the corresponding coefficient in D; if the coefficient in D at P is negative, then we require that f has a zero of at least that multiplicity at P, if the coefficient in D is positive, f can have a pole of at most that order.

Example 297.

(1) If $D = n[P]$ with $n \geq 0$ and $P \in C$, then $\mathscr{L}(D)$ is the space of rational functions that are regular over C except at P where the function is allowed to have a pole of order at most n. The sequence $(\ell(n[P]))_{n \geq 0}$ is clearly increasing.

For $n = 0$, $\mathscr{L}(0) = \bar{\mathbf{K}}$ (by Corollary 289), and, thus, $\ell(0) = 1$.

(2) If $\deg(D) < 0$ then $\mathscr{L}(D) = \{0\}$ and $\ell(D) = 0$. This is because an $f \in \bar{\mathbf{K}}(C)^\times \cap \mathscr{L}(D)$ would have degree $\geq -\deg(D) > 0$ (impossible since it has degree zero, see Corollary 287).

Let C be a smooth projective plane curve of genus g. Let $w \in \Omega^1(\bar{\mathbf{K}}(C))$ be a nonzero rational differential form and $P \in C$. If t is a uniformizing parameter at P, we have $w = f dt$ with $f \in \bar{\mathbf{K}}(C)^\times$ (Proposition 292). The vanishing order of f at P does not depend on the choice of t (if t' is another uniformizing parameter at P, then, by Definition-Proposition 293, we have $dt' = u dt$ with $u \in \mathscr{O}_{C,P}^\times$). We denote $\mathrm{Ord}_P(w) := \mathrm{Ord}_P(f)$, and we set:

$$(w) = \mathrm{div}(w) := \sum_{P \in C} \mathrm{Ord}_P(w).[P] \in \mathbb{Z}[C].$$

This is in fact a finite sum. As $\Omega^1(\bar{\mathbf{K}}(C))$ has dimension 1 over $\bar{\mathbf{K}}(C)$ and $\mathrm{div}(hw) = \mathrm{div}(w) + \mathrm{div}(h)$, the class of $\mathrm{div}(w)$ in $\mathrm{Pic}(C)$ depends only on C. We call it the *canonical divisor* and we denote it by K_C.

Definition 298. Let C be a smooth projective plane curve. If D is a divisor over C, we denote by $\mathscr{I}(D)$ the $\bar{\mathbf{K}}$-vector space of all rational differential forms w over C such that $(w) \geq D$. We also denote $i(D) := \dim_{\bar{\mathbf{K}}} \mathscr{I}(D)$.

Proposition 299. *Let C be a smooth projective plane curve and D a divisor over C. Then the $\bar{\mathbf{K}}$-vector space $\mathscr{I}(D)$ is isomorphic to $\mathscr{L}(K_C - D)$. Or, also, the $\bar{\mathbf{K}}$-vector space $\mathscr{L}(D)$ is isomorphic to $\mathscr{I}(K_C - D)$.*

Proof. It suffices to prove that $\mathscr{L}(D)$ is isomorphic to $\mathscr{I}(K_C - D)$. Suppose that $K_C = \mathrm{div}(w)$ for some $w \in \Omega^1(\bar{\mathbf{K}}(C))$. For $f \in \bar{\mathbf{K}}(C)^\times$, we have:

$$(f) + D = (fw) - (w) + D = (fw) - (K_C - D).$$

So, $f \in \mathscr{L}(D)$ if and only if $fw \in \mathscr{I}(K_C - D)$. Thus, we have a linear map $\mathscr{L}(D) \to \mathscr{I}(K_C - D)$ given by $f \mapsto fw$. Since every rational differential forms in $\mathscr{I}(K_C - D)$ can be presented as fw for some rational differential form f (by virtue of Proposition 292), this latter map is an isomorphism.

Corollary 300. *Let C be a smooth projective plane curve of genus g. Then $\mathscr{L}(K_C)$ is isomorphic as $\bar{K}$-vector space to $\Omega^1(C)$. In particular, $\ell(K_C) = g$.*

Proof. By Proposition 299, $\mathscr{L}(K_C)$ is isomorphic as $\bar{K}$-vector space to $\mathscr{I}(K_C - K_C) = \mathscr{I}(0) = \Omega^1(C)$.

Theorem 301. (Riemann-Roch)

Let C be a smooth projective plane curve of genus g and canonical divisor K_C. Then, for all $D \in \mathbb{Z}[C]$, we have:

$$\ell(D) - \ell(K_C - D) = \deg(D) - g + 1.$$

The number $\ell(D)$ is the one of interest, while $\ell(K_C - D) = i(D)$ is thought of as a correction term. So the theorem may be roughly paraphrased by saying:

$$\text{dimension } - \text{ correction } = \text{ degree } - \text{ genus } + 1.$$

The correction term $\ell(K_C - D)$ is always non-negative, so that $\ell(D) \geq \deg(D) - g + 1$. This is called Riemann's inequality. Roch's part of the statement is the description of the possible difference between the sides of the inequality.

Corollary 302. *Let C be a smooth projective plane curve of genus g and canonical divisor K_C. Then $\deg(K_C) = 2g - 2$, and, for all $D \in \mathbb{Z}[C]$, if $\deg(D) \geq 2g - 1$ then*

$$\ell(D) = \deg(D) - g + 1.$$

Proof. To obtain $\deg(K_C) = 2g - 2$, just put $D = K_C$ in Riemann-Roch Theorem 301. The second assertion follows from the fact that $\ell(K_C) = g$ (Corollary 300, or take $D = 0$ in Riemann-Roch Theorem 301) and if $\deg(D) < 0$ then $\ell(D) = 0$ (Example 297).

For example, for $P \in C$, the sequence $(\ell(n[P]))_{n \geq 0}$ is $(1, 2, 3, 4, 5, \ldots)$ in case the genus is 0 and $(1, 1, 2, 3, 4, \ldots)$ in case the genus is 1.

Remarks and Definitions 303.

1) Recall that a *Riemann surface* is a one-dimensional, connected, complex manifold. Every Riemann surface is a two-dimensional real manifold. A two-dimensional real manifold can be turned into a Riemann surface if and only if it is orientable and metrizable. So the sphere and torus admit complex structures, but the *Möbius strip*[2] does not.

Any connected open set in $\mathbb{C}$ is a noncompact Riemann surface. Any compact Riemann surface is homeomorphic to a sphere with $g \geq 0$ handles (g is called the

---

[2]The source of Fig. 5.5 is https://en.wikipedia.org/wiki (Möbius strip).

Figure 5.5: The Möbius strip

genus of the Riemann surface). A sphere without handles is simply a sphere. A sphere with one handle is a torus. The *Riemann sphere* is the set $\mathbb{P}^1(\mathbb{C}) = \mathbb{C} \cup \{\infty\}$ with the neighborhoods $U_0 = \mathbb{C}$ and $U_1 = (\mathbb{C} \setminus \{0\}) \cup \{\infty\}$ and corresponding charts

$$\phi_0(z) = z \text{ and } \phi_1(z) = \begin{cases} \frac{1}{z} \text{ if } z \neq \infty \\ 0 \text{ if } z = \infty. \end{cases}$$

2) Recall that a function f on a Riemann surface S is called *meromorphic* if f is holomorphic on $S \setminus \mathscr{P}$, where $\mathscr{P}$ is a discrete subset of S (when S is compact, $\mathscr{P}$ is finite), and f has poles at every point $x \in \mathscr{P}$. Of course, any holomorphic function is meromorphic and the sum, product, quotient of two meromorphic functions are meromorphic. We know by *"Riemann Existence Theorem"* that every compact Riemann surface admits a nonconstant meromorphic function.

3) Recall that a *holomorphic* 1-*form* w on a Riemann surface S is a complex differential form of degree 1 on S that can be written in local coordinates as $w = f\,dz$, for some f holomorphic function f of the local coordinate z (of course, we do require a compatibility condition when two charts have overlapping domains). We denote by $\Omega^1(S)$ the $\mathbb{C}$-vector space of holomorphic 1-forms on S and $g := \dim_{\mathbb{C}} \Omega^1(S)$ the *genus* of S.

In the same way, a *meromorphic* 1-*form* w on a Riemann surface S is an expression which locally of the form $f\,dz$, where f is meromorphic in the coordinate z.

4) Similarly to the curve case, a divisor $D = \sum_{P \in S} n_p.[P] \in \mathbb{Z}[S]$ (a finite sum) on a compact Riemann surface S is an element of the free abelian group generated by the points of S. We define $\deg(D) := \sum_{P \in S} n_p \in \mathbb{Z}$. We say that $\sum_{P \in S} n_p.[P] \geq \sum_{P \in S} m_p.[P]$ if $n_P \geq m_P$ for all $P \in S$. We define the $\mathbb{C}$-vector space $\mathscr{L}(D)$ to be the set of all meromorphic functions f such that $\operatorname{div}(f) + D \geq 0$, where $\operatorname{div}(f) := \sum_{P \in S} \operatorname{Ord}_P(f).[P] \in \mathbb{Z}[S]$ (it is in fact a finite sum since S is compact) is called a *principal divisor*. We denote $\ell(D) := \dim_{\mathbb{C}} \mathscr{L}(D)$. Similarly, we define the $\mathbb{C}$-vector space $\mathscr{I}(D)$ to be the set of all meromorphic 1-forms w for which

$\mathrm{div}(w) := \sum_{P \in S} \mathrm{Ord}_P(w).[P] \geq D$, where $\mathrm{Ord}_P(w) = \mathrm{Ord}_P(f_P)$, f_P being a local representation of w at P, i.e., w is represented locally as $f_P dz$ in a neighborhood of P. We denote $i(D) := \dim_{\mathbb{C}} \mathscr{I}(D)$. Since, any meromorphic 1-form can be presented as any other meromorphic 1-form multiplied by some meromorphic function, all the divisors of meromorphic 1-forms have the same class in the *Picard group* $\mathrm{Pic}(S)$ (the quotient of $\mathbb{Z}[S]$ by the subgroup of principal divisors). We call it *the canonical divisor* and we denote it by K_S. As in Proposition 299, the $\mathbb{C}$-vector space $\mathscr{I}(D)$ is isomorphic to $\mathscr{L}(K_S - D)$. In particular, if g is the genus of S, then $\mathscr{L}(K_S)$ is isomorphic as $\mathbb{C}$-vector space to $\mathscr{I}(0)$ (the linear space of all holomorphic 1-forms on S), and, thus, $\ell(K_S) = g$.

5) If S is a compact Riemann surface of genus g then $\deg(K_S) = 2g - 2 = -\chi$ (see Remark 323.1). To see this, let $w = df$ be a meromorphic 1-form on S, where f is a nonconstant meromorphic function on S. As explained in Remark 323.2, f can be seen as a holomorphic map $S \rightarrow \mathbb{P}^1(\mathbb{C})$. Suppose that f has degree n and is ramified (see Definition 311) at m points on S (we can suppose that they are not among the poles of f by "rotating" the Riemann sphere) with ramification indexes $e_1, \ldots, e_m$, correspondingly. The function f has exactly n simple poles, and at each of them the 1-form df has a double pole (since $d(\frac{1}{z}) = \frac{-1}{z^2}$). Moreover, the zeroes of df are the m ramifications points of the map f with multiplicities $e_1 - 1, \ldots, e_m - 1$, correspondingly. By virtue of Remark 323.3, we have

$$\deg(K_S) = \deg(\mathrm{div}(w)) = \sum_{i=1}^{m} (e_i - 1) - 2n = -\chi = 2g - 2.$$

6) Define *the principal part* at $z = a$ of a function $f = \sum_{k=-\infty}^{+\infty} a_k(z-a)^k$ as $\sum_{k=-\infty}^{-1} a_k z^k$ (the portion of the Laurent series consisting of terms of negative degree moved to 0).

The *residue* of a meromorphic function f on open subset of $\mathbb{C}$ at a pole a, denoted by $\mathrm{res}_a f$, is the coefficient a_{-1} of the Laurent series of f at a. The definition of a residue can be generalized to arbitrary Riemann surfaces. Suppose w is a meromorphic 1-form on a Riemann surface. Write $w = f(z)dz$ in local coordinates at x. Then the residue of w at x is defined to be the residue of $f(z)$ at the point corresponding to x.

By the *Stokes theorem*, we know that the sum of all the residues of a meromorphic 1-form on a Compact Riemann surface is 0. And conversely, the *Reverse Residue theorem* says:

Considering a set of points $\{a_1, \ldots, a_m\}$ on a compact Riemann surface S and a set of principal parts $\{f_1, \ldots, f_m\}$, then there exists a meromorphic function f on S that has principal part f_i at each point a_i and has no other poles if and only if $\sum_{i=1}^{m} \mathrm{res}_{a_i} f_i w = 0$ for all holomorphic 1-forms w on S.

7) It is worth pointing out that, to prove the Riemann-Roch Theorem 301 for algebraic curves over a field **K**, we can make use of the idea of residues. Let C be a smooth projective plane curve, $P \in C$, and t a uniformizing parameter at P. Hereafter we introduce the development of a rational function as a Laurent series. Consider $f \in \mathscr{O}_{C,P}$ and write it as $f = a_0 + t f_1$ with $a_0 = f(P) \in \bar{\mathbf{K}}$ and $f_1 \in \mathscr{O}_{C,P}$. By iterating this process, we see that there is a unique formal series

$$\tau(f) = \sum_{n \geq 0} a_n T^n \in \bar{\mathbf{K}}[[T]] \text{ such that } f \in \sum_{n=0}^{N} a_n t^n + \langle t^{N+1} \rangle \text{ for all } N \geq 0. \text{ The map}$$

$\tau : \mathscr{O}_{C,P} \to \bar{\mathbf{K}}[[T]]$ is a one-to-one (because $\cap_{n \geq 0} \langle t^n \rangle = \{0\}$) homomorphism of rings, and, thus, gives rise to a one-to-one homomorphism of fields $\tau : \bar{\mathbf{K}}(C) \to \bar{\mathbf{K}}((T))$, where $\mathscr{O}_{C,P}$ is the preimage of $\bar{\mathbf{K}}[[T]]$.

For $w = f dt \in \Omega^1(\bar{\mathbf{K}}(C))$, the *residue* of w at P, denoted by $\mathrm{res}_P w$, is the coefficient of T^{-1} in the Laurent development $\tau(f)$ of f at P. Then we have:

(i) $\mathrm{res}_P w$ doesn't depend on the choice of t,

(ii) $\mathrm{res}_P w$ is $\bar{\mathbf{K}}$-linear in w,

(iii) $\mathrm{res}_P w = 0$ if and only if w is regular at P,

(iv) $\forall w \in \Omega^1(\bar{\mathbf{K}}(C))$, we have $\sum_{P \in C} \mathrm{res}_P w = 0$.

8) Over $\mathbb{C}$, Riemann-Roch Theorem 301 can be reformulated as follows:
Let D be a divisor on a compact Riemann surface S of genus g. Then

$$\ell(D) - i(D) = \deg(D) - g + 1.$$

Hereafter a brief overview on how to prove Riemann-Roch Theorem 301 over $\mathbb{C}$. First, since $i(D) = \ell(K_S - D)$ and $\deg(K_S) = 2g - 2$, we only need to prove that $\ell(D) - i(D) \geq \deg(D) - g + 1$. Moreover, we can reduce to the case $D \geq 0$ (for this, one has only to prove that $\ell(D - [P]) - i(D - [P]) \geq \ell(D) - i(D) - 1$ for any divisor D and point P on S). Now, consider a divisor $D = \sum_{i=1}^{m} n_i \cdot [P_i] \in$ $\mathbb{N}[S]$. The idea is to consider the linear map $\varphi : \mathscr{L}(D) \to E := \prod_{i=1}^{m} \bigoplus_{j=1}^{n_i} \mathbb{C} \cdot \frac{1}{z^j}$, that sends $f \in \mathscr{L}(D)$ to the tuple of principal parts of f at the points P_i moved to 0. Note that E is a $\mathbb{C}$-vector space of dimension $\deg(D)$, and $\mathrm{Ker}(\varphi) = \mathbb{C}$. For example, if $D = 2P_1 + P_2$, then $\varphi(f) = (\frac{a}{z^2} + \frac{b}{z}, \frac{c}{z})$ for some $a, b, c \in \mathbb{C}$. We have $\ell(D) = \dim_{\mathbb{C}} \mathrm{Ker}(\varphi) + \dim_{\mathbb{C}} \mathrm{Im}(\varphi) = 1 + \dim_{\mathbb{C}} \mathrm{Im}(\varphi)$. But, by the Reverse Residue Theorem (see item 6), $\mathrm{Im}(\varphi) = \cap_{w \in \Omega^1(S)} \mathrm{Ker}(R_w)$, where R_w is the linear map $E \to \mathbb{C}$, $\{f_1, \ldots, f_m\} \mapsto \sum_{i=1}^{m} \mathrm{res}_{P_i} f_i w$. Thus, $\ell(D) = 1 + \deg(D) - \dim_{\mathbb{C}} F =$

$1 + \deg(D) - g + \dim_{\mathbb{C}} \text{Ker}(\psi)$, where F is the $\mathbb{C}$-vector space generated by the R_w's and $\psi : \Omega^1(S) \to F$, $w \mapsto R_w$. Since $\text{Ker}(\psi) = \mathscr{I}(D)$, we obtain $\ell(D) = 1 + \deg(D) - g + i(D)$.

In the following, we give an important application of Riemann-Roch Theorem 301. We give a necessary and sufficient condition for a projective plane curve to be rational in terms of its genus. Recall that a unirational curve is always irreducible. Note that, in order to obtain a parametrization over $\mathbf{K}$ (instead of $\bar{\mathbf{K}}$), we need to suppose that C has a rational point.

Corollary 304. *Let C be an irreducible projective plane curve over a field $\mathbf{K}$ with genus g. Then the following assertions are equivalent:*

(i) $g = 0$.

(ii) $\bar{\mathbf{K}}(C) = \bar{\mathbf{K}}(t)$ for a nonconstant rational function t, i.e., C is rational.

(iii) C has a rational parametrization, i.e., C is unirational.

Proof. We will give a proof in case C is smooth. Recall that an irreducible algebraic variety V is said to be *rational* if $\mathbf{K}(V)$ is isomorphic to $\mathbf{K}(u_1, \ldots, u_d)$, where u_i are algebraically independent over $\mathbf{K}$. Clearly, a rational variety is unirational, i.e., has a rational parametrization. Conversely, a rational parametrization of V gives rise to a morphism from $\mathbf{K}(V)$ to $\mathbf{K}(u_1, \ldots, u_d)$, which, however, cannot be surjective. *Lüroth's theorem* (every field between $\mathbf{K}$ and $\mathbf{K}(X)$ must be generated as an extension of $\mathbf{K}$ by an element of $\mathbf{K}(X)$) asserts that every algebraic unirational curve is rational. So, in our situation (dimension $d = 1$), the assertions (ii) and (iii) are equivalent.

"(i) $\Rightarrow$ (ii)" Let $P \in C$. By Corollary 302, we know that $\ell([P]) = 2$ and, thus, there exists $t \in (\bar{\mathbf{K}}(C) \setminus \bar{\mathbf{K}}) \cap \mathscr{L}([P])$. The point P is the only allowed pole of t. As $t \notin \bar{\mathbf{K}}$, t has a simple pole at P. Since $\deg(\text{div}(t)) = 0$, there exists $Q \in C$ such that $\text{div}(t) = [Q] - [P]$. By Theorem 286, we deduce that $\deg(t) = 1$, and, thus, $\bar{\mathbf{K}}(C) = \bar{\mathbf{K}}(t)$.

"(ii) $\Rightarrow$ (i)" If $\bar{\mathbf{K}}(C) = \bar{\mathbf{K}}(t)$ then $\deg(t) = 1$ and necessarily $\text{div}(t) = [Q] - [P]$ for two points $P \neq Q$ on C. Now, for $0 \leq i \leq n$, an $f \in (\bar{\mathbf{K}}(C))^{\times}$ is regular over C except at P where it has a pole of order i if and only if $f \in \bar{\mathbf{K}} t^i$ (this is because $f \in \bar{\mathbf{K}}(t)$; one can think of Q as 0 and P as ∞ in $\bar{\mathbf{K}} \cup \{\infty\}$). It follows that
$$\mathscr{L}(n[P]) = \bigoplus_{i=0}^{n} \bar{\mathbf{K}} t^i, \quad \ell(n[P]) = n + 1, \text{ and, thus, } g = 0.$$

Corollary 305. *The only unirational smooth projective plane curves are lines and irreducible conics.*

Proof. Use Corollary 304 and Proposition 295.

5.4 Rational maps between algebraic curves

Definition 306. Let C and C' be two affine plane curves. A *regular map* or a *morphism* $\phi = (\phi_1, \phi_2) : C \to C'$ is a map such that ϕ_1 and ϕ_2 are regular (polynomial) over C.

A morphism $\phi : C \to C'$ is called an *isomorphism* if there exists a morphism $\psi : C' \to C$ such that $\phi \circ \psi = \text{id}_{C'}$ and $\psi \circ \phi = \text{id}_C$. If $C = C'$, the isomorphism ϕ is called an *automorphism*.

Example 307.

(1) Let $f(X,Y) = X^n + Y^n - 1$, $g(X,Y) = X + Y - 1 \in \bar{K}[X,Y]$. The map $\phi : V(f) \to V(g)$ which sends (a,b) to (a^n, b^n) is a morphism of curves.

(2) Let $f(X,Y) \in \bar{K}[X,Y]$. The map $p_x : V(f) \to V(Y)$ which sends (a,b) to $(a,0)$ (projection onto the x-axis) and the map $p_y : V(f) \to V(X)$ which sends (a,b) to $(0,b)$ (projection onto the y-axis) are morphisms of curves.

(3) Let $f(X,Y) = Y^n - g(X) \in \bar{K}[X,Y]$. Assume that $\text{char}(\bar{K})$ is coprime with n, so that $\bar{K}$ contains n distinct solutions to the equation $z^n = 1$. Let $\xi \neq 1$ be such a solution. The map $\sigma : V(f) \to V(f)$ which sends (a,b) to $(a, \xi b)$ is an automorphism of the curve $V(f)$.

Let C and C' be two affine plane curves. If $\phi : C \to C'$ is a morphism and $f \in \bar{K}[C']$ is a regular function, then $\phi^* f := f \circ \phi$ is a regular function over C. Thus, we have a $\bar{K}$-algebras morphism $\phi^* : \bar{K}[C'] \to \bar{K}[C]$. Moreover, if C and C' are irreducible and ϕ is nonconstant, then ϕ^* is one-to-one (an $f \in \bar{K}[C']$ vanishing on the infinite set $\phi(C)$ is zero). In that case, ϕ induces a morphism between fields $\phi^* : \bar{K}(C') \to \bar{K}(C)$. Moreover, C and C' are regularly isomorphic via ϕ if and only if ϕ^* is an isomorphism of $\bar{K}$-algebras.

Definition 308. Let C and C' be two projective plane curves with C irreducible and $C' = V(H)$. A *rational map* $\phi : C \dashrightarrow C'$ is a map $\phi = (f_0 : f_1 : f_2)$ with $f_0, f_1, f_2 \in \bar{K}(C)$ not all zero satisfying $H(f_0, f_1, f_2) = 0$. We say that ϕ is *regular* at $P \in C$ if there exists $g \in \bar{K}(C)^\times$ such that the $f_i' = g f_i$ are regular and not all zero at P; we then set $\phi(P) = (f_0'(P) : f_1'(P) : f_2'(P)) \in C'$. A rational map that is regular at every point is called a *morphism*.

Note that a morphism of curves is either constant or surjective (see Proposition 317).

Proposition 309. *Let $\phi : C \dashrightarrow C'$ be a rational map between two projective plane curves with C smooth. Then ϕ is a morphism (regular).*

Proof. Let $P \in C$ and t a uniformizing parameter at P. Denoting $\phi = (f_0 : f_1 : f_2)$ and $m = \min(\text{Ord}_P(f_0), \text{Ord}_P(f_1), \text{Ord}_P(f_2)) \in \mathbb{Z}$, the rational functions $t^{-m} f_i$ are regular and not all zero at P. Thus, ϕ is regular at P.

Example 310.

(1) Let C be a smooth projective plane curve and $\mathbb{P}^1 = V(Z)$. An $f \in \bar{\mathbf{K}}(C)$
defines a rational map $\tilde{f} = (f : 1) : C \dashrightarrow \mathbb{P}^1$. By Proposition 309, $\tilde{f}$ is
everywhere regular with $\tilde{f}(P) = f(P)$ if f is regular at P, and $\tilde{f}(P) = \infty$
otherwise. Then, there is a one-to-one correspondence between $\bar{\mathbf{K}}(C) \cup$
$\{\infty\}$ (∞ denotes the constant map $\infty(P) = (1 : 0 : 0)$) and the set of regular
maps $C \dashrightarrow \mathbb{P}^1$. Note that since $\bar{\mathbf{K}}(C) \cup \{\infty\}$ is not a field, it does not make
sense to interpret the set of regular maps $C \dashrightarrow \mathbb{P}^1$ as a field.

(2) Let C be the smooth projective plane curve $V(XZ - Y^2)$ (the projective
closure of a parabola in $\bar{\mathbf{K}}^2$) and $\mathbb{P}^1 = V(Z)$. Let $\phi = (x^2 : xy : y^2) : \mathbb{P}^1 \dashrightarrow C$.
As x, y do not vanish simultaneously, one of x^2 and y^2 is nonzero, and, thus,
ϕ is a rational map. Moreover, let $\psi = (x : y) = (y : z) : C \dashrightarrow \mathbb{P}^1$. As x, y, z
cannot all be 0 simultaneously, so at least one of the forms $(x : y)$ and $(y : z)$
is nonzero, and, thus, ψ is a rational map.

Note that $\psi \circ \phi = \mathrm{id}_{\mathbb{P}^1}$ and $\phi \circ \psi = \mathrm{id}_C$. Therefore ϕ and ψ are inverses of
each other, and the projective line and the conic curve C are isomorphic.

Definition 311. The notion of degree of a mapping is fundamental in algebra and
topology; a degree d map is "d-to-one on most points". The degree of a rational
map $\phi : C \dashrightarrow C'$ between two irreducible curves measures how often the map
covers the image curve, i.e. the cardinality of a generic fibre $\phi^{-1}(Q)$, for $Q \in C'$.
Those points where the cardinality of the fibre does not equal the degree of the
mapping are called *ramification points* for the rational mapping (there is only a
finite number of such points). A point $Q \in C'$ is called a *branch point* for ϕ if it
is the image of a ramification point for ϕ.

Example 312. Let $C = V(f)$ (resp. $C = V(F)$) be a smooth affine (resp. projective) plane curve where $f \in \bar{\mathbf{K}}[X,Y]$ (resp. F is a homogeneous polynomial in
$\bar{\mathbf{K}}[X,Y,Z]$).

Define $\pi : C \to \mathbb{A}^1(\bar{\mathbf{K}})$ by $\pi(x,y) = x$ (resp. $\pi : C \dashrightarrow \mathbb{P}^1(\bar{\mathbf{K}})$ by $\pi(x : y : z) =$
$(x : z)$).

Then, by the Implicit Function theorem, π is ramified at $P \in C$ if and only if
$\frac{\partial f}{\partial Y}(P) = 0$ (resp. $\frac{\partial F}{\partial Y}(P) = 0$).

Example 313. Let $\mathbf{K}$ be a field of characteristic not equal to 2. The morphism
$\phi : \mathbb{A}^1(\bar{\mathbf{K}}) \to \mathbb{A}^1(\bar{\mathbf{K}})$ given by $\phi(x) = x^2$ is clearly two-to-one away from the point
$x = 0$. We say that ϕ has degree 2.

Example 313 suggests several possible definitions for degree: the first in
terms of the number of preimages of a general point in the image; the second
in terms of the degrees of the polynomials defining the map. A third definition
is to recall the injective field homomorphism $\phi^* : \bar{\mathbf{K}}(\mathbb{A}^1) \to \bar{\mathbf{K}}(\mathbb{A}^1)$. One sees that
$\phi^*(\bar{\mathbf{K}}(\mathbb{A}^1)) = \bar{\mathbf{K}}(x^2) \subseteq \bar{\mathbf{K}}(x)$ and that $[\bar{\mathbf{K}}(x) : \bar{\mathbf{K}}(x^2)] = 2$. This latter formulation
turns out to be a suitable definition for degree.

Definition 314. Let $\phi : C \dashrightarrow C'$ be a rational map between irreducible projective plane curves, and suppose that ϕ is nonconstant. If $f \in \bar{\mathbf{K}}(C')$, then it can be seen as a rational map $C' \dashrightarrow \mathbb{P}^1$ (Example 310). The composition $\phi^* f := f \circ \phi$ is well-defined as a rational map (the set of points where f is not regular is finite, and, thus, its preimage by ϕ is finite). Therefore, we obtain a morphism between fields $\phi^* : \bar{\mathbf{K}}(C') \to \bar{\mathbf{K}}(C)$. This extension is finite. As a matter of fact, since $\phi^*(\bar{\mathbf{K}}(C'))$ is isomorphic to $\bar{\mathbf{K}}(C')$, it has transcendence degree 1 over $\bar{\mathbf{K}}$. Since $\bar{\mathbf{K}}(C)$ also has transcendence degree 1 over $\bar{\mathbf{K}}$, we infer that $\bar{\mathbf{K}}(C)/\phi^*(\bar{\mathbf{K}}(C'))$ is an algebraic extension. Finally, $\bar{\mathbf{K}}(C)$ is a finite algebraic extension of $\phi^*(\bar{\mathbf{K}}(C'))$ since $\bar{\mathbf{K}}(C)$ is finitely-generated over $\bar{\mathbf{K}}$. The *degree* of ϕ is:

$$\deg(\phi) := [\bar{\mathbf{K}}(C) : \phi^*(\bar{\mathbf{K}}(C'))].$$

The same definition holds for nonconstant morphisms between irreducible affine plane curves (as explained above).

We say that ϕ is *separable* if the field extension $\bar{\mathbf{K}}(C)/\phi^*(\bar{\mathbf{K}}(C'))$ is separable. We also define the *separable degree* of ϕ, denoted by $\deg_s(\phi)$, as the separable degree of the field extension $\bar{\mathbf{K}}(C)/\phi^*(\bar{\mathbf{K}}(C'))$ (see Definition 223).

Remark 315.

1. The notion of degree can be used to characterize the birationality of rational maps since a noncontant rational map $\phi : C \dashrightarrow C'$ between two irreducible projective plane curves is birational if and only if its degree is one.

2. As the degree of algebraic field extensions is multiplicative, if $\phi_1 : C_1 \dashrightarrow C_2$ and $\phi_2 : C_2 \dashrightarrow C_3$ are noncontant rational maps between irreducible projective plane curves, then $\deg(\phi_2 \circ \phi_1) = \deg(\phi_2) \cdot \deg(\phi_1)$.

Example 316. Let $\phi : \mathbb{A}^1(\bar{\mathbf{K}}) \to \mathbb{A}^1(\bar{\mathbf{K}})$ be a nonconstant morphism given by $\phi(x) = a(x)$ for some polynomial $a(x) \in \bar{\mathbf{K}}[x]$. Let $\theta = a(x)$ and $d = \deg(\theta)$. We have $\phi^*(\bar{\mathbf{K}}(\mathbb{A}^1)) = \bar{\mathbf{K}}(\theta) \subseteq \bar{\mathbf{K}}(x)$. By successive Euclidean divisions by θ, it is easy to see that $1, x, \ldots, x^{d-1}$ is a basis for $\bar{\mathbf{K}}(x)$ as a $\bar{\mathbf{K}}(\theta)$-vector space. Or, also, it is immediate that the minimal polynomial of x over $\bar{\mathbf{K}}(\theta)$ is, up to a nonzero scalar, $F(T) = a(T) - a(x)$. So, $\deg(\phi) = \deg(a(x))$.

An important fact is that nonconstant rational maps between projective curves are always surjective.

Proposition 317. *Every nonconstant rational map $\phi : C \dashrightarrow C'$ between irreducible projective plane curves is surjective.*

Proof. Let $\xi = (\alpha_1, \alpha_2) \in C'$, denote by t_1, t_2 the coordinates functions on C', and set $\mathfrak{m}_\xi = \langle t_1 - \alpha_1, t_2 - \alpha_2 \rangle$ the ideal of $\bar{\mathbf{K}}[C']$ consisting of functions that vanish at ξ. Thus, the equations of the variety $\phi^{-1}(\{\xi\})$ are $\phi^*(t_1) = \alpha_1$ and $\phi^*(t_2) = \alpha_2$. By Theorem 133 (the weak Nullstellensatz), $\phi^{-1}(\{\xi\}) = \emptyset$ if and only if $\langle \phi^*(t_1) - \alpha_1, \phi^*(t_2) - \alpha_2 \rangle = \bar{\mathbf{K}}[C]$. Viewing $\bar{\mathbf{K}}[C']$ as a subring of $\bar{\mathbf{K}}[C]$ (i.e., confounding f with $\phi^*(f)$ for $f \in \bar{\mathbf{K}}[C']$), the above condition becomes $\mathfrak{m}_\xi \bar{\mathbf{K}}[C] = \bar{\mathbf{K}}[C]$. The desired conclusion follows from the classical algebraic result saying that if $\mathbf{T}$ is a ring containing a subring $\mathbf{R}$ and $\mathbf{T}$ is a finite $\mathbf{R}$-module, then for any ideal $\mathfrak{a} \subsetneq \mathbf{R}$, we have $\mathfrak{a}\mathbf{T} \subsetneq \mathbf{T}$.

Definition 318. Let $\phi : C \dashrightarrow C'$ be a nonconstant rational map between smooth projective plane curves. By Proposition 309, ϕ is everywhere regular. Let $P \in C$, $Q = \phi(P)$, and t_Q a uniformizing parameter at Q. The *ramification index* of ϕ at P is the nonnegative integer

$$e_\phi(P) = \mathrm{Ord}_P(\phi^* t_Q) = \mathrm{Ord}_P(t_Q \circ \phi).$$

It does not depend on the choice of t_Q. We say that ϕ is not ramified if $e_\phi(P) = 1$ for all $P \in C$, or, equivalently, if for every $Q \in C'$, $\sharp(\phi^{-1}(\{Q\})) = \deg(\phi)$ (see Theorem 319 below).

Examples on how to compute ramification points and ramifications indexes an be found in Exercises 368 and 369.

The following theorem is a generalization of Theorem 286 (this latter corresponds to $C' = \mathbb{P}^1$, $\phi = f$, and $Q = 0$).

Theorem 319. *Let $\phi : C \dashrightarrow C'$ be a nonconstant rational map between two smooth projective plane curves. Then:*

1. For every $Q \in C'$, we have:

$$\deg(\phi) = \sum_{\substack{P \in C \\ \phi(P) = Q}} e_\phi(P).$$

2. For all but finitely many $Q \in C'$, we have:

$$\sharp(\phi^{-1}(\{Q\})) = \deg_s(\phi).$$

Theorem 320. *(Riemann-Hurwitz)*

Let $\phi : C \dashrightarrow C'$ be a nonconstant separable rational map between two smooth projective plane curves C, C' of genera (plural of genus) g and g', respectively. Then:

$$2g - 2 \geq \deg(\phi)(2g' - 2) + \sum_{P \in C}(e_\phi(P) - 1).$$

In addition, equality holds if and only if ϕ is tamely ramified, that is, either $\mathrm{char}(\mathbf{K}) = 0$ *or* $\mathrm{char}(\mathbf{K}) = p > 0$ *and none of the $e_\phi(P)$ is divisible by p.*

Example 321. Let **K** be a field of characteristic $p > 0$. Set $\mathbb{P}^1 = V(Z)$ and consider the morphism $\mathbb{P}^1 \dashrightarrow \mathbb{P}^1$ given on $\mathbb{A}^1$ by $\psi(x) = x^p - x$. Then one checks that this is unramified on $\mathbb{A}^1$, and ramified to order p at ∞. Thus, we see that the inequality in Theorem 320 is strict without the tame ramification hypothesis since $-2 > -2p + (p-1) = -(p+1)$.

For additional examples, see Exercises 368 and 369.

Corollary 322. *Let $\phi : C \dashrightarrow C'$ be a nonconstant separable rational map between two smooth projective plane curves C, C' of genera g and g', respectively. Then $g \geq g'$, and equality is only possible if $g' = 0, 1$ or $\deg(\phi) = 1$.*

Proof. We have $g \geq g'$ because ϕ induces a one-to-one linear map $\phi^* : \Omega^1(C') \to \Omega^1(C)$, or simply, because $g - 1 \geq \deg(\phi)(g' - 1) \geq g' - 1$. By virtue of this latter inequality, if $g' > 1$ and $\deg(\phi) > 1$, then $g > g'$.

Remark 323.

1) Over $\mathbb{C}$, Riemann-Hurwitz Theorem 320 can be reformulated as follows:

Let $\phi : X \to X'$ be a nonconstant holomorphic map between compact Riemann Surfaces of genera g and g', respectively. Suppose that ϕ has degree n and is ramified at m points on X with ramification indexes $e_1, \ldots, e_m$, correspondingly. Then

$$2 - 2g = n(2 - 2g') - \sum_{i=1}^{m}(e_i - 1).$$

To see this, first recall that the *Euler characteristic* for a Riemann surface X is $\chi = V - E + F$, where V, E, and F are respectively the numbers of vertices, edges and faces of a triangulation of X. The Euler characteristic is a topological invariant, and, in particular, it does not depend on the considered triangulation. When we see a compact Riemann surface as a sphere with g handles (g being the genus), then, after triangulation, we obtain that

$$\chi = 2 - 2g.$$

Now triangulate the surface Y in such a way that all branch points (i.e., the corresponding values $\phi(a_i)$ of ramification points a_i) are among the vertices of the triangulation. If V', E', and F' are the numbers of vertices, edges and faces of the considered triangulation, respectively, then the Euler characteristic of X' is $\chi' = V' - E' + F'$. Considering the preimage by ϕ of our triangulation on Y, obtain a triangulation of X whose number of edges is $E = nE'$, number of faces is $F = nF'$, and number of vertices is $V = nV' - \sum_{i=1}^{m}(e_i - 1)$. So, the Euler characteristic of X is $\chi = 2 - 2g = V - E + F = nV' - \sum_{i=1}^{m}(e_i - 1) - nE' + nF' =$

$n\chi' - \sum_{i=1}^{m}(e_i - 1) = n(2 - 2g') - \sum_{i=1}^{m}(e_i - 1).$

For example, the Riemann sphere maps to itself by the function z^n, which has ramification index n at 0, for any integer $n > 1$. So, necessarily, there is a ramification at the point at ∞. In order to balance the equation $2 = 2 \cdot n - (n-1) - (e_\infty - 1)$, we must have ramification index n at ∞.

2) If f is a meromorphic function on a Riemann surface X and $\mathscr{P}$ is the set of its poles, we can define a map $\bar{f} : X \to \mathbb{P}^1(\mathbb{C}) = \mathbb{C} \cup \{\infty\}$ by $\bar{f}(x) = \infty$ if $x \in \mathscr{P}$, and $\bar{f}(x) = f(x)$, otherwise. Then $\bar{f}$ is holomorphic since for $x \in \mathscr{P}$, the composition of f with the chart $\phi_1 : U_1 = (\mathbb{C} \setminus \{0\}) \cup \{\infty\} \to \mathbb{C}$, $z \mapsto \frac{1}{z}$, is bounded in a neighborhood of x and, thus, can be made holomorphic by the removable singularity theorem. Conversely, if we have a holomorphic map $g : X \to \mathbb{P}^1(\mathbb{C})$, we obtain a meromorphic function f by restriction of g to the preimage of $\mathbb{C}$ (f has poles at the point of $g^{-1}(\{\infty\})$). So, we see that there is a natural bijection between meromorphic functions on a Riemann surface X and holomorphic maps from X to $\mathbb{P}^1(\mathbb{C})$.

3) Let f be a nonconstant meromorphic function on a compact Riemann surface X of genus g and Euler characteristic $\chi = 2 - 2g$. Then, as explained above, it can be seen as a holomorphic map $X \to \mathbb{P}^1(\mathbb{C})$. Suppose that f has degree n and is ramified at m points on X with ramification indexes $e_1, \ldots, e_m$, correspondingly. Then, by Riemann-Hurwitz Theorem 320, we have:

$$\chi = 2 - 2g = 2n - \sum_{i=1}^{m}(e_i - 1).$$

5.5   Resultants and Bézout's theorem

Borrowing words from [23], Bézout's theorem is concerned with the intersections of algebraic plane curves. For instance, Fig. 5.6 illustrates some different types of intersections. The ellipse and the quartic illustrate a polynomial of degree 2 and a polynomial of degree 4 that have 6 points in common. The other examples show two quadratic polynomials (a circle and a parabola) which appear to intersect 3 times, two parabolas that intersect twice, and two cubics that intersect 9 times. Counting points of intersection in these examples suggests that, for two plane curves C and D, defined by polynomials $f(X,Y)$ and $g(X,Y)$, respectively,

$$\sharp(\text{points in } C \cap D) \leq \deg(f)\deg(g).$$

As usual, in order to replace the inequality with an equality (this is exactly what Bézout's theorem claims), we need to find the right hypotheses. In our earlier discussion, we could replace the inequality 4 with an equality provided

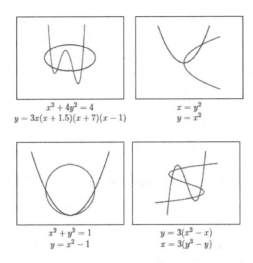

$$x^2 + 4y^2 = 4$$
$$y = 3x(x + 1.5)(x + 7)(x - 1)$$

$$x = y^2$$
$$y = x^2$$

$$x^2 + y^2 = 1$$
$$y = x^2 - 1$$

$$y = 3(x^3 - x)$$
$$x = 3(y^3 - y)$$

Figure 5.6: Number of points in the intersection of two curves

we allowed nonreal roots, and counted roots (points of intersection) with multiplicity. While these extra conditions do give equality in all examples above, unfortunately, even with these provisions, we cannot replace the inequality with an equality in the case of the intersection of any two algebraic plane curves. To see why, consider the case of two parallel lines. No matter how carefully we count intersections, two parallel lines simply do not intersect. So to get an equality in our equation, we need to see our curves in the projective plane (as another example, see Fig. 5.7).

Bézout's theorem is one of the "great theorems" in algebraic geometry. It explains the intersections of polynomial curves in the projective plane. More precisely, it states that the number of common points of two plane algebraic curves which do not share a common component (that is, which do not have infinitely many common points) is at most equal to the product of their degrees, and equality holds if one counts points at infinity and points with complex coordinates (or more generally, coordinates from the algebraic closure of the ground field), and if each point is counted with its intersection multiplicity. For exploring Bézout theorem, the resultant will be significantly helpful.

5.5.1 Resultants

We will give a brief outline of *resultant*: an important idea in constructive algebra whose development owes considerably to famous pioneers such as Bézout, Cayley, Euler, Herman, Hurwitz, Kronecker, Macaulay, Noether, and Sylvester, among others. The resultant of two univariate polynomials over a field is a

Figure 5.7: Intersection of two hyperbolas $x^2 - y^2 = 1$ and $y^2 - x^2 = 1$

polynomial expression of their coefficients, which is equal to zero if and only if the polynomials have a common root.

Lemma 324. *Let $\mathbf{K}$ be a field and $f, g \in \mathbf{K}[X] \setminus \{0\}$. The following assertions are equivalent:*

(i) *f and g have a common nonconstant factor (i.e., $\gcd(f, g) \neq 1$).*

(ii) *$\exists F, G \in \mathbf{K}[X] \setminus \{0\}$ such that $Gf = Fg$ with $\deg(G) < \deg(g)$ and $\deg(F) < \deg(f)$.*

Proof. "(i) $\Rightarrow$ (ii)" Take $G = \frac{g}{\gcd(f,g)}$ and $F = \frac{f}{\gcd(f,g)}$.

"(ii) $\Rightarrow$ (i)" Use Gauss's lemma.

Definition 325. For a field $\mathbf{K}$ and $\ell \in \mathbb{N}$, we denote by $\mathbf{K}_\ell[X]$ the $\mathbf{K}$-vector space formed by polynomials in $\mathbf{K}[X]$ of degree $\leq \ell$. For $f = a_n X^n + \cdots + a_1 X + a_0, g = b_m X^m + \cdots + b_1 X + b_0 \in \mathbf{K}[X]$ with $a_n, b_m \neq 0$, we consider the $\mathbf{K}$-linear map:

$$\varphi_{f,g} : \quad \begin{array}{ccc} \mathbf{K}_{m-1}[X] \times \mathbf{K}_{n-1}[X] & \longrightarrow & \mathbf{K}_{m+n-1}[X] \\ (G, F) & \longmapsto & Gf + Fg \end{array}$$

Considering a basis $\mathscr{B} = ((X^{m-1}, 0), (X^{m-2}, 0), \ldots, (1, 0), (0, X^{n-1}), (0, X^{n-2}), \ldots, (0, 1))$ for $\mathbf{K}_{m-1}[X] \times \mathbf{K}_{n-1}[X]$, and a basis $\mathscr{B}' = (X^{m+n-1}, \ldots, 1)$ for $\mathbf{K}_{m+n-1}[X]$, we have

$$\text{Mat}(\varphi_{f,g}, \mathscr{B}, \mathscr{B}') = \begin{pmatrix} a_n & & & & b_m & & & \\ a_{n-1} & a_n & & & b_{m-1} & b_m & & \\ a_{n-2} & a_{n-1} & \ddots & & b_{m-2} & b_{m-1} & \ddots & \\ \vdots & & \ddots & a_n & \vdots & & \ddots & b_m \\ \vdots & & & a_{n-1} & \vdots & & & b_{m-1} \\ a_0 & & & & b_0 & & & \\ & a_0 & & \vdots & & b_0 & & \vdots \\ & & \ddots & & & & \ddots & \\ & & & a_0 & & & & b_0 \end{pmatrix}$$

$$\underbrace{}_{m \text{ columns}} \quad \underbrace{}_{n \text{ columns}}$$

This matrix is called *the Sylvester matrix of f and g with respect to X* and denoted by $\text{Syl}(f, g, X)$. The same definition applies when replacing $\mathbf{K}$ with a ring $\mathbf{R}$. The resultant of f and g, denoted by $\text{Res}_X(f, g)$, or simply $\text{Res}(f, g)$ if there is no risk of ambiguity, is the determinant of $\text{Syl}(f, g, X)$. It is a homogeneous polynomial at $a_n, \ldots, a_0$ of degree m, a homogeneous polynomial at $b_m, \ldots, b_0$ of degree n, and contains the monomial $a_n^m b_0^n$ with coefficient 1.

Note that if g is constant, say $g = b \in \mathbf{R} \setminus \{0\}$, then $\text{Res}(f, b) = b^{\deg f}$. Moreover, it is easy to see that

$$\text{Res}(g, f) = (-1)^{mn} \text{Res}(f, g)$$

$((-1)^{mn}$ being the signature of the corresponding columns permutation).

Theorem 326. *Let $f, g \in \mathbf{K}[X] \setminus \{0\}$ where $\mathbf{K}$ is a field. Then*

(i) $1 \in \langle f, g \rangle \Leftrightarrow \gcd(f, g)$ *is constant* $\Leftrightarrow \text{Res}(f, g) \neq 0$.

(ii) f *and g have a common factor* $\Leftrightarrow \gcd(f, g)$ *is nonconstant* $\Leftrightarrow \text{Res}(f, g) = 0$.

Proof. $\gcd(f, g) = 1 \Leftrightarrow \varphi_{f,g}$ is an isomorphism $\Leftrightarrow \text{Res}(f, g) \neq 0$.

The resultant is an efficient tool for eliminating variables as can be seen by the following proposition. Applying this proposition in the particular case where $\mathbf{R}[X] = \mathbf{K}[X_1, \ldots, X_n]$, $\mathbf{K}$ a field, $\text{Res}_{X_n}(f, g)$ is in the first elimination ideal $\langle f, g \rangle \cap \mathbf{K}[X_1, \ldots, X_{n-1}]$.

Theorem 327. *Let $\mathbf{R}$ be a ring. Then, for any $f, g \in \mathbf{R}[X]$, there exist $h_1, h_2 \in \mathbf{R}[X]$ such that*

$$h_1 f + h_2 g = \text{Res}_X(f, g) \in \mathbf{R},$$

with $\deg(h_1) \leq m - 1$ and $\deg(h_2) \leq \ell - 1$.

Proof. First notice that

$$(X^{\ell+m-1},\dots,X,1)\,\mathrm{Syl}(f,\,g,\,X) = (X^{m-1}f,\dots,f,X^{\ell-1}g,\dots,g).$$

Thus, by Cramer's rule, considering 1 as the $(\ell+m-1)^{\mathrm{th}}$ unknown of the linear system whose matrix is $\mathrm{Syl}(f,g,\,X)$, $\mathrm{Res}_X(f,g)$ is the determinant of the Sylvester matrix of f and g in which the last row is replaced by $(X^{m-1}f,\dots,f,$ $X^{\ell-1}g,\dots,g)$.

Example 328. Let $f = XY - 1, g = X^2 + Y^2 - 4 \in \mathbb{Z}[X,Y]$. Then

$$\mathrm{Res}_X(f,g) = \begin{vmatrix} Y & 0 & 1 \\ -1 & Y & 0 \\ Xf & f & g \end{vmatrix} = Y^4 - 4Y^2 + 1 = -(XY+1)f + Y^2 g$$

$$\in \langle f,g \rangle \cap \mathbb{Z}[Y] \setminus \{0\}.$$

Proposition 329. *If $f = a_n(X - z_1)\cdots(X - z_n) = a_n X^n + \cdots + a_0$ and $g = b_m(X - \xi_1)\cdots(X - \xi_m) = b_m X^m + \cdots + b_0$, then we have the universal formula:*

$$\mathrm{Res}(f,g) = a_n^m b_m^n \prod_{i=1}^{n}\prod_{j=1}^{m}(z_i - \xi_j) = a_n^m \prod_{i=1}^{n} g(z_i) = (-1)^{mn} b_m^n \prod_{j=1}^{m} f(\xi_j).$$

Proof. Consider $a_n, z_1, \dots, z_n, b_m, \xi_1, \dots, \xi_m$ as indeterminates over $\mathbb{Z}$. As $\mathrm{Res}(f,g)$ vanishes when substituting z_i by ξ_j (by Theorem 327), it is divisible by $\prod_{i=1}^{n}\prod_{j=1}^{m}(z_i - \xi_j)$. Moreover, since a_n divides all the a_i's and b_m divides all the b_j's, then $a_n^m b_m^n$ divides $\mathrm{Res}(f,g)$. It follows that $S := a_n^m b_m^n \prod_{i=1}^{n}\prod_{j=1}^{m}(z_i - \xi_j)$ divides $\mathrm{Res}(f,g)$. Note that

$$S = a_n^m \prod_{i=1}^{n} g(z_i) = (-1)^{mn} b_m^n \prod_{j=1}^{m} f(\xi_j).$$

From these expressions, we infer that S is homogeneous at $a_n,\dots,a_0$ of degree m and homogeneous at $b_m,\dots,b_0$ of degree n. As $\mathrm{Res}(f,g)$ is divisible by S and shares the same homogeneity properties with S, there exists $a \in \mathbb{Q}$ such that $\mathrm{Res}(f,g) = aS$. When seen as sums of monomials, both $\mathrm{Res}(f,g)$ and S contain the monomial $a_n^m b_0^n$ with coefficient 1, and therefore $a = 1$.

As an immediate consequence, we obtain:

Corollary 330. *Let $f,g,h \in \mathbf{R}[X] \setminus \{0\}$ and $a \in \mathbf{R}$, where $\mathbf{R}$ is a ring. Then:*

(i) $\mathrm{Res}(f, X - a) = (-1)^{\deg f} f(a)$.

(ii) $\mathrm{Res}(f, gh) = \mathrm{Res}(f, g)\,\mathrm{Res}(f, h)$.

(iii) $\mathrm{Res}(f, g + hf) = \mathrm{Res}(f, g)$, with the additional hypothesis that f is monic (i.e., with leading coefficient 1) or that $\deg(g) = \deg(g + hf)$.

Definition 331. For $f = a_n X^n + \cdots + a_1 X + a_0 \in K[X]$ (**K** a field) with $a_n \neq 0$, we define the *discriminant* of f as

$$\mathrm{disc}(f) := (-1)^{\frac{n(n-1)}{2}} a_n^{-1} \mathrm{Res}(f, g).$$

From the discussion above, f has a multiple root in some extension field exactly when its discriminant vanishes. In characteristic 0, this is equivalent to saying that f is not square-free. In nonzero characteristic p, $\mathrm{disc}(f) = 0$ if and only if f is not square-free or it has an irreducible factor which not separable (that is, the irreducible factor is a polynomial in X^p).

Example 332. Let $f = aX^2 + bX + c \in K[X]$ (**K** a field with $\mathrm{char}(K) \neq 2$) with $a \neq 0$. The discriminant of f is

$$\mathrm{disc}(f) := -a^{-1} \begin{vmatrix} a & b & c \\ 2a & b & 0 \\ 0 & 2a & b \end{vmatrix} = b^2 - 4ac.$$

Proposition 333. If $f = a_n(X - z_1) \cdots (X - z_n) = a_n X^n + \cdots + a_0$, then we have the universal formula:

$$\mathrm{disc}(f) = (-1)^{\frac{n(n-1)}{2}} a_n^{2n-2} \prod_{i \neq j}(z_i - z_j) = \left(a_n^{n-1} \prod_{i<j}(z_i - z_j) \right)^2.$$

Proof. This follows from Proposition 329 and the fact that

$$f'(z_j) = a_n(z_j - z_1) \cdots (z_j - z_{j-1})(z_j - z_{j+1}) \cdots (z_j - z_n).$$

In particular, over the reals, if all the roots are real, then the discriminant is positive.

5.5.2 Bézout's theorem

The following result is a cornerstone of our study of Bézout's theorem.

Lemma 334. Let $f, g \in K[X_0, \ldots, X_n]$ be homogeneous polynomials of total degree m, n respectively, where **K** is a field. If $f(1, 0, \ldots, 0)$ and $g(1, 0, \ldots, 0)$ are nonzero then $\mathrm{Res}_{X_0}(f, g)$ is either zero or homogeneous in $X_1, \ldots, X_n$ of degree mn.

Proof. First notice that as $f(1,0,\ldots,0)$ and $g(1,0,\ldots,0)$ are nonzero, then the Sylvester matrix $\mathrm{Syl}(f,g,X_0)$ has size $(m+n)\times(m+n)$. Moreover, its $(i,j)^{\mathrm{th}}$ entry is homogeneous of degree $m-j+i$ if $i \leq n$, and $-j+i$ otherwise. Writing $\mathrm{Res}_{X_0}(f,g)$ as the determinant of $\mathrm{Syl}(f,g,X_0)$, it is a sum of certain products over permutations. Some of these products are zero and others are not; we have to prove that the ones that are nonzero have degree mn. The product for a permutation $\sigma \in S_{n+m}$ has degree

$$\sum_{i=1}^{n}(m-\sigma(i)+i)+\sum_{i=n+1}^{n+m}(-\sigma(i)+i)=mn+\sum_{i=1}^{n+m}(-\sigma(i)+i)=mn.$$

In case $n=2$, Lemma 334 shows that $\mathrm{Res}_{X_0}(f,g)$ is a homogeneous bivariate polynomial. By the fundamental theorem of algebra, every univariate polynomial over an algebraically closed field can be written as the product of linear terms. As a consequence, every bivariate homogeneous polynomial over an algebraically closed field does factor into linear homogeneous factors in an essentially unique way, that is up to permutation of the factors and multiplication of the factors by nonzero constants.

Lemma 335. *Let $h \in \bar{\mathbf{K}}[X,Y]$ be a nonzero homogeneous polynomial of degree δ, where $\mathbf{K}$ is a field. Then h can be written in the form*

$$h = c(s_1X-r_1Y)^{m_1}\cdots(s_\ell X-r_\ell Y)^{m_\ell},$$

where $c \neq 0$ in $\bar{\mathbf{K}}$ and $(r_1,s_1),\ldots,(r_\ell,s_\ell)$ are distinct point of $\mathbb{P}^1(\bar{\mathbf{K}})$. Furthermore,

$$V(h)=\{(r_1,s_1),\ldots,(r_\ell,s_\ell)\}\subseteq\mathbb{P}^1(\bar{\mathbf{K}}),\ \text{and}\ \sharp(V(h))\leq\delta.$$

Proof. Write $h = X^r g$ with $g = \sum_{i=0}^{d}a_iX^iY^{d-i}$, $a_i \in \bar{\mathbf{K}}$, and $a_0 \neq 0$. From the factorization of a univariate polynomial over $\bar{\mathbf{K}}$ into degree one factors ensues the factorization:

$$g = Y^d\sum_{i=0}^{d}a_i\left(\frac{X}{Y}\right)^i = Y^d\prod_{j=1}^{d}\left(s_j\left(\frac{X}{Y}\right)-r_j\right)=\prod_{j=1}^{d}(s_jX-r_jY).$$

The following theorem is a more precise formulation of Lemma 273. It bounds the number of points in the intersection of two curves using their degrees (recall that the degree of a curve is the degree of a square-free polynomial defining it, see Definition 271).

Theorem 336. (Weak Bézout's Theorem)

Let C, D be two projective curves in $\mathbb{P}^2(\bar{\mathbf{K}})$ of degrees m, n respectively, and without common irreducible components, where $\mathbf{K}$ is a field. Then $C \cap D$ is finite and has at most mn points.

Proof. Suppose the curves have $mn + 1$ points $P_0, \ldots, P_{mn}$ in common, and let us prove that they have a common component. As $\bar{\mathbf{K}}$ is infinite, we can pick a point which is not in $C \cup D \cup \bigcup_{0 \leq i < j \leq mn} (P_i P_j)$, where $(P_i P_j)$ denotes the line passing through P_i and P_j. By a change of coordinates, we can suppose that this point is $(0 : 0 : 1)$. In these coordinates, the curves have equations:

$$F(x, y, z) = z^m + a_1 z^{m-1} + \cdots + a_m = 0,$$

$$G(x, y, z) = z^n + b_1 z^{n-1} + \cdots + b_n = 0,$$

where the a_i's and b_i's are homogeneous polynomials in $\bar{\mathbf{K}}[x, y]$ of degree i. Denoting by $P_i = (x_i : y_i : z_i)$, the points $(x_i : y_i)$ are pairwise distinct in $\mathbb{P}^1(\bar{\mathbf{K}})$ by our choice of coordinates (for $0 \leq i < j \leq mn$, $(x_i : y_i) = (x_j : y_j) \Rightarrow (0 : 0 : 1) \in (P_i P_j) = V(y_i X - x_i Y))$. By virtue of Lemmas 334 and 335, we infer that $\text{Res}_Z(F(X, Y, Z), G(X, Y, Z)) = 0$, and, thus, F and G have a common factor. $\blacksquare$

Hereafter an important consequence of Bézout's Theorem 336.

Corollary 337. *If two projective curves C, D in $\mathbb{P}^2(\bar{\mathbf{K}})$ ($\mathbf{K}$ a field) of degree n intersect in n^2 distinct points, and if exactly mn of these points lie on an irreducible curve of degree m, then the remaining $n^2 - mn$ points lie on a curve of degree $n - m$.*

Proof. Denote by $C = V(F), D = V(G)$, where F, G are homogeneous polynomials in $\bar{\mathbf{K}}[X, Y, Z]$ of degree n, and $E = V(H)$ the irreducible curve containing the mn points, where H is a homogeneous polynomials in $\bar{\mathbf{K}}[X, Y, Z]$ of degree m.

If E is a component of C (idem for D) then $F = HL$ for some polynomial L, and the curve $V(L)$ has degree $n - m$ and contains the $n^2 - mn$ points of $C \cap D$ which are not on E.

Now, suppose that E is not a component of $C \cup D$, and pick a point $P = (\alpha : \beta : \gamma) \in E$ with $a := F(\alpha, \beta, \gamma) \neq 0$ and $b := G(\alpha, \beta, \gamma) \neq 0$. Then, $P \in E' := V(bF - aG)$. As E' and E have at least $mn + 1$ points of intersection and E is irreducible, then, by virtue of Bézout's Theorem 336, we infer that $E \subseteq E'$, that is, $bF - aG = HL'$ for some homogeneous polynomial $L' \in \bar{\mathbf{K}}[X, Y, Z]$. The curve $V(L')$ has degree $n - m$ and contains the $n^2 - mn$ points of $C \cap D$ which are not on E. $\blacksquare$

As an application of Corollary 337, we will prove the following result of Pascal (see Fig. 5.8).

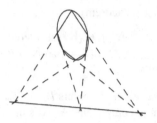

Figure 5.8: Pascal's Theorem

Corollary 338. (Pascal's theorem)

Let A, B, C, A', B', C' *be a hexagon on an irreducible conic. Then the points of intersection* $(AB') \cap (A'B)$, $(AC') \cap (A'C)$, *and* $(BC') \cap (B'C)$ *lie on the same line.*

Proof. The product of the linear polynomials defining the lines (AC'), (BA'), (CB') and that of (AB'), (BC'), (CA') define two cubics. As they intersect in 9 points and 6 among them are on the irreducible conic, then, by virtue of Corollary 337, the other 3 points are on a curve of degree $3 - 2 = 1$, that is, are collinear.

Hereafter we give a definition of intersection multiplicity. This notion will allow us to replace the inequality in Bézout's Theorem 336 with an equality.

Definition 339. Let $C = V(F)$, $D = V(G)$ be two projective curves in $\mathbb{P}^2(\bar{\mathbf{K}})$ of degrees $m = \deg(F)$, $n = \deg(G)$, and without common irreducible components, where F, G are two square-free homogeneous polynomials in $\bar{\mathbf{K}}[X, Y, Z]$, $\mathbf{K}$ being a field. As in the proof of Bézout's Theorem 336, by a change of coordinates, we can suppose that

$$(0 : 0 : 1) \notin C \cup D \cup \bigcup_{P \neq Q \text{ in } C \cap D} (PQ). \tag{5.1}$$

Here, it is worth pointing out that the hypothesis "$(0 : 0 : 1) \notin C \cup D$" is so that $\mathrm{Res}_Z(F(X,Y,Z), G(X,Y,Z))$ has the "good" degree $\deg(C) \cdot \deg(D)$ (see Example 341), while the hypothesis "$(0 : 0 : 1) \notin \bigcup_{P \neq Q \text{ in } C \cap D} (PQ)$" is so that from the point $(0 : 0 : 1)$ one can distinguish between the points in $C \cap D$ (see Example 342).

Then, by virtue of the proof of Bézout's Theorem 336 and Lemma 335, there is a bijection between points P of intersection and linear factors of $\mathrm{Res}_Z(F(X,Y,Z), G(X,Y,Z))$. The *intersection multiplicity* $I_P(C, D)$ is defined to be the multiplicity of this factor. This definition does not depend on the choice of coordinate system satisfying (5.1).

With Definition 339, the proof of Bézout's Theorem 336 immediately implies:

Theorem 340. (Strong Bézout's Theorem)

Let C, D be two projective curves in $\mathbb{P}^2(\bar{\mathbf{K}})$ of degrees m, n respectively, and without common irreducible components, where $\mathbf{K}$ is a field. Then

$$\sum_{P \in C \cap D} I_P(C,D) = mn.$$

Example 341. (Cusp and 4 leaf clover)

Figure 5.9: Cusp and 4 leaf clover

Consider the two affine plane curves $C : y^2 = x^3$, and $D : (x^2 + y^2)^3 = 4x^2y^2$ over the complex numbers. Strong Bézout's Theorem 340 says that there are exactly 18 intersection points counted with multiplicities. Substituting $y^2 = x^3$ into $(x^2 + y^2)^3 = 4x^2y^2$, one gets $x^6(x+1)^3 = 4x^5$. Hence $x^5(x(x+1)^3 - 4) = 0$. There are 4 distinct roots of $(x(x+1)^3 - 4)$ and for each root we get two points from $y^2 = x^3$. So there are 8 complex points other than the origin. In particular, this shows that C and D have no common components. But, what is the intersection multiplicity of $P = (0,0)$?

When seen in $\mathbb{P}^2(\mathbb{C})$, the curves become $\tilde{C} = V(F)$ and $\tilde{D} = V(G)$, where $F = X^3 - Y^2Z$ and $G = (X^2 + Y^2)^3 - 4X^2Y^2Z^2$. We have $\mathrm{Res}_Z(F,G) = Y^2(-4X^8 + X^6Y^2 + 3X^4Y^4 + 3X^2Y^6 + Y^8)$ but this resultant does not give the correct intersection multiplicities since $(0:0:1) \in \tilde{C} \cap \tilde{D}$. It is easy to see that the point $(1:0:0)$ satisfies Condition 5.1. So, the information on intersection multiplicities is coded in

$$\mathrm{Res}_X(F,G) = Y^{10}(Y^8 + 3Y^6Z^2 - 69Y^4Z^4 + 181Y^2Z^6 - 64Z^8).$$

We read that the intersection multiplicity of $P = (0,0)$ is 10 while the intersection multiplicity of all the other 8 complex points is 1 since $\mathrm{disc}(1 + 3t^2 - 69t^4 + 181t^6 - 64t^8) = -8683048622142521344 \neq 0$. We see that

$$10 + 8 = 18 = 3 \cdot 6 = \deg(C) \cdot \deg(D).$$

Example 342. Consider the two affine plane curves $C : (x^2 + y^2)^2 - (x^2 - y^2) = 0$ and $D : (x^2 + y^2)^2 - (y^2 - x^2) = 0$ over the complex numbers (see Fig. 5.10). Strong Bézout's Theorem 340 says that there are exactly 16 intersection points counted with multiplicities.

The only point of intersection in the affine plane $\mathbb{A}^2(\mathbb{C})$ is $(0,0)$. When seen in $\mathbb{P}^2(\mathbb{C})$, the curves become $\tilde{C} = V(F)$ and $\tilde{D} = V(G)$, where $F = (X^2 + Y^2)^2 - (X^2 - Y^2)Z^2$ and $G = (X^2 + Y^2)^2 - (Y^2 - X^2)Z^2$. There are only two points of intersection at ∞, $(1 : \pm i : 0)$. In total, there are 3 intersection points in $\mathbb{P}^2(\mathbb{C})$. In particular, this shows that C and D have no common components.

Figure 5.10: Lemniscates of Bernoulli

As in Example 341, $(0 : 0 : 1) \in \tilde{C} \cap \tilde{D}$. But, contrary to Example 341, the point $(1 : 0 : 0)$ does not satisfy Condition 5.1 since it is on the same line with the two points $(1 : \pm i : 0)$ (the line at ∞). So, only the information on intersection multiplicity of $(0 : 0 : 1)$ (because $(1 : 0 : 0)$ is not collinear with $(0 : 0 : 1)$ and a point of intersection at ∞) is coded in

$$\operatorname{Res}_X(F, G) = 256 Y^8 Z^8.$$

We read that the intersection multiplicity of $(0 : 0 : 1)$ is 8 (this corresponds to the factor Y^8 in $\operatorname{Res}_X(F,G)$). Nevertheless, despite that from the point $(0 : 0 : 1)$ we can not distinguish between the two points $(1 : \pm i : 0)$ (this corresponds to the factor Z^8 in $\operatorname{Res}_X(F,G)$), we can also extract the information that the intersection multiplicity of the two conjugate points at ∞ is 4. We see that

$$8 + 4 + 4 = 16 = 4 \cdot 4 = \deg(C) \cdot \deg(D).$$

But if we want to do the job in good and due form, we have to make a coordinates change. We see that the point $(1 : 1 : 1)$ satisfies Condition 5.1, so we can consider the coordinates change $\varphi(x, y, z) = (x - z, y - z, z)$ transforming $(1 : 1 : 1)$ into $(0 : 0 : 1)$, and $\tilde{C}, \tilde{D}$ into $\varphi(\tilde{C}) = V(F \circ \varphi^{-1}) = V(F(X + Z, Y + Z, Z))$, $\varphi(\tilde{D}) = V(G \circ \varphi^{-1}) = V(G(X + Z, Y + Z, Z))$. Now, the whole information on intersection multiplicities is coded in

$$\operatorname{Res}_Z(F(X + Z, Y + Z, Z), G(X + Z, Y + Z, Z))$$

$$= 256X^{16} - 2048X^{15}Y + 8192X^{14}Y^2 - 22528X^{13}Y^3 + 48128X^{12}Y^4 - 83968X^{11}Y^5$$

$$+ 122880X^{10}Y^6 - 153600X^9Y^7 + 165376X^8Y^8 - 153600X^7Y^9 + 122880X^6Y^{10}$$

$$- 83968X^5Y^{11} + 48128X^4Y^{12} - 22528X^3Y^{13} + 8192X^2Y^{14} - 2048XY^{15} + 256Y^{16}$$

$$= 256(X-Y)^8(X^2+Y^2)^4 = 256(X-Y)^8(X-iY)^4(X+iY)^4.$$

We read that the intersection multiplicity at the point $\varphi(0:0:1) = (-1:-1:1)$ is 8 while it is 4 at each of the points $\varphi(1:\pm i:0) = (1:\pm i:0)$.

It is worth pointing out that, for studying intersection multiplicities, the approach we adopted so far is global, that is, we obtain all the intersection multiplicities at once. A different way of approaching the same problem, is to see what happens locally as explained in the remark below.

Remark 343. Let $C = V(F), D = V(G)$ be two curves, where $F, G \in \mathbf{K}[X,Y]$, $\mathbf{K}$ being a field. We do not suppose that C and D have no common irreducible components. Let $P = (a,b) \in \mathbb{A}^2(\mathbf{K})$.

1. The intersection multiplicity $I_P(C,D)$ can also be defined as

$$I_P(C,D) = \dim_{\mathbf{K}}(\mathbf{K}[X,Y]_{\langle X-a, Y-b \rangle}/\langle F, G \rangle) \in \mathbb{N} \cup \{\infty\}.$$

2. For a rational function $f = \frac{g(x,y)}{h(x,y)}$ on the curve C, with $g, h \in \mathbf{K}[X,Y]$ such that $F \nmid g$ and $F \nmid h$, considering a nonsingular point P on C, the order of f at P is

$$\mathrm{Ord}_P(f) = I_P(C, V(g)) - I_P(C, V(h)).$$

3. $I_P(C,D)$ is a nonnegative integer if C and D share no common component which passes through P (in which case we say C and D *intersect properly* at P). Moreover, $I_P(C,D) = \infty$ if C and D do not intersect properly at P.

4. $I_P(C,D) = 0$ if and only if $P \notin C \cap D$. In fact, $I_P(C,D)$ depend only on the components of C and D passing through P.

5. We say that C and D *intersect transversally* at P if C and D are both smooth at P and the tangent lines at P are distinct (see Fig. 5.11). Then, C and D intersect transversally at P if and only if $I_P(C,D) = 1$ if and only if $\langle F, G \rangle = \mathfrak{m}$, where $\mathfrak{m}$ is the maximal ideal of the local ring $\mathbf{K}[X,Y]_{\langle X-a, Y-b \rangle}$ (see Exercise 388 and its solution).

6. $I_P(C,D)$ is invariant under coordinates change, and $I_P(C,D) = I_P(D,C)$.

7. If $F = \prod F_i^{r_i}$ and $G = \prod G_j^{s_j}$, then $I_P(C,D) = \sum_{i,j} r_i s_j I_P(F_i, G_j)$, that is, intersection number is additive over unions.

8. $I_P(C,D) = I_P(C, V(G+RF))$ for any $R \in \mathbf{K}[X,Y]$ (we convene that $I_P(C, \emptyset) = 0$).

9. $I_P(C,D) \geq \mathrm{mult}_C(P) \cdot \mathrm{mult}_D(P)$, with equality if and only if C and D have no tangents in common at P (see Definition 345).

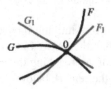

Figure 5.11: Transversal intersection

Example 344. Consider the two affine plane curves $C = V(F)$ and $D = V(G)$ with $F = Y^2 - X^3$, $G = X^2 - Y^3 \in \mathbb{R}[X,Y]$. Let us compute the intersection multiplicity of the two curves at the origin $P = (0,0)$ (see Fig. 5.12) using the Remark 343. We have:

$$
\begin{aligned}
I_P(C,D) &= I_P(V(Y^2 - X^3), V(X^2 - Y^3)) \\
&= I_P(V(Y^2 - X^3 + X(X^2 - Y^3)), V(X^2 - Y^3)) \\
&= I_P(V(Y^2 - XY^3), V(X^2 - Y^3)) \\
&= 2 \cdot I_P(Y, V(X^2 - Y^3)) + I_P(V(1 - XY), V(X^2 - Y^3)) \\
&= 2 \cdot \dim_{\mathbb{R}}(\mathbb{R}[X]/\langle X^2 \rangle) + 0 = 4.
\end{aligned}
$$

This result was predictable since, at the origin, C has a double tangent $y = 0$ while D has a double tangent $x = 0$ (by Remark 343, different tangents implies that $I_P(C,D) = \mathrm{mult}_C(P) \cdot \mathrm{mult}_D(P) = 2 \times 2 = 4$).

With an immediate computation, we see that there is no intersection point at infinity. If we look at Fig. 5.12, we see that there is only one other intersection point $Q = (1,1)$ in the affine plane. At that point, the curves intersect transversally and so we should have $I_Q(C,D) = 1$.

But where are the other $4 = 9 - 4 - 1$ intersection points promised by Strong Bézout's Theorem 340?

In fact, these points are $R_1 = (z_1, z_1^4), \ldots, R_4 = (z_4, z_4^4)$ in the complex plane, where the z_k's correspond to the four complex roots of $t^4 + t^3 + t^2 + t + 1 = \frac{t^5 - 1}{t - 1}$. Necessarily, we have $I_{R_k}(C,D) = 1$ for $1 \leq k \leq 4$.

Of course, we can retrieve this result in one shot with the resultant

$$\text{Res}_X(Y^2Z - X^3, X^2Z - Y^3) = Y^4(Z^5 - Y^5).$$

We read that the intersection multiplicity of $P = (0,0)$ is 4 while the intersection multiplicity of all the other 5 points (1 real and 4 complex) is 1 (since disc$(t^5 - 1) \neq 0$).

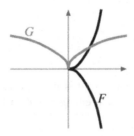

Figure 5.12: Intersection of the curves $y^2 = x^3$ and $x^2 = y^3$

5.6 Genus computation via quadratic transformations

Definition 345. Let $C : f(x,y) = 0$ be an affine curve defined over a field $\mathbf{K}$ and $P = (a,b)$ a singular point of multiplicity $m = \text{mult}_C(P) \geq 2$ (see Definition 162). In such case, m is the minimum of the degrees of the nonzero homogeneous components of $f(X - a, Y - b)$.

The polynomial $G_P(X,Y) := \displaystyle\sum_{i+j=m} \frac{\partial^{i+j} f}{\partial X^i \partial Y^j}(P) \frac{X^i Y^j}{i! \, j!}$ being homogeneous of degree m, by virtue of Lemma 335, it factors in $\bar{\mathbf{K}}[X,Y]$ into m linear polynomials. Since the number of factors of a polynomial is invariant under linear changes of coordinates, also the bivariate polynomial

$$H_P(X,Y) := \sum_{i+j=m} \frac{\partial^{i+j} f}{\partial X^i \partial Y^j}(P) \frac{(X - a)^i (Y - b)^j}{i! \, j!}$$

factors into m linear polynomials. These linear factors are the tangents of C at the singular point P. They are not always distinct, and the multiplicity of a tangent is the multiplicity of the corresponding factor.

A multiple point of a curve with multiplicity $m \geq 1$ that has exactly m distinct tangents (i.e., all its tangents are simple) is called *ordinary*. Otherwise it is called

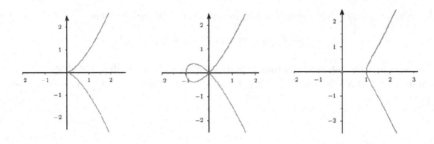

Figure 5.13: $y^2 = x^3$ Figure 5.14: $y^2 = x^3 + x^2$ Figure 5.15: $y^2 = x^3 - x^2$

nonordinary. So, a nonordinary point is necessarily singular. These properties are obviously invariant under linear changes of coordinates.

All the definitions above can be naturally generalized to a projective curve using affine charts.

Example 346.

- The unique singular point $(0,0)$ of the curve in Fig. 5.13 (over the reals) is nonordinary of multiplicity 2 with a double tangent $y = 0$ since $H_{(0,0)}(X,Y) = Y^2$.

- The unique singular point $(0,0)$ of the curve in Fig. 5.14 (over the reals) is ordinary of multiplicity 2 with two tangents $y = x$ and $y = -x$ since $H_{(0,0)}(X,Y) = Y^2 - X^2 = (Y - X)(Y + X)$.

- The unique singular point $(0,0)$ of the curve in Fig. 5.15 (over the reals) is ordinary of multiplicity 2 with two tangents $y = ix$ and $y = -ix$ (over the complexes) since $H_{(0,0)}(X,Y) = Y^2 + X^2 = (Y - iX)(Y + iX)$.

Example 347. We consider the irreducible curve $C = V(f)$, where $f(X,Y) = X^2Y^2 + 36X + 24Y + 108 \in \mathbb{R}[X,Y]$ (see Fig. 5.16).

In order to find the singular points of C in the affine plane $(z \neq 0)$, we need to solve the following polynomial system:

$$\begin{cases} x^2y^2 + 36x + 24y + 108 &= 0 \\ 2xy^2 + 36 &= 0 \\ 2x^2y + 24 &= 0. \end{cases} \qquad (5.2)$$

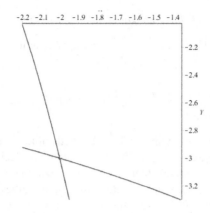

Figure 5.16: $x^2y^2 + 36x + 24y + 108 = 0$

We calculate the reduced Gröbner basis G of

$$\langle x^2y^2 + 36x + 24y + 108,\ 2xy^2 + 36,\ 2x^2y + 24 \rangle$$

with respect to the lexicographic order with $x > y$, using the computer algebra software MAPLE. We obtain $G = \{y + 3,\ x + 2\}$. Therefore, the unique singular point of this curve in the affine plane $(z \neq 0)$ is $P = (-2, -3)$. Let us see whether it is ordinary. We have

$$f(X - 2, Y - 3) = X^2Y^2 - 6X^2Y - 4XY^2 + 9X^2 + 24XY + 4Y^2,$$

whose nonzero homogeneous component of lowest degree is $G_P(X,Y) = 9X^2 + 24XY + 4Y^2$. As $\mathrm{disc}(9 + 24Z + 4Z^2) = 423 \neq 0$, we infer that P is an ordinary singular point of C of multiplicity 2.

Let us see whether C has singular points at ∞. The projective completion of C is $\tilde{C} = V(H)$, where $H(X,Y,Z) = X^2Y^2 + 36XZ^3 + 24YZ^3 + 108Z^4$. The system $H = \frac{\partial H}{\partial X} = \frac{\partial H}{\partial Y} = \frac{\partial H}{\partial Z} = Z = 0$ has two solutions $Q = (1:0:0)$ and $R = (0:1:0)$, which are the singular points of C at ∞. As the nonzero homogeneous component of $H(1,Y,Z)$ (resp. $H(X,1,Z)$) of lowest degree is Y^2 (resp. X^2), we deduce that Q (resp. R) is a nonordinary singular point of multiplicity 2.

Proposition 348. *Let C be an irreducible projective plane curve of degree d, then*

$$\frac{(d-1)(d-2)}{2} \geq \sum_{P \in \mathrm{Sing}(C)} \frac{\mathrm{mult}_C(P)(\mathrm{mult}_C(P)-1)}{2}.$$

In particular, $\sharp(\mathrm{Sing}(C)) \leq \frac{(d-1)(d-2)}{2}$ *(the number of singular points is at most the arithmetic genus).*

Proof. Denote $C = V(F)$ with $F \in K[X,Y]$ irreducible of degree d. As the curve $D := V(\frac{\partial F}{\partial X})$ has degree $d-1$ and has no common irreducible components with C (because C is irreducible), then, by Strong Bézout's Theorem 340, it shares with C exactly $d(d-1)$ points counted with multiplicities. Now, each point P of multiplicity m on $V(F)$ has multiplicity at least $m-1$ on $V(\frac{\partial F}{\partial X})$, and so its intersection multiplicity $I_P(C,D) \geq m(m-1)$ (see Remark 343). It follows that $d(d-1) \geq \sum_{P \in \mathrm{Sing}(C)} \mathrm{mult}_C(P)(\mathrm{mult}_C(P)-1)$.

   The projective plane curves of degree d live in a projective space of dimension $\frac{d(d+3)}{2}$ (see Exercise 379). Also, from Exercise 379, we know that the projective plane curves of degree $d-1$ which have multiplicity at least $\mathrm{mult}_C(P)-1$ at the multiple points P of C form a K-vector space V of dimension $r := \frac{d(d+3)}{2} - \sum_{P \in \mathrm{Sing}(C)} \frac{\mathrm{mult}_C(P)(\mathrm{mult}_C(P)-1)}{2}$. As passing through a point is one linear condition on the coordinates, there is a projective plane curve E in V that has r more common points with C. By Strong Bézout's Theorem 340 applied to C and E, we get $d(d-1) \geq \sum_{P \in \mathrm{Sing}(C)} \mathrm{mult}_C(P)(\mathrm{mult}_C(P)-1) + r$, and, thus, $\frac{(d-1)(d-2)}{2} \geq \sum_{P \in \mathrm{Sing}(C)} \frac{\mathrm{mult}_C(P)(\mathrm{mult}_C(P)-1)}{2}$.

   The following theorem gives an explicit formula for computing the genus of an algebraic curve. For a proof, see [24]. It generalizes Theorem 295 to the singular case with only ordinary singularities. It says that in case C has only ordinary singularities, the difference between $\sum_{P \in \mathrm{Sing}(C)} \frac{\mathrm{mult}_C(P)(\mathrm{mult}_C(P)-1)}{2}$ and $\frac{(d-1)(d-2)}{2}$ is exactly equal to the genus of C.

Theorem 349. *Let C be an irreducible projective plane curve with only ordinary singularities, and let d be the degree of C. Then the genus of C is*

$$\frac{(d-1)(d-2)}{2} - \sum_{P \in \mathrm{Sing}(C)} \frac{\mathrm{mult}_C(P)(\mathrm{mult}_C(P)-1)}{2}.$$

Example 350. We consider the irreducible curve $C = V(f)$, where $f(X,Y) = X^4 - XY + Y^4 \in \mathbb{R}[X,Y]$ (see Fig. 5.17).

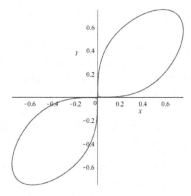

Figure 5.17: $x^4 - xy + y^4 = 0$

In order to find the singular points of C in the affine plane $(z \neq 0)$, we need to solve the following polynomial system:

$$\begin{cases} x^4 - xy + y^4 &= 0 \\ 4x^3 - y &= 0 \\ 4y^3 - x &= 0. \end{cases} \tag{5.3}$$

We calculate the reduced Gröbner basis G of $\langle x^4 - xy + y^4, 4x^3 - y, 4y^3 - x \rangle$ with respect to the lexicographic order with $x > y$, using the computer algebra software MAPLE. We obtain $G = \{x, y\}$. The unique singular point of this curve in the affine plane $(z \neq 0)$ is then $P = (0,0)$. As the nonzero homogeneous component of f of lowest degree is XY, we infer that P is an ordinary singular point of C of multiplicity 2. It is easy to check that there is no singular point at ∞.

We conclude that the genus of C is $\frac{(4-1)(4-2)}{2} - \frac{2(2-1)}{2} = 3 - 1 = 2$. In particular, C is not rational by virtue of Corollary 304.

It is worth mentioning that the nonsingular model of the curve in Example 350 is not plane since there is no degree d such that $\frac{(d-1)(d-2)}{2} = 2$. However, it is always possible, given an irreducible plane curve, to find a plane curve with only ordinary multiple points that is birationally equivalent, so that the formula in Theorem 349 gives the genus. This will be done using quadratic transformations (blow-up) of homogeneous curves.

Definition 351. The transformation $\mathfrak{q}$ of the projective plane $\mathbb{P}^2(\mathbf{K})$ defined by $\mathfrak{q}(x : y : z) = (yz : xz : xy)$, is called the *standard quadratic transformation* or *standard Cremona transformation*. It induces an isomorphism from U onto itself, where $U = \mathbb{P}^2(\mathbf{K}) \setminus V(XYZ)$ (points not lying on one of the lines $x = 0$ or $y = 0$ or $z = 0$). So the quadratic transformation is a birational map between $\mathbb{P}^2(\mathbf{K})$ and itself. In fact, $\mathfrak{q}$ is its own inverse, as can be seen from

$$\mathsf{q}(\mathsf{q}(x:y:z)) = (xzxy:yzxy:yzxz) = xyz(x:y:z) = (x:y:z). \qquad (5.4)$$

For the special points $(1:0:0)$, $(0:1:0)$ and $(0:0:1)$, the quadratic transformation is not defined. These points are called *the fundamental points* of the transformation. Every point lying on one of the lines $x = 0$, $y = 0$ or $z = 0$ is sent to the point $(1:0:0)$, $(0:1:0)$ or $(0:0:1)$, respectively. These lines are called *the irregular lines* of the transformation. When we will speak of *nonfundamental intersections* of a curve with an irregular line, we mean intersection points different from the fundamental points.

Let $P = (0:0:1)$. Then q sends the line $z = 0$ to the point P, and hence $\mathsf{q}^{-1} = \mathsf{q}$ blows up the point P to the line $z = 0$.

Proposition and Definition 352. *Let $H(X,Y,Z)$ be a homogeneous polynomial of degree d and not divisible by any of X, Y or Z. Denote by m, n, r the multiplicities on the curve $V(H)$ of the fundamental points $P = (0:0:1)$, $Q = (0:1:0)$, and $R = (1:0:0)$, respectively. Set*

$$\overline{H}(X,Y,Z) = \frac{H(YZ,XZ,XY)}{X^r Y^n Z^m}.$$

Then:

1. *$\overline{H}$ is a homogeneous polynomial of degree $2d - m - n - r$, and $\overline{\overline{H}} = H$.*

2. *The fundamental points P, Q, R have multiplicities $d - n - r$, $d - m - r$, $d - m - n$, respectively, on the curve $V(\overline{H})$.*

3. *If no tangent at any of the fundamental points P, Q, R is an irregular line, then there is a one-to-one correspondence, preserving multiplicities (multiplicities of the tangents from one side and intersection multiplicities on the other side) between the tangents to $V(H)$ at P, Q, R and the nonfundamental intersections of $V(\overline{H})$ with the irregular lines $z = 0$, $y = 0$, and $x = 0$, respectively. The tangents to $V(\overline{H})$ at the fundamental points P, Q, R are distinct from the irregular lines and correspond to the nonfundamental intersections of C with $z = 0$, $y = 0$, $x = 0$, respectively.*

4. *H is irreducible if and only if so is $\overline{H}$.*

We say that $\overline{H}$ is *the quadratic transform* of H. Denoting by $C = V(H)$ and $\overline{C} = V(\overline{H})$, we will also say that the curve $\overline{C}$ is the quadratic transform of C.

Proof.

1. Write $H(X,Y,Z) = H_m(X,Y)Z^{d-m} + H_{m+1}(X,Y)Z^{d-m-1} + \cdots + H_{d-1}(X,Y)Z + H_d(X,Y)$, where $H_i(X,Y)$ is either 0 or homogeneous degree i, and $H_m(X,Y) \neq 0$ (since P has multiplicity m). We immediately see that $H(YZ,XZ,XY)$ is divisible by Z^m. Likewise, it is divisible by Y^n and X^r.

Since $\overline{H}$ can be seen as $H(YZ, XZ, XY)$ with all factors X, Y, Z removed, and since $q(q(H)) = (XYZ)^d H$ (by virtue of 5.4), doing this twice, we retrieve H, since H is not divisible by any of X, Y or Z.

2. Note that we have the relation $X^r Y^n Z^m = \frac{(XYZ)^d H(X^{-1}, Y^{-1}, Z^{-1})}{\overline{H}(X,Y,Z)}$ in which we can read the multiplicities of P, Q, R on the curve $V(H)$. So, from the equality

$$\frac{(XYZ)^{2d-m-n-r}\overline{\overline{H}}(X^{-1}, Y^{-1}, Z^{-1})}{\overline{\overline{H}}(X,Y,Z)} = \frac{(XYZ)^{2d-m-n-r}\overline{H}(X^{-1}, Y^{-1}, Z^{-1})}{H(X,Y,Z)}$$

$$= \frac{(XYZ)^{2d-m-n-r}X^r Y^n Z^m}{(XYZ)^d} = X^{d-m-n}Y^{d-m-r}Z^{d-n-r},$$

we can read the multiplicities of P, Q, R on the curve $V(\overline{H})$.

3. It suffices to prove the result for P. As above, write $H(X,Y,Z) = H_m(X,Y)Z^{d-m} + H_{m+1}(X,Y)Z^{d-m-1} + \cdots + H_{d-1}(X,Y)Z + H_d(X,Y)$, where $H_i(X,Y)$ is either 0 or homogeneous degree i, and $H_m(X,Y) \neq 0$ and is not divisible by X or Y. We have $\overline{H}(X,Y,0) = X^{d-m-r}Y^{d-m-n}H_m(X,Y)$. It follows that the nonfundamental intersections of $V(\overline{H})$ with $z = 0$ are given by the factors of $H_m(X,Y)$. In more details, by Lemma 335, $H_m(X,Y)$ factorizes linearly as $H_m(X,Y) = (a_1X - b_1Y)^{\ell_1} \cdots (a_sX - b_sY)^{\ell_s}$, and, thus, the nonfundamental intersections of $V(\overline{H})$ with the irregular line $z = 0$ are the $(a_i : b_i : 0)$, $1 \leq i \leq s$.

As $\overline{\overline{H}} = H$, the tangents to $V(\overline{H})$ at P correspond to the nonfundamental intersections of $V(H)$ with $z = 0$. These tangents are different from $x = 0$ or $y = 0$ because for all $1 \leq i \leq s$, we have $a_i b_i \neq 0$ (since, by hypothesis, $x = 0$ and $y = 0$ are not tangent to $V(H)$ at P).

4. Any factorization of H or $\overline{H}$ yields to a factorization of the other.

The third assertion in the previous Proposition is no longer true if we drop the condition that no tangent at any of the fundamental points is an irregular line. Hereafter a counterexample.

Example 353. Consider the irreducible projective plane curve $C = V(H)$, where $H(X,Y,Z) = -X^3Z + Y^4 + Y^2Z^2 \in \mathbb{R}[X,Y,Z]$.

The only singular point of C is $P = (0:0:1)$. Examining $H(X,Y,1), H(X,1,Z)$, and $H(1,Y,Z)$, we see that $P = (0:0:1)$ is a nonordinary double point of C (with a double tangent $y = 0$), $Q = (0:1:0)$ is not a point of C, and $R = (1:0:0)$ is a nonsingular point of C (with tangent $z = 0$). Thus,

$$\bar{H} = \frac{H(YZ, XZ, XY)}{Z^2 X} = X^3Y^2 + X^3Z^2 - Y^4Z.$$

Note that there is no one-to-one correspondence between the tangents to C at P and the nonfundamental intersections of $\bar{C}$ with $z = 0$ since $\bar{C} \cap (z = 0) = \{Q, R\}$.

Proposition 354. *Let $C = V(H)$ be a projective plane curve, where $H \in \mathbf{K}[X, Y, Z]$ is an irreducible homogeneous polynomial of degree d. Denote by m the multiplicity on the curve C of the point $P = (0 : 0 : 1)$. Suppose that lines $x = 0$ and $y = 0$ are not tangent to C, and intersect C in $d - m$ points other than P, and that $z = 0$ intersects this curve in d distinct points different from $Q = (0 : 1 : 0)$ and $R = (1 : 0 : 0)$ (in particular, $Q, R \notin C$). Then $\bar{C}$ has degree $2d - m$ and:*

1. *Inside $U = \mathbb{P}^2(\mathbf{K}) \setminus V(XYZ)$, the curves C and $\bar{C}$ are isomorphic, and, thus, they have the same multiplicities.*

2. *All the points P, Q, R are ordinary on $\bar{C}$ of multiplicities $d, d - m, d - m$, respectively.*

3. *Apart from P, Q, R, the curve $\bar{C}$ does not intersect the lines $x = 0$ and $y = 0$, and has total intersection multiplicities m with the line $z = 0$.*

Proof. For the first item, see Definition 351. The second and third items follow immediately from Proposition and Definition 352.3.

Note that, with the hypotheses of Proposition 354, new ordinary singularities might be created on $\bar{C}$ at the fundamental points, and also new singularities (ordinary or not) might be created on the line $z = 0$.

For an irreducible projective plane curve C of degree d, we set

$$g^*(C) := \binom{d-1}{2} - \sum_{M \in C} \binom{\text{mult}_C(M)}{2}.$$

Proposition 348 affirms that $g^*(C) \geq 0$ while Theorem 349 says that if C has only ordinary singularities then $g^*(C)$ is the genus of C.

Proposition 355. *With the hypotheses of Proposition 354, we have:*

$$g^*(\bar{C}) = g^*(C) - \sum_{M = (a:b:0) \in \bar{C};\, ab \neq 0} \binom{\text{mult}_{\bar{C}}(M)}{2}.$$

Proof. Setting $e := \displaystyle\sum_{M = (a:b:0) \in \bar{C};\, ab \neq 0} \binom{\text{mult}_{\bar{C}}(M)}{2}$, by virtue of Proposition 354, we have

$$g^*(C) - g^*(\bar{C}) = \left(\binom{d-1}{2} - \binom{m}{2} \right) - \left(\binom{2d-m-1}{2} - \binom{d}{2} - 2\binom{d-m}{2} - e \right)$$

$$= e + \binom{d}{2} + \binom{d-1}{2} + 2\binom{d-m}{2} - \binom{2d-m-1}{2} - \binom{m}{2} = e.$$

From Proposition 355, we see that $g^*(C)$ decreases at each quadratic transformation of the curve (at each iteration, we make a coordinates change so that the curve fulfills the hypotheses of Proposition 354), unless $\text{mult}_{\overline{C}}(M) = 1$ for each $M \in \overline{C} \cap V(Z)$, in which case it stays the same but the number of nonordinary multiple points has decreased. This provides an algorithm transforming in a finite number of steps via birational maps (a finite number N of quadratic transformations with $N \leq g^*(C) + \sharp\{\text{nonordinary singular points of } C\}$) an irreducible projective plane curve C into an irreducible projective plane curve having only ordinary singularities. As the genus is a birational invariant and as we have a formula (Theorem 349) for computing the genus of irreducible curve having only ordinary singularities, this algorithm allows genus computation.

Note that the condition in Proposition 354 that $P = (0 : 0 : 1) \in C$ is easily achieved by translating in the affine space (say $z = 1$) the point $P = (a, b)$ to $(0, 0)$. This amounts to changing the defining polynomial $F(X, Y)$ to $F(X + a, Y + b)$. The remaining conditions are usually satisfied for a "random" curve, so they are easily achieved via a linear coordinates change.

Example 356. Consider the curve $C = V(F)$ with $F(X, Y) = Y^2 - X^3 \in \mathbb{R}[X, Y]$ (see Fig. 5.18).

The cubic C has a unique nonordinary singular double point $P = (0 : 0 : 1)$. The projective completion of C is $\tilde{C} = V(G)$ with $G(X, Y, Z) = Y^2 Z - X^3$. The polynomial G does not fulfill the hypotheses of Proposition 354 (for example, because $(0 : 1 : 0) \in \tilde{C}$). Note that $g^*(\tilde{C}) = 1 - 1 = 0$ (so, we will obtain a birational equivalent curve without nonordinary singularities with only one quadratic transformation). If we send (X, Y, Z) to $(X - Y, X + Y, Y + Z)$, the equation becomes $H(x, y, z) = 0$ with

$$H(X, Y, Z) = (X + Y)^2 (Y + Z) - (X - Y)^3$$
$$= -X^3 + 4X^2 Y + X^2 Z - XY^2 + 2XYZ + 2Y^3 + Y^2 Z.$$

Let us check that H fulfills the hypotheses of Proposition 354:

For $z = 0$, the equation becomes $-x^3 + 4x^2 y - xy^2 + 2y^3 = 0$, with $\text{disc}(-t^3 + 4t^2 - t + 2) = -464 \neq 0$ (3 simple points distinct from $(0 : 1 : 0)$ and $(1 : 0 : 0)$).

For $y = 0$, the equation becomes $-x^3 + x^2 z = x^2(z - x) = 0$. So, we have besides P only one simple point $(1 : 0 : 1)$.

For $x = 0$, the equation becomes $2y^3 + y^2 z = y^2(2y + z) = 0$. So, we have besides P only one simple point $(0 : 1 : -2)$.

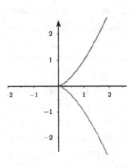

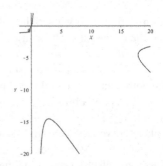

Figure 5.18: $y^2 = x^3$

Figure 5.19: $x^3y + 2x^3 + 2x^2y^2 - x^2y + xy^3 + 4xy^2 - y^3 = 0$

We find

$$\bar{H} = \frac{H(YZ, XZ, XY)}{Z^2} = X^3Y + 2X^3Z + 2X^2Y^2 - X^2YZ + XY^3 + 4XY^2Z - Y^3Z.$$

We see that the origin (in the affine chart $z = 1$, see Fig. 5.19) is an ordinary triple singular point as disc$(2t^3 - t^2 + 4t - 1) = -464 \neq 0$. So, $P = (0 : 0 : 1)$ is an ordinary triple point on $V(\bar{H})$. The points $Q = (0 : 1 : 0)$ and $R = (1 : 0 : 0)$ are nonsingular points on $V(\bar{H})$.

At $z = 0$, the equation of $V(\bar{H})$ becomes $xy(x + y)^2 = 0$. We see that besides $Q = (0 : 1 : 0)$ and $R = (1 : 0 : 0)$, $V(\bar{H})$ intersects $V(Z)$ at the point $M = (1 : -1 : 0)$ with intersection multiplicity 2. This point is nonsingular as the component of degree one of $\bar{H}(1, Y - 1, Z)$ is $8Z$.

As a conclusion, $V(\bar{H})$ is birationally equivalent to C and has only nonordinary singularities (one ordinary triple point at P). Its genus (which is also the genus of C) is $\binom{3}{2} - \binom{3}{2} = 0$, as expected.

Example 357. (Example 353 continued)

Consider the irreducible projective plane curve $C = V(H)$, where $H(X,Y,Z) = -X^3Z + Y^4 + Y^2Z^2 \in \mathbb{R}[X,Y,Z]$ (see Fig. 5.20).

As already seen, C does not satisfy the hypotheses of Proposition 354. If we send (X,Y,Z) to $(X,X+Y,X+Z)$, the equation becomes $G(x,y,z) = 0$ with

$$G(X,Y,Z) = X^4 + 6X^3Y + X^3Z + 7X^2Y^2 + 4X^2YZ + X^2Z^2 + 4XY^3 + 2XY^2Z + 2XYZ^2 + Y^4 + Y^2Z^2.$$

Let us check that G fulfills the hypotheses of Proposition 354:

For $z = 0$, the equation becomes $x^4 + 6x^3y + 7x^2y^2 + 4xy^3 + y^4 = 0$, with disc$(t^4 + 6t^3 + 7t^2 + 4t + 1) = -400 \neq 0$ (4 simple points distinct from $(1 : 0 : 0)$).

For $y = 0$, the equation becomes $x^4 + x^3z + x^2z^2 = x^2(x^2 + xz + z^2) = 0$. So, we have besides P two simple points since disc$(t^2 + t + 1) \neq 0$.

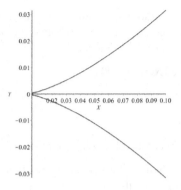

Figure 5.20: $-x^3 + y^4 + y^2 = 0$

For $x = 0$, the equation becomes $y^2(y^2 + z^2) = 0$. So, we have besides P two simple points since $\text{disc}(t^2 + 1) \neq 0$.

We find

$$\bar{G} = \frac{G(YZ, XZ, XY)}{Z^2}$$
$$= X^4Y^2 + X^4Z^2 + 2X^3Y^3 + 2X^3Y^2Z + 4X^3YZ^2 + X^2Y^4$$
$$+ 4X^2Y^3Z + 7X^2Y^2Z^2 + XY^4Z + 6XY^3Z^2 + Y^4Z^2$$

(see Fig. 5.21). At $z = 0$, the equation of $V(\bar{G})$ becomes $x^2y^2(x+y)^2 = 0$. We see that besides $Q = (0 : 1 : 0)$ and $R = (1 : 0 : 0)$, $V(\bar{G})$ intersects $V(Z)$ at the point $M = (1 : -1 : 0)$ with intersection multiplicity 2. This point is nonsingular as the component of degree one of $\bar{G}(1, Y - 1, Z)$ is $-Z$.

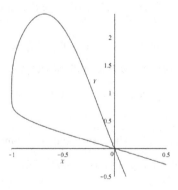

Figure 5.21: $x^4y^2 + x^4 + 2x^3y^3 + 2x^3y^2 + 4x^3y + x^2y^4 + 4x^2y^3 + 7x^2y^2 + xy^4 + 6xy^3 + y^4 = 0$

As a conclusion, $V(\bar{G})$ is birationally equivalent to C and has only nonordinary singularities (an ordinary quadruple point at P, two ordinary double points at Q and R). Its genus (which is also the genus of C) is $\binom{5}{2} - \binom{4}{2} - 2\binom{2}{2} = 2$.

For another example, see Exercise 390.

5.7   Exercises

Exercise 358. Show that a reducible projective plane curve is singular.

Exercise 359. Show (without using neither Proposition 348 nor Exercise 380) that an irreducible conic is smooth.

Exercise 360. Let $C = V(H)$ be a projective plane curve over a field $\mathbf{K}$ with $\mathrm{char}(\mathbf{K}) = 0$, where $H \in \mathbf{K}[X,Y,Z]$ is homogeneous. Show that the set of singular points of C is $V\left(\frac{\partial H}{\partial X}, \frac{\partial H}{\partial Y}, \frac{\partial H}{\partial Z}\right)$.

Exercise 361. Show that if P is a nonsingular point of an irreducible plane curve C, there is no rational function f over C having a pole at P and such that $f(1 - f)$ is regular at P.

Exercise 362. Let $C \subseteq \mathbb{A}^2(\bar{\mathbf{K}})$ ($\mathbf{K}$ a field) be an irreducible affine plane curve and $P \in C$. We denote by $\mathfrak{M}_P$ the maximal ideal of $\bar{\mathbf{K}}[C] = \bar{\mathbf{K}}[X,Y]/\mathscr{I}(C)$ formed by functions vanishing at P. Recall that writing $C = V(F)$, where F is a square-free polynomial in $\bar{\mathbf{K}}[X,Y]$, we have $\mathscr{I}(C) = \langle F \rangle$. With a translation, we can, without loss of generality, suppose that $P = (0,0)$ and, thus, $\mathfrak{M}_P = I/\langle F \rangle = \langle x, y \rangle$, where $I = \langle X, Y \rangle \subseteq \bar{\mathbf{K}}[X,Y]$.

The goal of this exercise is to prove that if C is nonsingular at P then $\dim_{\bar{\mathbf{K}}} \mathfrak{M}_P/\mathfrak{M}_P^2 = 1$, and if C is singular at P then $\dim_{\bar{\mathbf{K}}} \mathfrak{M}_P/\mathfrak{M}_P^2 = 2$.

1) Show that $\mathfrak{M}_P/\mathfrak{M}_P^2 = \langle \bar{x}, \bar{y} \rangle$, where $\bar{x}, \bar{y}$ denote the classes of x, y modulo $\mathfrak{M}_P^2$, respectively.

2) Show that if C is singular at P then $\dim_{\bar{\mathbf{K}}} \mathfrak{M}_P/\mathfrak{M}_P^2 = 2$.

3) Show that if C is nonsingular at P then $\dim_{\bar{\mathbf{K}}} \mathfrak{M}_P/\mathfrak{M}_P^2 = 1$.

Exercise 363. We consider the affine plane curve $C: y^2 = x^4 + 1$ defined over $\mathbb{C}$.

1) Show that C is irreducible.

2) Show that C is smooth. What about its projective completion?

3) Show that x is a uniformizing parameter at $P = (0,1) \in C$.

4) Compute $\mathrm{Ord}_P(y - 1)$.

Exercise 364. We consider the affine plane curve $C: y^2 + x^2y + x = 0$ defined over an algebraically closed field $\mathbf{K}$. We denote by $\bar{C}$ the projective completion of C.

1) Show that C is irreducible and give $\mathscr{I}(C)$.

2) Determine the singular points of $\bar{C}$.

3) Give a uniformizing parameter t at $Q = (0,0) \in C$. Compute $\mathrm{Ord}_Q(x)$, $\mathrm{Ord}_Q(y)$, and $\mathrm{Ord}_Q(x+y^2)$.

Exercise 365. Show that for a smooth projective plane curve C over a field $\mathbf{K}$ with genus 0, we have $\mathrm{Pic}^0(C) = 0$, that is, every degree-zero divisor is principal.

Exercise 366. Consider the curve $C : x+y+x^4+y^3 = 0$ over $\mathbf{K} = \mathbb{C}$. Compute the principal divisor of the rational function $f = \frac{y}{x^2-y} \in \mathbf{K}(C)^\times$.

Exercise 367. Let $\phi : \mathbb{P}^1(\bar{\mathbf{K}}) = V(Z) \to \mathbb{P}^1(\bar{\mathbf{K}}) = V(Z)$ be a nonconstant rational map given by $\phi(x:1) = (\frac{a(x)}{b(x)} : 1)$ with $a(x), b(x) \in \bar{\mathbf{K}}[x]$ and $\gcd(a(x),b(x)) = 1$. Show that $\deg(\phi) = \max(\deg(a(x)), \deg(b(x)))$.

Exercise 368. We suppose that the base field has characteristic zero.

1) Let $\mathbb{P}^1 = V(Z)$ and consider the rational map $\phi = (x^2(x+y)^4 : y^6) : \mathbb{P}^1 \dashrightarrow \mathbb{P}^1$. Compute the ramification points of ϕ as well as their ramification indexes and check the Riemann-Hurwitz Formula 320.

2) The same question for the morphism $\mathbb{P}^1 \dashrightarrow \mathbb{P}^1$ given on $\mathbb{A}^1$ by $\psi(x) = x^n - x$ ($n \geq 2$).

Exercise 369. Use the Riemann-Hurwitz Formula 320 to compute the genus of the Fermat curve $C = \{(x:y:z) \in \mathbb{P}^2(\mathbb{C}) \mid x^d + y^d + z^d = 0\}$ for $d \geq 2$.

Exercise 370. Using the properties of the resultant, show that

$$\mathrm{Res}_X(aX^2 + bX + c,\ cX^2 + bX + a) = (a+b+c)(a-b+c)(a-c)^2.$$

Exercise 371. Let $f,g \in \mathbf{R}[X,Y] \setminus \{0\}$ ($\mathbf{R}$ a ring) with $n = \deg_X(f)$, $m = \deg_X(g)$, and $\deg_Y(f), \deg_Y(g) \leq d$. Show that

$$\deg_Y(\mathrm{Res}_X(f,g)) \leq (m+n)d.$$

Exercise 372. (Specialization of the resultant)

Let $f,g \in \mathbf{K}[X_1,\ldots,X_n]$, where $\mathbf{K}$ is a field, and set $h = \mathrm{Res}_{X_1}(f,g)$. Let $c = (c_2,\ldots,c_n) \in \mathbf{K}^{n-1}$, denote by

$$f(X_1,c) = f(X_1,c_2,\ldots,c_n) \text{ and } g(X_1,c) = g(X_1,c_2,\ldots,c_n).$$

Denote by $f = a_0 X_1^\ell + \cdots + a_\ell$, and $g = b_0 X_1^m + \cdots + b_m$, with $a_0 \neq 0, b_0 \neq 0, a_i, b_j \in \mathbf{K}[X_2,\ldots,X_n]$.

We suppose that $a_0(c) \neq 0$ and $b_0(c) = 0$.

1) We begin by supposing that the degree of g at X_1 decreases by 1 when replacing $(X_2,\dots,X_n)$ with $(c_2,\dots,c_n)$, i.e., $b_1(c) \neq 0$. Show that

$$h(c) = a_0(c)\operatorname{Res}_{X_1}(f(X_1,c),g(X_1,c)).$$

2) Now, we deal with the general case. Suppose that the degree of g at X_1 is $m - p$ with $p \geq 1$. Show that

$$h(c) = a_0(c)^p \operatorname{Res}_{X_1}(f(X_1,c),g(X_1,c)).$$

3) Check this formula with $f = X_1^2 X_2 + 3X_1 - 1$, $g = 6X_1^2 + X_2^2 - 4$, $h = \operatorname{Res}_{X_1}(f,g)$, $c = 0$.

Exercise 373. 1) Let m, n be integers ≥ 1 and $\alpha, \beta \in \mathbf{K}$ ($\mathbf{K}$ a field). We set $\delta = m \wedge n$. Show that

$$\operatorname{Res}(X^m - \alpha, X^n - \beta) = (-1)^m (\beta^{\frac{m}{\delta}} - \alpha^{\frac{n}{\delta}})^\delta.$$

2) Deduce that if p is a prime number $\neq 2$ then for all $a \in \mathbb{F}_p$, we have:

$$\operatorname{Res}(X^2 - a, X^{p-1} - 1) = (a^{\frac{p-1}{2}} - 1)^2.$$

3) Deduce that that if p is a prime number $\neq 2$, we have (see Definition 410):

$$a^{\frac{p-1}{2}} = \begin{cases} 0 & \text{if } a = 0, \\ 1 & \text{if } a \neq 0 \text{ and } a \text{ is a square in } \mathbb{F}_p, \\ -1 & \text{if } a \neq 0 \text{ and } a \text{ is not a square in } \mathbb{F}_p. \end{cases}$$

Exercise 374. (A "fast" algorithm for computing the resultant)

Let $f = a_0 X^\ell + \cdots + a_\ell$, $g = b_0 X^m + \cdots + b_m \in \mathbf{K}[X]$ ($\mathbf{K}$ a field) and suppose that $a_0 \neq 0$, $b_0 \neq 0$, and $\ell \geq m$.

a) Set $\tilde{f} = f - \frac{a_0}{b_0} X^{\ell-m} g$ (note that $\deg \tilde{f} \leq \ell - 1$). We suppose that $\deg \tilde{f} = \ell - 1$. Show that

$$\operatorname{Res}(f,g) = (-1)^m b_0 \operatorname{Res}(\tilde{f},g).$$

b) Now, we don't exclude the possibility that the degree of $\tilde{f}$ is $< \ell - 1$. Show that

$$\operatorname{Res}(f,g) = (-1)^{m(\ell-\deg \tilde{f})} b_0^{\ell-\deg \tilde{f}} \operatorname{Res}(\tilde{f},g).$$

c) We perform an Euclidean division of f by g in $\mathbf{K}[X]$: $f = qg + r$ with $\deg r < \deg g$. Show that

$$\operatorname{Res}(f,g) = (-1)^{m(\ell-\deg r)} b_0^{\ell-\deg r} \operatorname{Res}(r,g).$$

d) Deduce an algorithm for computing $\operatorname{Res}(f,g)$. Do the computations by hand for $f = X^4 + 1$ and $g = X^2 - X + 2$ in $\mathbb{Q}[X]$.

Exercise 375. (Suslin's lemma, particular case)

1) Let $\mathbf{R}$ be a ring and $f, g \in \mathbf{R}[X] \setminus \{0\}$ with f monic. Show that

$$1 \in \langle f, g \rangle \text{ in } \mathbf{R}[X] \quad \Longleftrightarrow \quad \mathrm{Res}(f, g) \in \mathbf{R}^{\times}.$$

2) Let $\mathbf{A}$ be a ring containing an infinite field $\mathbf{K}$ and let us fix a sequence $(y_i)_{i \in \mathbb{N}}$ of pairwise distinct elements in $\mathbf{K}$. Let $v_1, \ldots, v_n \in \mathbf{A}[X]$ such that v_1 is monic of degree d and $n \geq 2$. Show that

$$1 \in \langle v_1, \ldots, v_n \rangle \Leftrightarrow 1 \in \langle \mathrm{Res}_X(v_1, v_2 + y_i v_3 + \cdots + y_i^{n-2} v_n), 0 \leq i \leq (n-2)d \rangle.$$

3) Let $f_1, \ldots, f_n \in \mathbb{Q}[X]$ ($n \geq 2$) and suppose that $\deg f_1 = \min\{\deg f_i\} = d$. Show that

$$1 \in \langle f_1, \ldots, f_n \rangle \Leftrightarrow \exists \ 0 \leq i \leq (n-2)(d+1) \ | \\ \mathrm{Res}_X(f_1, f_2 + if_3 + \cdots + i^{n-2} f_n) \neq 0.$$

Take $f_1 = X^5 - X^4 + 3X^2 - 3$, $f_2 = 2X^3 + 3X^2 - X - 4$, $f_3 = 3X^5 + 2X^4 - X^3 - X^2 - 3$, and check that $1 \notin \langle f_1, f_2, f_3 \rangle$.

Exercise 376. (Dedekind-Mertens, by Th. Coquand)

If U, V are two sub-$\mathbb{Z}$-modules of a ring $\mathbf{A}$, we denote by UV the submodules generated by the uv's, with $u \in U$ and $v \in V$. For $f \in \mathbf{A}[T]$, we denote by $[f]$ the sub-$\mathbb{Z}$-module of $\mathbf{A}$ generated by the coefficients of f.

Let $f = \sum_{i \geq 0} a_i T^i$, $g = \sum_{j=0}^{m} b_j T^j$, $h = fg \in \mathbf{A}[T]$ and denote by $F = [f]$, $G = [g]$ and $H = [h]$.

Denote also $\tilde{g} = \sum_{j=0}^{m-1} b_j T^j = g - b_m T_m$, and $\tilde{G} = [\tilde{g}]$. By induction on m (using $\tilde{G}$), show that

$$F^{m+1} G = F^m H.$$

Exercise 377. (Suslin's Lemma, particular case, by C. Quitté)

Let $\mathbf{A}$ be a ring, $d \geq 1$, $v = (X - x_1) \cdots (X - x_d) \in \mathbf{A}[X]$, $u, w \in \mathbf{A}[X]$, and take $d + 1$ elements $u_0, \ldots, y_d$ in $\mathbf{A}$.

For $0 \leq i \leq d$, set

$$r_i = \mathrm{Res}_X(v, u + y_i w) = \prod_{j=1}^{d} (u_j + y_i w_j),$$

with $u_j = u(x_j)$ and $w_j = w(x_j)$. Moreover, set

$$\pi = \prod_{i < j} (y_i - y_j)$$

and

$$c_0 + c_1 Y + \cdots + c_d Y^d = (u_1 + w_1 Y) \cdots (u_d + w_d Y).$$

1) Show that $\langle u_1, w_1 \rangle \langle u_2, w_2 \rangle^2 \cdots \langle u_d, w_d \rangle^d \subseteq \langle c_0, \ldots, c_d \rangle$.

2) Show that $\pi\langle c_0, \ldots, c_d \rangle \subseteq \langle r_0, \ldots, r_d \rangle$.

3) Deduce that $\pi\langle u_1, w_1 \rangle \langle u_2, w_2 \rangle^2 \cdots \langle u_d, w_d \rangle^d \subseteq \langle r_0, \ldots, r_d \rangle$.

4) Deduce that if $1 \in \langle u, v, w \rangle$ then $1 \in \langle r_0, \ldots, r_d \rangle$.

Exercise 378. We consider the complex plane curve $C: x^3 + y^3 - 3xy = 0$.

1) Show that C is irreducible.

2) Show that C has a unique singular point.

3) What is the nature of this singular point?

Exercise 379. In this exercise, by a projective plane curve with multiple components, we mean a formal linear combination $C = n_1 C_1 + \cdots + n_r C_r$, where the C_i's are pairwise distinct irreducible projective plane curves and $n_i \in \mathbb{N}$. The number n_i gives the multiplicity with which we count the component C_i. If $F_i = 0$ is the equation of C_i, then the equation of C is naturally $F_1^{n_1} \ldots F_r^{n_r} = 0$, and the order of C is the degree of this homogeneous polynomial.

1) Show that the set of all curves of order m in $\mathbb{P}^2(\bar{\mathbf{K}})$ forms a projective space over $\bar{\mathbf{K}}$ of dimension $\frac{m(m+3)}{2}$ (**K** a field).

2) Show that the condition for a curve of order m in $\mathbb{P}^2(\bar{\mathbf{K}})$ to have multiplicity at least s at a particular point P is equivalent to $\frac{s(s+1)}{2}$ linearly independent conditions.

3) Deduce that the curves of order m in $\mathbb{P}^2(\bar{\mathbf{K}})$ which have multiplicity at least s_j at $P_j \in \mathbb{P}^2(\bar{\mathbf{K}})$ form a $\bar{\mathbf{K}}$-linear system of dimension $\geq \frac{m(m+3)}{2} - \sum_j \frac{s_j(s_j+1)}{2}$.

Exercise 380. Let **K** be a field, $d \geq 1$, and denote by $\mathbf{K}[X,Y,Z]_d$ the subspace of the **K**-vector space $\mathbf{K}[X,Y,Z]$ formed by 0 and homogeneous polynomials of degree d at X, Y, Z.

1) Determine $\dim_{\mathbf{K}} \mathbf{K}[X,Y,Z]_d$ and deduce that by $\binom{d+2}{2} - 1$ pairwise distinct points in $\mathbb{P}^2(\mathbf{K})$ passes at least one curve of degree d in $\mathbb{P}^2(\mathbf{K})$.

2) Let $C \subseteq \mathbb{P}^2(\mathbf{K})$ be an irreducible curve of degree d. Our goal is to prove that C has at most $\binom{d-1}{2} = \frac{(d-1)(d-2)}{2}$ singular points (retrieving a consequence of Proposition 348). By way of contradiction, suppose that C has $N = \binom{d-1}{2} + 1$ singular points $P_1, \ldots, P_N$. Choose $3d - 3$ points $Q_1, \ldots, Q_{3d-3}$ on $C \setminus \{P_1, \ldots, P_N\}$ and a point $Q \notin C$. Show that there exists a curve $C' \subseteq \mathbb{P}^2(\mathbf{K})$ of degree d passing through $P_1, \ldots, P_N, Q_1, \ldots, Q_{3d-3}, Q$. Show that $I_P(C, C') \geq 2$ for $P \in \{P_1, \ldots, P_N\}$. Deduce a contradiction.

Exercise 381. Check Strong Bézout's Theorem 340 on the intersection points of the curves $C: x^2 + y^2 = 1$, and $D: y^2 = x^3 - x^2 - x - 1$ over the complex numbers (see Fig. 5.22).

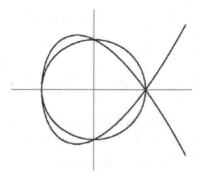

Figure 5.22: Intersection of a conic with a cubic

Exercise 382. Show (without using neither Proposition 348 nor Exercise 380) that an irreducible projective plane curve of degree 3 has at most one singular point.

Exercise 383. Let C be a projective plane curve of degree 4 with 4 singular points of order 2. Show that C is reducible.

Exercise 384. Let C be the affine curve $V(Y^2 - X^3 + X) \subseteq \mathbb{A}^2(\mathbb{C})$, $P = (a,b)$ a point of C, and $\Delta : x = a$ the vertical line passing through P. Show that if Δ is not tangent to C at P, then Δ intersects C at another point P' and Δ is not tangent to C at P'.

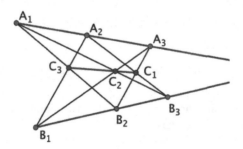

Figure 5.23: Pappus's theorem

Exercise 385. (Pappus's theorem)

Let D_1, D_2 be two distinct lines in $\mathbb{P}^2(\mathbb{C})$. Let A_1, A_2, A_3 be points on D_1 and B_1, B_2, B_3 points on D_2 such that $\sharp\{A_1,A_2,A_3,B_1,B_2,B_3\} = 6$. Denote by L_{ij} the unique line in $\mathbb{P}^2(\mathbb{C})$ passing through A_i and B_j. Let $\{C_k\} = L_{ij} \cap L_{ji}$ for each triplet (i,j,k) such that $\{i,j,k\} = \{1,2,3\}$ (see Fig. 5.23). Show that the points C_1, C_2, C_3 are collinear.

Exercise 386.

1) Show that a line L is tangent to a projective plane curve C at a smooth point P if and only if $I_P(C,L) \geq 2$.

2) Let E be a smooth projective plane curve of degree 3 and $P \in E$. Could the tangent line L at P intersect the curve in three distinct points?

Exercise 387. (Points of inflexion and Hessian curve)

Let C be a plane curve over $\mathbb{C}$. A nonsingular point $P \in C$ with tangent line L is called an *r-tuple point of inflexion* if L is not a component of C and $I_P(C,L) = r+2 \geq 3$ (see Fig. 5.24).

1) Show that a nonsingular point P on a cubic C with tangent line L is a point of inflexion if and only if $I_P(C,L) = 3$ if and only if L does not meet C at another point.

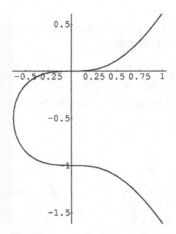

Figure 5.24: The elliptic curve $y^2 + y = x^3$ has two points of inflexion at $(0,0)$ and $(0,-1)$ in the real affine plane

2) Consider a projective plane curve $C\colon F(x_0,x_1,x_2) = 0$, where $F \in \mathbb{C}[X_0,X_1,X_2]$ is a homogeneous polynomial of degree $d \geq 3$. The *Hessian* of C is defined as

$$H_f := \det\left(\frac{\partial^2 F}{\partial X_{i-1}\partial X_{j-1}}\right)_{1\leq i,j\leq 3}.$$

2.a) Show that if $H_f \neq 0$ (Hesse proved that this amounts to saying that C dos not decompose into lines) then H_f is a homogeneous polynomial of degree $3(d-2)$.

The projective plane curve $\mathscr{H} := V(H_f)$ is called the *Hessian curve* of C. In particular, the Hessian curve of a cubic is again a cubic.

2.b) Show that $\mathscr{H}$ goes through each singular point P of C.

2.c) Show that a nonsingular point P on C is a point of inflexion if and only if $P \in C \cap \mathscr{H}$, and that if C is smooth then the number N of inflexion points $\in [\![1, 3d(d-2)]\!]$.

More generally, we can prove that a nonsingular point P on C is an r-tuple point of inflexion if and only if $\mathscr{H}$ meets C at P with multiplicity r, and a smooth curve of degree d has exactly $3d(d-2)$ inflexion points counting multiplicities.

2.d) Find the points of inflexion of the cubic $y^2 + y = x^3$ defined over $\mathbb{C}$.

2.e) (Plücker) Show that a smooth cubic has exactly 9 distinct points of inflexion and these are all simple.

Exercise 388. Let $C = V(F)$, $D = V(G)$ be two projective plane curve, where $F, G \in \mathbf{K}[X, Y]$. We say that they *intersect transversally* at a point $P \in C \cap D$ if C and D are both smooth at P and the tangent lines at P are distinct (see Fig. 5.25).

Show that C and D intersect transversally at P if and only if $\langle F, G \rangle = \mathfrak{m}_P$ in $\mathscr{O}_P$ if and only if $I_P(C, D) = 1$, where $\mathfrak{m}_P$ denotes the maximal ideal of the local ring $\mathscr{O}_P := \mathbf{K}[X, Y]_{\langle X-a, Y-b \rangle}$.

Hint: use a consequence of *Nakayama's lemma* saying that for a local ring **R** with maximal ideal $\mathfrak{m}$ and $x_1, \ldots, x_r \in \mathfrak{m}$, $\mathfrak{m} = \langle x_1, \ldots, x_r \rangle$ if and only if $\bar{x}_1, \ldots, \bar{x}_r$ generate $\mathfrak{m}/\mathfrak{m}^2$ as an **R**-module.

Figure 5.25: Curves intersecting transversally

Exercise 389. Let $P = (a, b)$ be a nonsingular point on a curve $C = V(F)$, $F \in \mathbf{K}[X, Y]$. Let $D = V(G)$ and $E = V(H)$ ($G, H \in \mathbf{K}[X, Y]$) that do not have a common component with C through P.

1) Show that

$$\langle F, G \rangle \subseteq \langle F, H \rangle \text{ in } \mathscr{O}_P \implies I_P(C, D) \geq I_P(C, E),$$

where $\mathscr{O}_P := \mathbf{K}[X, Y]_{\langle X-a, Y-b \rangle}$ is the local ring at P with maximal ideal $\mathfrak{m}_P$.

2) Let $L = V(\ell)$ be a line through P different from the tangent to C at P. Show that

$$\langle F, G \rangle = \langle F, \ell^{I_P(C,D)} \rangle.$$

3) Show that

$$\langle F, G \rangle \subseteq \langle F, H \rangle \text{ in } \mathscr{O}_P \quad \Leftrightarrow \quad I_P(C,D) \geq I_P(C,E).$$

4) Does 3) remain true if we relax the smoothness hypothesis by only assuming the curve C to be irreducible?

Exercise 390. Consider the irreducible projective plane curve $C = V(H)$, where $H(X,Y,Z) = X^2Z^2 - Y^3Z + Y^4 \in \mathbb{R}[X,Y,Z]$. Find a projective plane curve birationally equivalent to C without nonordinary singular points. Deduce the genus of C.

5.8   Solutions to the exercises

Exercise 358:

Let $C: F(x_0,x_1,x_2) = 0$ with $F(x_0,x_1,x_2) = p(x_0,x_1,x_2)q(x_0,x_1,x_2)$, $\deg(p) \geq 1$, $\deg(q) \geq 1$, and $\gcd(p,q) = 1$. We have $\frac{\partial F}{\partial x_i} = (\frac{\partial p}{\partial x_i})q + (\frac{\partial q}{\partial x_i})p$, and, thus, all the points in $V(p) \cap V(q)$ are singular (we have $\deg(p)\deg(q)$ points counted with their multiplicities by Strong Bézout's Theorem 340).

Exercise 359:

Let P be a point on an irreducible conic C. By considering a local affine chart, we can suppose that $C = V(F)$ is in the affine plane. Via a change of variables, we can also suppose that $P = (0,0)$, and $F(0,0) = 0$. If P were a nonsingular point of C, then F would not have a homogeneous component of degree 1, and, thus, F would be a homogeneous polynomial of degree 2. By Lemma 335, F would be the product of two linear factor, in contradiction with the fact that C is irreducible.

Exercise 360:

Let P be a point on C. By a coordinates change, we can suppose that P is in the affine plane $(z \neq 0)$. Setting $P = (a : b : 1)$, then P is a singular point of C if and only if (a,b) is a singular point of $D := V(F(X,Y))$, where $F(X,Y) = H(X,Y,1)$. The desired result ensues from the relations:

$$\begin{cases} \frac{\partial F}{\partial X}(a,b) = \frac{\partial H}{\partial X}(P) \\ \frac{\partial F}{\partial Y}(a,b) = \frac{\partial H}{\partial Y}(P) \\ X\frac{\partial H}{\partial X} + Y\frac{\partial H}{\partial Y} + Z\frac{\partial H}{\partial Z}(P) = \deg(H)H, \end{cases}$$

(the third of which, called *Euler's Formula*, being true since H is homogeneous).

Exercise 361:

As f is not regular at P, we have $\mathrm{Ord}_P(f) < 0$, and, thus, we have also $\mathrm{Ord}_P(1 - f) < 0$. It follows that

$$\mathrm{Ord}_P(f(1-f)) = \mathrm{Ord}_P(f) + \mathrm{Ord}_P(1-f) < 0,$$

that is, $f(1 - f)$ is not regular at P.

Exercise 362:

1) This follows from the fact that $I/I^2 = \langle \bar{X}, \bar{Y} \rangle$, where $\bar{X}, \bar{Y}$ denote the classes of X, Y modulo I^2, respectively.

2) Let $a, b \in \bar{\mathbf{K}}$ such that $a\bar{x} + b\bar{y} = 0$, or, equivalently, such that $aX + bY \in \langle F \rangle$. But, as C is singular at $(0,0)$, $F \in I^2 = \langle X^2, XY, Y^2 \rangle$, and this would imply that $a = b = 0$. It follow that $(\bar{x}, \bar{y})$ is a basis for $\mathfrak{M}_P/\mathfrak{M}_P^2$ as $\bar{\mathbf{K}}$-vector space.

3) Since C is nonsingular at $(0,0)$ then the term $\frac{\partial F}{\partial X}(0,0) \cdot X + \frac{\partial f}{\partial Y}(0,0) \cdot Y$ of degree one of the Taylor expansion of F at $(0,0)$ is nonzero. It follows that we have a nontrivial linear relation $\frac{\partial F}{\partial X}(0,0)\bar{x} + \frac{\partial F}{\partial Y}(0,0)\bar{y} = 0$ and, thus, $\dim_{\bar{\mathbf{K}}} \mathfrak{M}_P/\mathfrak{M}_P^2 \leq 1$. On the other hand, $\bar{x} = \bar{y} = 0$ would imply that $X, Y \in \langle F \rangle + I^2$, and, hence, the 2-dimensional $\bar{\mathbf{K}}$-vector space I/I^2 (with basis $(\bar{X}, \bar{Y})$) would be generated by $\bar{F}$ only, a contradiction. We conclude that $\dim_{\bar{\mathbf{K}}} \mathfrak{M}_P/\mathfrak{M}_P^2 = 1$.

Exercise 363:

1) C is irreducible since $F = Y^2 - X^4 - 1$ is irreducible in $\mathbb{C}[X, Y]$. To see this, by way of contradiction, supposing that $Y^2 - X^4 - 1 = (Y + a(X))(Y + b(X))$ with $a(X), b(X) \in \mathbb{C}[X]$, we would have $a(X) + b(X) = 0$ and $X^4 + 1 = -a(X)b(X) = a(X)^2$, in contradiction with the fact that $G = X^4 + 1$ is square-free ($G \wedge G' = 1$).

2) The only point where $\frac{\partial F}{\partial x}(x, y)$ and $\frac{\partial F}{\partial y}(x, y)$ vanish simultaneously is $(0,0) \notin C$. So, C is smooth. The projective completion of C is $\bar{C} = V(H)$, where $H = Y^2Z^2 - X^4 - Z^4$. Putting $Z = 0$, we see that $Q = (0 : 1 : 0)$ is the only point of C at ∞. In the affine chart $\{y \neq 0\}$, the equation of $\bar{C}$ is $z^2 - x^4 - z^4 = 0$, and we see that the point $(0,0)$ is singular. We conclude that $\bar{C}$ is singular with a unique singular point $Q = (0 : 1 : 0)$.

3) By Proposition 283 (keeping its notation), we know that the ideal $\mathfrak{M}_{C,P}$ is principal generated by any $t \in \mathfrak{M}_P \setminus \mathfrak{M}_P^2$. We know also that as $\mathbb{C}$-vector space, $\mathfrak{M}_P/\mathfrak{M}_P^2$ is generated by $\bar{x}$ and $\bar{y} - \bar{1}$, where the classes are modulo $\mathfrak{M}_P^2$. But, as $(y - 1)(y + 1) = x^4 \in \mathfrak{M}_P^2$ and $y + 1 \notin \mathfrak{M}_P$, we conclude that $y - 1 \in \mathfrak{M}_{C,P}^2$. As, a generator for $\mathfrak{M}_{C,P}$ is either x or $y - 1$ and this latter is zero modulo $\mathfrak{M}_{C,P}^2$, we infer that $\mathfrak{M}_{C,P}$ is generated by x, that is, x is a uniformizing parameter at $P = (0,1) \in C$.

Or, alternatively, since $\frac{\partial F}{\partial Y}(0,1) = 2 \neq 0$, we deduce that x is a uniformizing parameter at $P = (0,1)$.

4) $\mathrm{Ord}_P(y-1) = \mathrm{Ord}_P(\frac{x^4}{y+1}) = 4$ since $\mathrm{Ord}_P(x) = 1$ and $\mathrm{Ord}_P(y+1) = 0$.

Exercise 364:

1) C is irreducible since $F = Y^2 + X^2Y + X$ is irreducible in $\mathbf{K}[X,Y]$. To see this, by way of contradiction, supposing that $Y^2 + X^2Y + X = (Y + a(X))(Y + b(X))$ with $a(X), b(X) \in \mathbf{K}[X]$, we would have $a(X) + b(X) = X^2$ and $a(X)b(X) = X$, a contradiction since this would imply that $\lambda^{-1} + \lambda X = X^2$ for some $\lambda \in \mathbf{K}^\times$. One can also use Eisenstein's criterion below with irreducible element $X \in \mathbf{K}[X]$.

Eisenstein's criterion. *Let* $\mathbf{D}$ *be a unique factorization domain with field of fractions* $\mathbf{F}$, *and* $f = a_0 + a_1Y + \cdots + a_nY^n \in \mathbf{D}[Y]$. *Suppose that there exists an irreducible element* $a \in \mathbf{D}$ *such that*

- *a divides each a_i for $i \neq n$,*
- *a does not divide each a_n, and*
- *a^2 does not divide each a_0,*

then f is irreducible in $\mathbf{F}[Y]$. *If, in addition, f is primitive (i.e., it has no nontrivial constant divisors), then it is irreducible in* $\mathbf{D}[Y]$.

As F is irreducible, we have $\mathscr{I}(C) = \langle F \rangle$.

2) $\bar{C} = V(H)$, with $H = X^2Y + Y^2Z + Z^2X$. The points at ∞ of C (i.e., the points of $\bar{C} \cap V(Z)$) are $(1:0:0)$ and $(0:1:0)$.

Let $P = (x:y:z)$ be a singular of $\bar{C}$. We have $H(x,y,z) = \frac{\partial H}{\partial X}(x,y,z) = \frac{\partial H}{\partial Y}(x,y,z) = \frac{\partial H}{\partial Z}(x,y,z) = 0$, that is, $x^2y + y^2z + z^2x = 2xy + z^2 = x^2 + 2yz = y^2 + 2zx = 0$. If $xyz = 0$, then we would have $x = y = z = 0$, which is impossible. So, $xyz \neq 0$. We have $x^2y = (-2yz)y = -2y^2z$, and, by circular symmetry, $y^2z = -2z^2x$ and $z^2x = -2x^2y$. It follows that $x^2y = -8x^2y$, that is, $9x^2y = 0$. If $\mathrm{char}(\mathbf{K}) \neq 3$, we get a contradiction. If $\mathrm{char}(\mathbf{K}) = 3$, we obtain $x^2 = yz$, $y^2 = xz$, $z^2 = yz$, and, thus, $x^3 = y^3 = z^3 = xyz$. As the map $\mathbf{K} \to \mathbf{K}$; $u \mapsto u^3$, is one-to-one, we infer that $x = y = z = 1$, and, thus, $P = (1:1:1) \in \bar{C}$. We conclude that $\bar{C}$ is smooth if $\mathrm{char}(\mathbf{K}) \neq 3$, and has a unique singular point $(1:1:1)$ if $\mathrm{char}(\mathbf{K}) = 3$.

3) By Proposition 283 (keeping its notation), we know that the ideal $\mathfrak{M}_{C,Q}$ is principal generated by any $t \in \mathfrak{M}_Q \setminus \mathfrak{M}_Q^2$. We know also that as $\mathbf{K}$-vector space, $\mathfrak{M}_Q/\mathfrak{M}_Q^2$ is generated by $\bar{x}$ and $\bar{y}$, where the classes are modulo $\mathfrak{M}_Q^2$. But, as $x(xy + 1) = -y^2 \in \mathfrak{M}_Q^2$ and $xy + 1 \notin \mathfrak{M}_Q$, we conclude that $x \in \mathfrak{M}_{C,Q}^2$. As, a generator for $\mathfrak{M}_{C,Q}$ is either y or x and this latter is zero modulo $\mathfrak{M}_{C,Q}^2$, we infer that $\mathfrak{M}_{C,Q}$ is generated by y, that is, $t = y$ is a uniformizing parameter at $Q = (0,0) \in C$.

Or, alternatively, since $\frac{\partial F}{\partial X}(0,0) = 1 \neq 0$, we deduce that y is a uniformizing parameter at Q.

$\mathrm{Ord}_Q(x) = 2$ since $x = -\frac{y^2}{xy+1}$; $\mathrm{Ord}_Q(y) = 1$; $\mathrm{Ord}_Q(x+y^2) = 2 \times 2 + 1 = 5$ since $x + y^2 = -x^2y$.

Exercise 365:

Let $D = \sum_{P \in C} n_P.[P] \in \mathbb{Z}[C]$ with $\deg(D) = \sum_{P \in C} n_P = 0$. By Corollary 302, we know that $\ell(D) = 1$ and, thus, there exists $f \in \bar{\mathbf{K}}(C)^\times \cap \mathscr{L}(D)$. By definition, $\mathrm{Ord}_P(f) \geq -n_P$ for all $P \in C$. Because $\deg(D) = \deg(\mathrm{div}(f)) = 0$, we should have $\mathrm{Ord}_P(f) = -n_P$ for all $P \in C$, that is $D = \mathrm{div}(\frac{1}{f})$.

Exercise 366:

Denote by $F = X + Y + X^4 + Y^3$, so that $C = V(F)$. We have $\mathrm{div}(f) = \mathrm{div}(y) - \mathrm{div}(x^2 - y)$. Consider the projective completion $\tilde{C} = V(H)$ of C with $H = XZ^3 + YZ^3 + X^4 + Y^3Z$. It has a single point $(0:1:0)$ at ∞.

Let us compute $\mathrm{div}(y)$. The function y has 4 simple zeroes corresponding to the zeroes of $x + x^4$, namely, the points $(0,0)$, $(-1,0)$, $(e^{\frac{i\pi}{3}}, 0)$ and $(e^{\frac{5i\pi}{3}}, 0)$. All the orders of y at these points are 1 since $x + x^4$ has only simple roots. In more details, at the point $(\alpha, 0)$ with $\alpha \in \{0, -1, e^{\frac{i\pi}{3}}, e^{\frac{5i\pi}{3}}\}$, $x - \alpha$ is a uniformizing parameter since $\frac{\partial F}{\partial Y}(\alpha, 0) = 1 \neq 0$. As $y = \frac{-x(1+x^3)}{1+y^2} = \frac{-x(x+1)(x-e^{\frac{i\pi}{3}})(x-e^{\frac{5i\pi}{3}})}{1+y^2}$, we see that $\mathrm{Ord}_{(\alpha,0)}(y) = \mathrm{Ord}_{(\alpha,0)}(x - \alpha) = 1$. Since $\deg(\mathrm{div}(y)) = 0$, we deduce that

$$\mathrm{div}(y) = [(0:0:1)] + [(-1:0:1)] + [(e^{\frac{i\pi}{3}}:0:1)] + [(e^{\frac{5i\pi}{3}}:0:1)] - 4[(0:1:0)].$$

Let us compute $\mathrm{div}(x^2 - y)$. The function $x^2 - y$ has 6 simple zeroes corresponding to the 6 roots of $x^6 + x^4 + x^2 + x$, namely, the points $(0,0)$, $(\beta_1, \beta_1^2), \ldots,$ (β_5, β_5^2), where $\beta_1, \ldots, \beta_5$ are the roots of $x^5 + x^3 + x + 1$ (this quintic is not solvable by radicals since it has Galois group S_5). All the orders of $x^2 - y$ at these points are 1 since $x^5 + x^3 + x + 1$ has only simple roots (for example, because its discriminant is nonzero). In more details, at the point (β, β^2) with $(\beta \in \{0, \beta_1, \ldots, \beta_5\})$, $x - \beta$ is a uniformizing parameter since $\frac{\partial F}{\partial Y}(\beta, \beta^2) = 1 + 3\beta^4 \neq 0$. As $x^2 - y = \frac{x^6 + x^4 + x^2 + x}{x^4 + x^2 y^2 + y^2 + 1}$, we see that $\mathrm{Ord}_{(\beta, \beta^2)}(x^2 - y) = \mathrm{Ord}_{(\beta, \beta^2)}(x - \beta) = 1$. Since $\deg(\mathrm{div}(x^2 - y)) = 0$, we deduce that

$$\mathrm{div}(x^2 - y) = [(0:0:1)] + [(\beta_1:\beta_1^2:1)] + \cdots + [(\beta_5:\beta_5^2:1)] - 6[(0:1:0)].$$

We conclude that

$$\mathrm{div}(f) = [(-1:0:1)] + [(e^{\frac{i\pi}{3}}:0:1)] + [(e^{\frac{5i\pi}{3}}:0:1)]$$
$$- [(\beta_1:\beta_1^2:1)] - \cdots - [(\beta_5:\beta_5^2:1)] + 2[(0:1:0)].$$

One more question: how to retrieve that $\mathrm{Ord}_\infty(x^2 - y) = -6$ directly without using the fact that $\deg(\mathrm{div}(x^2 - y)) = 0$, with $\infty = (0:1:0)$?

Answer: We have $\mathrm{Ord}_\infty(x^2 - y) := \mathrm{Ord}_\infty(\frac{x^2 - yz}{z^2}) = I_\infty(\tilde{C}, V(X^2 - YZ)) - I_\infty(\tilde{C}, V(Z^2))$. But $I_\infty(\tilde{C}, V(X^2 - YZ)) = 2 \times 4 - 6 = 2$ by Strong Bézout's Theorem 340, while $I_\infty(\tilde{C}, V(Z^2))$ is the power of Z in $\text{Res}_X(XZ^3 + YZ^3 + X^4 + Y^3Z, Z^2) = (Z^2)^4 = Z^8$, that is, 8. It follows that $\text{Ord}_\infty(x^2 - y) = 2 - 8 = -6$.

Exercise 367:

Denoting $\theta = \frac{a(x)}{b(x)}$, we have $\phi^*(\bar{K}(\mathbb{P}^1)) = \bar{K}(\theta) \subseteq \bar{K}(x)$. Since $\bar{K}(\theta) = \bar{K}(\frac{1}{\theta})$, we may assume that $\deg(a(x)) \geq \deg(b(x))$. Moreover, replacing $b(x)$ by its remainder on division by $a(x)$, we can assume that $\deg(a(x)) > \deg(b(x))$. We claim that the minimal polynomial of x over $\bar{K}(\theta)$ is, up to a nonzero scalar, $F(T) = a(T) - \theta b(T)$ (whose degree at T is $\deg(a(x))$ as desired).

To prove the claim, first, we have $F(x) = 0$. Second, $a(T) - \theta b(T)$ is irreducible in $\bar{K}[\theta, T]$ since it is linear in θ. We conclude that $a(T) - \theta b(T)$ is irreducible in $\bar{K}(\theta)[T]$ using the following version of Gauss's lemma:

Let $\mathbf{R}$ be a gcd domain and $\mathbf{F}$ its field of fractions. A nonconstant polynomial in $\mathbf{R}[X]$ is irreducible in $\mathbf{R}[X]$ if and only if it is both irreducible in $\mathbf{F}[X]$ and primitive in $\mathbf{R}[X]$ (that is, the gcd of its coefficients is 1).

Exercise 368:

1) The fibre of ϕ at the point $(1 : 0)$ (the point at ∞) is the point $(1 : 0)$. So, the point at ∞ is a ramification point of index 6. We can also see this by restricting ϕ to the affine chart $(x = 1)$. Restricting ϕ to the affine chart $(y = 1)$ gives a map $\phi = x^2(x+1)^4 : \mathbb{A}^1 \to \mathbb{A}^1$. We see that ϕ is ramified at $(0 : 1)$ with index 2, at $(-1 : 1)$ with index 4, and at $(\frac{1}{3} : 1)$ with index 2 (this latter point can be found by setting the derivative of $x^2(x+1)^4$ equal to 0 and solving for x). Note that we have

$$\sum_{\substack{P \in \mathbb{P}^1 \\ \phi(P)=(0:1)}} e_\phi(P) = e_\phi(0 : 1) + e_\phi(-1 : 1) = 2 + 4 = 6 = \deg(\phi).$$

Riemann-Hurwitz Formula 320 says: $2 \times 0 - 2 = 6 \times (2 \times 0 - 2) + (5 + 3 + 1 + 1)$.

2) The map $\psi(x)$ has a ramification point at ∞ of index n plus $n - 1$ ramification points (corresponding to the zeroes of $nx^{n-1} - 1$) of index 2 each. Riemann-Hurwitz Formula 320 says: $2 \times 0 - 2 = n \times (2 \times 0 - 2) + (n - 1 + n - 1)$.

Exercise 369:

The Fermat Curve $C = V(F)$, with $F = X^d + Y^d + Z^d$, is a smooth projective plane curve. Let us denote its genus by g. Define $\pi : C \dashrightarrow \mathbb{P}^1(\mathbb{C})$ by $\pi(x : y : z) = (x : z)$. We know (see Example 312) that π is ramified at $P \in C$ if and only if $\frac{\partial F}{\partial Y}(P) = 0$. So, we have d ramification points $(\xi_i : 0 : 1)$ with $\xi_i^d = -1$. Now, as $\pi(x : y : z) = (1 : 0) \Leftrightarrow x = 1, y^d = -1, z = 1$, we infer that π has degree d. Moreover, since $\pi^{-1}(\{(\xi_i : 1)\}) = \{(\xi_i : 0 : 1)\}$, we deduce that the ramification index at $(\xi_i : 0 : 1)$ is d.

From the Riemann-Hurwitz Formula 320, we have: $2g - 2 = d \cdot (2 \times 0 - 2) +$ $d(d-1)$, and, thus, $g = \frac{(d-1)(d-2)}{2}$ as expected.

Exercise 370:

By Corollary 330 and Proposition 329:

$$\begin{aligned}
\mathrm{Res}_X(aX^2 + bX + c, \ cX^2 + bX + a) &= \mathrm{Res}_X(aX^2 + bX + c, \ (c-a)X^2 + a - c) \\
&= \mathrm{Res}_X(aX^2 + bX + c, \ (c-a)(X^2 - 1)) \\
&= \mathrm{Res}_X(aX^2 + bX + c, \ c - a)\mathrm{Res}_X(aX^2 + bX + c, \ X^2 - 1) \\
&= (c-a)^2(a+b+c)(a-b+c).
\end{aligned}$$

Exercise 371:

When we write $\mathrm{Res}_X(f,g)$ as a sum of $(m+n)!$ terms, each nonzero term has m factors which are coefficients of f and n factors which are coefficients of g. Thus, the degree of each term is at most $md + nd$.

Exercise 372:

1)

$$h(c) = \mathrm{Res}(f,g) = \begin{vmatrix}
a_0(c) & & & & 0 & & & \\
a_1(c) & a_0(c) & & & b_1(c) & 0 & & \\
& a_1(c) & \ddots & & b_1(c) & & \ddots & \\
\vdots & & \ddots & a_0(c) & \vdots & & \ddots & 0 \\
\vdots & & & a_1(c) & & & & b_1(c) \\
a_\ell(c) & & & & b_m(c) & & & \vdots \\
& a_\ell(c) & & \vdots & & b_m(c) & & \vdots \\
& & & & & & \ddots & \\
& & & a_\ell(c) & & & & b_m(c)
\end{vmatrix}$$

Now, one has only to expand down the first row.

2) It suffices to expand down the first row, then the second row and so on until the pth row of the Sylvester matrix above.

3) $f = X_1^2 X_2 + 3X_1 - 1$, $g = 6X_1^2 + X_2^2 - 4$,

$$\begin{aligned}
h = \mathrm{Res}_{X_1}(f,g) &= (-1)^{2 \times 2}\mathrm{Res}_{X_1}(g,f) = \mathrm{Res}_{X_1}(g,f) \\
&= X_2^6 - 8X_2^4 + 12X_2^3 + 70X_2^2 - 48X_2 - 180, \\
\mathrm{Res}_{X_1}(f(X_1,0), g(X_1,0)) &= \mathrm{Res}_{X_1}(3X_1 - 1, 6X_1^2 - 4) = -30, \\
h(0) &= -180 = 6^1 \times (-30).
\end{aligned}$$

Exercise 373:

1) We induct on $m+n$. We can suppose that $m \geq n$. We have:

$$\begin{aligned}
\text{Res}(X^m - \alpha, X^n - \beta) &= \text{Res}(X^{m-n}(X^n - \beta) + \beta X^{m-n} - \alpha, X^n - \beta) \\
&= \text{Res}(\beta X^{m-n} - \alpha, X^n - \beta) \\
&= \beta^{m-n}\text{Res}(X^{m-n} - \alpha\beta^{-1}, X^n - \beta) \\
&= \beta^{m-n}(-1)^{m-n}(\beta^{\frac{m-n}{\delta}} - (\alpha\beta^{-1})^{\frac{n}{\delta}})^{\delta} \\
&= (-1)^m(\beta^{\frac{m}{\delta}} - \alpha^{\frac{n}{\delta}})^{\delta}.
\end{aligned}$$

2) Take $\alpha = a$, $\beta = 1$, $m = 2$, $n = p-1$. We have $\delta = m \wedge n = 2 \wedge (p-1) = 2$.

3) If $a \neq 0$ then $a^{p-1} = 1$, and, thus, $a^{\frac{p-1}{2}} = \pm 1$. Moreover,

$$\text{Res}(X^2 - a, X^{p-1} - 1) = (a^{\frac{p-1}{2}} - 1)^2 = \prod_{x \in \mathbb{F}_p^\times} \text{Res}(X^2 - a, X - x) = \prod_{x \in \mathbb{F}_p^\times}(x^2 - a).$$

It follows that an $a \in \mathbb{F}_p^\times$ is a square in $\mathbb{F}_p$ if and only if $a^{\frac{p-1}{2}} = 1$.

Exercise 374:

a) $\tilde{f} = f - \frac{a_0}{b_0}X^{\ell-m}g = (a_1 - \frac{a_0}{b_0}b_1)X^{\ell-1} + \cdots + (a_m - \frac{a_0}{b_0}b_m)X^{\ell-m} + a_{m+1}X^{\ell-m-1} + \cdots + a_\ell$.

$$\text{Res}(f,g) = \begin{vmatrix}
a_0 & & & b_0 & & \\
\vdots & \ddots & & \vdots & \ddots & \\
 & & a_0 & & & b_0 \\
a_\ell & & \vdots & b_m & & \\
 & \ddots & & & \ddots & \vdots \\
 & & a_\ell & & & b_m
\end{vmatrix}$$

We perform the elementary columns operations C_j: $C_1 \leftarrow C_1 - a_0 b_0^{-1}C_{m+1}$, $C_2 \leftarrow C_2 - a_0 b_0^{-1}C_{m+2}, \ldots, C_\ell \leftarrow C_\ell - a_0 b_0^{-1}C_{m+\ell}$, and then we develop accordingly to the minors of the first row $L_1 = (0, \ldots, 0, b_0, 0, \ldots, 0)$; b_0 being at the $(m+1)^{\text{th}}$ position.

It follows that $\text{Res}(f,g) = (-1)^{m+2}b_0\text{Res}(\tilde{f},g) = (-1)^m b_0\text{Res}(\tilde{f},g)$.

b) If $a_1 - a_0 b_0^{-1}b_1 \neq 0$, then $\deg \tilde{f} = \ell - 1$ and we can use a).

If $a_1 - a_0 b_0^{-1}b_1 = 0$, then $\deg \tilde{f} < \ell - 1$ and we continue to develop $\text{Res}(f,g)$ accordingly to the first row. If $a_2 - a_0 b_0^{-1}b_2 \neq 0$, then $\deg \tilde{f} = \ell - 2$ and $\text{Res}(f,g) = (-1)^m(-1)^m b_0^2\text{Res}(\tilde{f},g) = (-1)^{2m}b_0^2\text{Res}(\tilde{f},g)$.

In general, we develop $\text{Res}(f,g)$ accordingly to rows $L_1,\ldots,l_{\ell-\deg \tilde{f}}$ and we find $\text{Res}(f,g) = (-1)^{m(\ell-\deg \tilde{f})}b_0^{\ell-\deg \tilde{f}}\text{Res}(\tilde{f},g)$.

c) When performing an Euclidean division of f by g in $\mathbf{K}[X]$, after each intermediary computational step, we find a remainder of the same type as $\tilde{f}$. Setting $r_1 = \tilde{f},\ldots,r_n = r$, we have

$$\text{Res}(f,g) = (-1)^{m(\ell-\deg r_1)}b_0^{\ell-\deg r_1}\text{Res}(r_1,g)$$

$$= (-1)^{m(\ell-\deg r_1)}b_0^{\ell-\deg r_1}(-1)^{m(\deg r_1-\deg r_2)}b_0^{\deg r_1-\deg r_2}\text{Res}(r_2,g)$$

$$= (-1)^{m(\ell-\deg r_2)}b_0^{\ell-\deg r_2}\text{Res}(r_2,g)$$

$$\vdots$$

$$= (-1)^{m(\ell-\deg r_n)}b_0^{\ell-\deg r_n}\text{Res}(r_n,g) = (-1)^{m(\ell-\deg r)}b_0^{\ell-\deg r}\text{Res}(r,g).$$

d) The Euclidean algorithm allows us to compute the gcd of f and g by successive Euclidean divisions:

$$f = q_1 g + R_1, \text{ with } \deg R_1 < \deg R_2,$$

$$g = q_2 R_1 + R_2, \text{ with } \deg R_2 < \deg R_1,$$

$$R_1 = q_3 R_2 + R_3, \text{ with } \deg R_3 < \deg R_2,$$

$$\vdots$$

$$R_N = q_{N+2}R_{N+1} + R_{N+2}, \text{ with } \deg R_{N+2} < \deg R_{N+1}.$$

Let R_N be the first constant remainder found: $R_N = 0 \Rightarrow \gcd(f,g) \neq 1$ and $\text{Res}(f,g) = 0$; $R_N \neq 0 \Rightarrow \gcd(f,g) = 1$ and $\text{Res}(f,g) \neq 0$. The algorithm's steps are the same as those of the Euclidean division algorithm until finding a first constant remainder R_N and we have $\text{Res}(R_{N-1},R_N) = R_N^{\deg R_{N-1}}$. The calculation is as follows:

$$\text{Res}(f,g) = (-1)^{\deg g(\deg f-\deg R_1)}b_0^{\deg f-\deg R_1}\text{Res}(R_1,g)$$

$$= (-1)^{\deg g(\deg f-\deg R_1)}b_0^{\deg f-\deg R_1}(-1)^{\deg g\deg R_1}\text{Res}(g,R_1)$$

$$= (-1)^{\deg g\deg f}b_0^{\deg f-\deg R_1}\text{Res}(g,R_1).$$

If for a polynomial h in $\mathbf{K}[X]$, we denote by $\text{lead}(h)$ its leading of h, we have:
$\text{Res}(f,g)$

$$= (-1)^{\deg g\deg f+\deg g\deg R_1+\cdots+\deg R_{N-1}\deg R_N}\text{lead}(g)^{\deg f-\deg R_1}\text{lead}(R_1)^{\deg g-\deg R_2}$$

$$\cdots \text{lead}(R_{N-1})^{\deg R_{N-2}-\deg R_N}\text{Res}(R_{N-1},R_N).$$

If we also introduce a function remainder$(f,g) = $ the remainder on Euclidean division of f by g, the algorithm is:

Input: f, g
Output: Res

$h := f$
$s := g$
$\text{Res} := 1$

While $\deg s > 0$ DO

  $r := \text{remainder}(h,s)$

  $\text{Res} := (-1)^{\deg h \deg s}\text{lead}(s)^{\deg h - \deg r}\text{Res}$

  $h := s$

  $s := r$

If $h = 0$ or $s = 0$ THEN $\text{Res} := 0$ ELSE

If $\deg h > 0$ THEN $\text{Res} := s^{\deg h}\text{Res}$

The example:

$f = (X^2 + X - 1)g - 3X + 3 \Rightarrow \text{Res}(f,g) = (-1)^{4 \times 2}(1)^{4-1}\text{Res}(g,-3X+3) = \text{Res}(g,-3X+3)$.

$g = (-\frac{1}{3}X)(-3X+3) + 2 \Rightarrow \text{Res}(g,-3X+3) = (-1)^{2 \times 1}(-3)^{2-0}\text{Res}(-3X+3,2) = 9 \times 2^1 = 18$.

It follows that $\text{Res}(f,g) = 18$.

Exercise 375:

1) "$\Leftarrow$" This follows from Theorem 327.

"$\Rightarrow$" Let $h_1, h_2 \in \mathbf{R}[X]$ such that $h_1 f + h_2 g = 1$. Since f is monic, we have $\text{Res}(f,h_2 g) = \text{Res}(f,h_2)\text{Res}(f,g)$ and $\text{Res}(f,h_2 g) = \text{Res}(f,h_1 f + h_2 g) = \text{Res}(f,1) = 1$.

2) "$\Leftarrow$" This follows from Theorem 327.

"$\Rightarrow$" Let us denote by $w_i = v_2 + y_i v_3 + \cdots + y_i^{n-2} v_n$ and $r_i = \text{Res}_X(v_1, w_i)$ for $0 \leq i \leq s = (n-2)d$. To prove that $\langle r_0, \ldots, r_s \rangle = \mathbf{A}$ it suffices to prove that for each maximal ideal $\mathfrak{M}$ of $\mathbf{A}$ there exists $0 \leq i \leq s$ such that $r_i \notin \mathfrak{M}$. For this, let $\mathfrak{M}$ be a maximal ideal of $\mathbf{A}$ and by way of contradiction suppose that $\overline{r_0}, \ldots, \overline{r_s} = 0$ in the residue field $\mathbf{K} := \mathbf{A}/\mathfrak{M}$. It is worth pointing out that $\overline{\text{Res}_X(v_1, w_i)} = \text{Res}_X(\overline{v_1}, \overline{w_i})$ since v_1 is monic.

This means that for each i there exists $\xi_i \in \overline{\mathbf{K}}$ such that $\overline{v_1}(\xi_i) = \overline{w_i}(\xi_i) = \overline{0}$. But since $\deg_X v_1 = d$, $\overline{v_1}$ has at most d distinct roots and hence there exists at least one root among the ξ_i repeated $n - 1$ times. We can suppose that $\xi_1 = \xi_2 = \cdots = \xi_{n-1} := \xi$. Thus, we have:

$$\begin{pmatrix} 1 & y_1 & \cdots & y_1^{n-2} \\ 1 & y_2 & \cdots & y_2^{n-2} \\ \vdots & \vdots & \vdots & \vdots \\ 1 & y_{n-1} & \cdots & y_{n-1}^{n-2} \end{pmatrix} \begin{pmatrix} v_2(\xi) \\ v_3(\xi) \\ \vdots \\ v_n(\xi) \end{pmatrix} = \begin{pmatrix} 0 \\ 0 \\ \vdots \\ 0 \end{pmatrix}.$$

Since the matrix above is a Vandermonde matrix, its determinant is equal to

$$\prod_{1 \leq i < j \leq n-1} (y_j - y_i),$$

which is invertible in $\mathbf{A}$. Thus, $\overline{v_1}(\xi) = \overline{v_2}(\xi) = \cdots = \overline{v_n}(\xi) = 0$, in contradiction with the fact that $1 \in \langle v_1, \ldots, v_n \rangle$.

3) This is a direct consequence of 2).

Taking $f_1 = X^5 - X^4 + 3X^2 - 3$, $f_2 = 2X^3 + 3X^2 - X - 4$, $f_3 = 3X^5 + 2X^4 - X^3 - X^2 - 3$ ($n = 3$, $d = 3$), we obtain:

$$\begin{cases} \operatorname{Res}_X(f_2, f_1) = 0 \\ \operatorname{Res}_X(f_2, f_1 + f_3) = 0 \\ \operatorname{Res}_X(f_2, f_1 + 2f_3) = 0 & \Rightarrow 1 \notin \langle f_1, f_2, f_3 \rangle. \\ \operatorname{Res}_X(f_2, f_1 + 3f_3) = 0 \\ \operatorname{Res}_X(f_2, f_1 + 4f_3) = 0 \end{cases}$$

Exercise 376:

Clearly, we have $F^m H \subseteq F^{m+1} G$. For the converse, since $f\tilde{g} = h - b_m f T^m$, we have $[f\tilde{g}] \subseteq H + b_m F$. By the induction hypothesis, we have

$$F^m \tilde{G} \subseteq F^{m-1}[f\tilde{g}] \subseteq F^{m-1} H + b_m F^m. \tag{5.5}$$

Now, as $a_i b_m = c_{i+m} - (a_{i+1} b_{m-1} + a_{i+2} b_{m-2} + \cdots)$, we have

$$a_i b_m \in c_{i+m} + \sum_{j>i} a_j \tilde{G}. \tag{5.6}$$

Multiplying (5.6) by F^m and using (5.5), we obtain

$$a_i b_m F^m \subseteq c_{i+m} F^m + \sum_{j>i} a_j F^m \tilde{G} \subseteq \left(c_{i+m} F^m + \sum_{j>i} a_j F^{m-1} H \right) + \sum_{j>i} a_j b_m F^m.$$

Setting $E_i = a_i b_m F^m$, we get

$$E_i \subseteq F^m H + \sum_{j>i+1} E_j.$$

As $E_j = 0$ for $j \gg 0$, the above containment gives $E_i \subseteq F^m H$, that is,

$$a_i b_m F^m \subseteq F^m H. \tag{5.7}$$

Multiplying (5.5) by a_i and using (5.7), we get

$$a_i F^m \tilde{G} \subseteq a_i F^{m-1} H + a_i b_m F^m \subseteq a_i F^{m-1} H + F^m H \subseteq F^m H. \tag{5.8}$$

Combining (5.7) and (5.8), we infer that $a_i F^m G \subseteq F^m H$, and, thus, $F^{m+1} G \subseteq F^m H$, as desired.

Exercise 377:

1) Use Dedekind-Mertens, Exercise 376.

2) The vectors ${}^t(r_0, \ldots, r_d)$ and ${}^t(c_0, \ldots, c_d)$ are relied by the Vandermonde matrix as follows:

$$\begin{pmatrix} 1 & y_0 & \cdots & y_0^d \\ 1 & y_1 & \cdots & y_1^d \\ \vdots & \vdots & \vdots & \vdots \\ 1 & y_d & \cdots & y_d^d \end{pmatrix} \begin{pmatrix} c_0 \\ c_1 \\ \vdots \\ c_d \end{pmatrix} = \begin{pmatrix} r_0 \\ r_1 \\ \vdots \\ r_d \end{pmatrix}.$$

It follows that

$$\pi \langle c_0, \ldots, c_d \rangle \subseteq \langle r_0, \ldots, r_d \rangle.$$

3) Use 1) and 2).

4) There exist $\tilde{u}, \tilde{v}, \tilde{w} \in \mathbf{A}[X]$ such that $u\tilde{u} + v\tilde{v} + w\tilde{w} = 1$. Replacing X by x_i, we obtain

$$1 \in \langle u_1, w_1 \rangle \langle u_2, w_2 \rangle^2 \cdots \langle u_d, w_d \rangle^d \subseteq \langle r_0, \ldots, r_d \rangle.$$

Exercise 378:

1) Suppose that we have $F(X,Y) = Y^3 - 3XY + X^3 = (Y+a)(Y^2 + bY + c) = Y^3 + (a+b)Y^2 + (ab+c)Y + ac$, with $a, b, c \in \mathbb{C}[X]$. Then, by identification, we would have $a + b = 0$, $ab + c = -3X$, and $ac = X^3$. This latter implies that $a = \lambda X^i$ and $c = \frac{1}{\lambda} X^j$ with $i + j = 3$ and $\lambda \in \mathbb{C}^\times$. In particular, we would have $c - a^2 = \frac{1}{\lambda} X^j - \lambda^2 X^{2i} = -3X$, which is impossible.

2) The singular points of C are determined by the following equations:

$$F(x,y) = \frac{\partial f}{\partial x}(x,y) = \frac{\partial f}{\partial y}(x,y) = 0.$$

So, in order to determine the singular points, we need to solve the following system of equations:

$$\begin{cases} x^3 + y^3 - 3xy &= 0 \\ x^2 - y &= 0 \\ y^2 - x &= 0. \end{cases} \tag{5.9}$$

We calculate the reduced Gröbner basis G of $\langle x^3 + y^3 - 3xy, x^2 - y, y^2 - x \rangle$ with regard to the lexicographic order with $x > y$ using the computing software MAPLE. We obtain $G = \{y, x\}$. So, $(0,0)$ is the unique singular point of C.

3) $(0,0)$ is double singular point of C with exactly 2 distinct tangents $x = 0$ and $y = 0$. So, it is ordinary.

Exercise 379:

1) The set of all homogeneous polynomials of degree m in the indeterminates $X_0, \ldots, X_k$ forms a $\bar{\mathbf{K}}$-vector space of dimension $N = \binom{m+k}{k}$. If one regards two homogeneous polynomials as equivalent when they are equal up to a nonzero constant, i.e., when they define the same curve, then the set of these equivalence classes forms a projective space $\mathbb{P}_{N-1}(\bar{\mathbf{K}})$. Thus, the set of all curves of order m in $\mathbb{P}^2(\bar{\mathbf{K}})$ forms a projective space over $\bar{\mathbf{K}}$ of dimension $N - 1 = \binom{m+2}{2} - 1 = \frac{m(m+3)}{2}$.

2) Via a coordinates change, we can suppose that P is the origin. Having multiplicity at least s at P gives $\binom{s+1}{2} = \frac{s(s+1)}{2}$ linearly independent conditions (the coefficients of monomials $X^i Y^j$ with $i + j < s$ must be zero).

3) This is an immediate consequence of 1) and 2).

Exercise 380:

1) Recall that the number of monomials at k variables of total degree at most d is $\binom{d+k}{d}$. So, $\dim_{\mathbf{K}} \mathbf{K}[X,Y,Z]_d = \binom{d+2}{d} = \binom{d+2}{2}$.

For $P \in \mathbb{P}^2(\mathbf{K})$, the subspace of all $f \in \mathbf{K}[X,Y,Z]_d$ such that $f(P) = 0$ is a hyperplane in $\mathbf{K}[X,Y,Z]_d$. It is folklore that for a family $(H_k)_{1 \leq k \leq n}$ of hyperplanes in a $\mathbf{K}$-vector space of finite dimension $n \geq 1$, we have

$$\dim_{\mathbf{K}}(\cap_{k=1}^{p} H_k) \geq n - p$$

for $p \in [\![1, n]\!]$. So, the intersection of $\binom{d+2}{2} - 1$ hyperplanes in $\mathbf{K}[X,Y,Z]_d$ has dimension ≥ 1, and, thus, nonzero.

2) We have $\binom{d-1}{2} + 1 + 3d - 3 + 1 = \binom{d+2}{2} - 1$. So, by 1), there exists a curve $C' \subseteq \mathbb{P}^2(\mathbf{K})$ of degree d passing through $P_1, \ldots, P_N, Q_1, \ldots, Q_{3d-3}, Q$. As pointed out in Remark 343, we have $I_P(C,C') \geq \text{mult}_C(P) \cdot \text{mult}_{C'}(P)$ for any $P \in \mathbb{P}^2(\mathbf{K})$. Hence, $I_P(C,C') \geq 2$ for $P \in \{P_1, \ldots, P_N\}$ (since $\text{mult}_C(P_i) \geq 2$). It follows that

$$\sum_{P \in C \cap C'} I_P(C,C') \geq (d-1)(d-2) + 2 + 3d - 3 = d^2 + 1 > d^2.$$

By Bézout's Theorem 336, we deduce that $C = C'$ (since C is irreducible). This contradicts the fact that $Q \in C' \setminus C$.

Exercise 381:

Strong Bézout's Theorem 340 says that there are exactly 6 intersection points counted with multiplicities. We see that there are 4 intersection points, namely $(\pm 1, 0)$ and $(0, \pm 1)$. In particular, this shows that C and D have no common components. But, what are the intersection multiplicities of these intersection points?

When seen in $\mathbb{P}^2(\mathbb{C})$, the curves become $\tilde{C} = V(F)$ and $\tilde{D} = V(G)$, where $F = X^2 + Y^2 - Z^2$ and $G = X^3 - X^2 Z - XZ^2 - Y^2 Z + Z^3$. There are no intersection points at ∞. We see that the point $(0:0:1)$ does not satisfy Condition 5.1 since it is collinear with the two intersection points $(\pm 1 : 0 : 1)$ (or $(0 : \pm 1 : 1)$). We see that the point $(1 : 1 : 1)$ satisfies Condition 5.1, so we can consider the coordinates change $\varphi(x,y,z) = (x - z, y - z, z)$ transforming $(1 : 1 : 1)$ into $(0 : 0 : 1)$, and $\tilde{C}, \tilde{D}$ into $\varphi(\tilde{C}) = V(F \circ \varphi^{-1}) = V(F(X+Z,Y+Z,Z))$, $\varphi(\tilde{D}) = V(G \circ \varphi^{-1}) = V(G(X+Z,Y+Z,Z))$. Now, the whole information on intersection multiplicities is coded in

$$\text{Res}_Z(F(X+Z,Y+Z,Z), G(X+Z,Y+Z,Z))$$
$$= -2X^5 Y + 9X^4 Y^2 - 12X^3 Y^3 + 4X^2 Y^4 = X^2 Y (2Y - X)^2 (Y - 2X).$$

We read that the intersection multiplicity is 2 at both of the points $\varphi(1:0:1) = (0:-1:1)$ and $\varphi(-1:0:1) = (-2:-1:1)$ while it is 1 at both of the points $\varphi(0:1:1) = (-1:0:1)$ and $\varphi(0:-1:1) = (-1:-2:1)$. We see that

$$2 + 2 + 1 + 1 = 6 = 2 \cdot 3 = \deg(C) \cdot \deg(D).$$

Exercise 382:

Let C be an irreducible projective plane curve of degree 3 and suppose, by way of contradiction, that it has two singular points P and Q. We can suppose that it is defined over an algebraically closed field. Then, necessarily the whole line (PQ) will be contained in C (contradicting the irreducibility of C) because, otherwise, by Strong Bézout's Theorem 340, we would have

$$\deg(C) \cdot \deg(PQ) = 3 = \sum_{R \in C \cap (PQ)} I_R(C, (PQ)) \geq 4$$

since $I_P(C, (PQ)) \geq 2$ and $I_Q(C, (PQ)) \geq 2$ (P and Q being singular).

Exercise 383:

By Proposition 348, we know that number of singular points of an irreducible projective plane curve of degree d is at most its arithmetic genus $\frac{(d-1)(d-2)}{2}$. For $d = 4$, $\frac{(d-1)(d-2)}{2} = 3$, and so, C is reducible.

Exercise 384:

Denote by $F = Y^2 - X^3 + X$ so that $C = V(F)$. Consider the projective completion $\tilde{C} = V(H)$ of C with $H = Y^2 Z - X^3 + XZ^2$. The curve $\tilde{C}$ is smooth and has a single point $O = (0:1:0)$ at ∞ (where the tangent line has equation $z = 0$). Denote by $\tilde{\Delta} : x = az$ the projective completion of Δ, and $\tilde{P} = (a:b:1)$.

We know that $\tilde{\Delta}$ is not tangent to $\tilde{C}$ neither at $\tilde{P}$ nor at O. Thus,

$$I_{\tilde{P}}(\tilde{C}, \tilde{\Delta}) + I_O(\tilde{C}, \tilde{\Delta}) = 1 + 1 = 2 < \deg(\tilde{C}) \cdot \deg(\tilde{\Delta}) = 3 \cdot 1 = 3.$$

By Strong Bézout's Theorem 340, $\tilde{C}$ and $\tilde{\Delta}$ intersect at another point P' with $I_{P'}(\tilde{C}, \tilde{\Delta}) = 1$, and, thus, $\tilde{\Delta}$ is not tangent to $\tilde{C}$ at P'. Since C has a unique point ∞, the point P' is in the affine plane $z \neq 0$.

Exercise 385:

As the cubics $L_{12} \cup L_{23} \cup L_{31}$ and $L_{21} \cup L_{32} \cup L_{13}$ intersect in 9 points and 3 among them are on a line (the P_i's), then, by virtue of Corollary 337, the other 6 points are on a curve of degree $3 - 1 = 2$, that is, on a conic. But, as 3 of these 6 points are on a line (the Q_i's), then, again by virtue of Corollary 337, the remaining 3 points (the C_i's) are on a curve of degree $2 - 1 = 1$, that is, are collinear.

Exercise 386:

1) By a coordinates change, we can suppose that $P = (0,0)$ and $L : x - y = 0$. Now write $C : F(x,y) = F_0(x,y) + F_1(x,y) + F_2(x,y) + \cdots = 0$, where $F_i \in \mathbf{K}[X,Y]$ is homogeneous of degree i. Then, L is tangent to C at P if and only if $F_0 = 0$ & $F_1(X,Y) = \lambda(X - Y)$ for some $\lambda \in \mathbf{K}^\times$, or, in other terms, if and only if X^2 divides $F(X,X)$. This latter condition is nothing but $\mathrm{Ord}_{P,L}(F(x,y)) \geq 2$. The desired result follows since $\mathrm{Ord}_{P,L}(F(x,y)) = I_P(C,L)$.

2) The answer is NO because this would contradict Strong Bézout's Theorem 340: $2 + 1 + 1 = 4 > \deg(E) \cdot \deg(L) = 3$.

Exercise 387:

1) This is a direct consequence of Strong Bézout's Theorem 340.

2.a) This is straightforward.

2.b) Let $P = (p_0, p_1, p_2)$ be a singular point of C. It is immediate that $P \in \mathcal{H}$ when $\mathrm{mult}_C(P) > 2$ because then all the $\frac{\partial^2 F}{\partial X_i \partial X_j}(P)$'s vanish by definition. If $\mathrm{mult}_C(P) = 2$ then, since $\frac{\partial F}{\partial X_i}$ is either zero or homogeneous of degree $d - 1$, we have (Euler's Formula, see the solution of Exercise 360):

$$\sum_j p_j \cdot \frac{\partial^2 F}{\partial X_i \partial X_j}(P) = (d-1)\frac{\partial F}{\partial X_i}(P) = 0.$$

It follows that $\det(\frac{\partial^2 F}{\partial X_{i-1} \partial X_{j-1}}(P))_{1 \leq i,j \leq 3} = 0$.

2.c) Set $x_0 = z$, $x_1 = x$, and $x_2 = y$. First, notice that after a coordinates change $\begin{pmatrix} x \\ y \\ z \end{pmatrix} \mapsto A \begin{pmatrix} x \\ y \\ z \end{pmatrix}$, with $A \in \mathrm{GL}_3(\mathbb{C})$, a polynomial $f \in \mathbb{C}[x,y,z]$ is transformed to $\hat{f} = f \circ A$, and it is immediate that $H_{\hat{f}} = \det(A)^2 \cdot H_f$. So, our hypotheses are stable under coordinates change.

By a coordinates change, we can work with affine coordinates (x,y) and suppose that $P = (0,0)$ with a tangent line $L : y = 0$. Now write $C : F(x,y) = y + (ax^2 + cxy + cy^2) + F_3(x,y) + \cdots = 0$, where $F_i \in \mathbb{C}[X,Y]$ is homogeneous of degree i, and $a,b,c \in \mathbb{C}$. The projective completion of C is $\tilde{C} : \tilde{F}(x,y,z) = yz^{d-1} + (ax^2 + cxy + cy^2)z^{d-2} + F_3(x,y)z^{d-3} + \cdots = 0$. We have

$$H_{\tilde{F}}(0:0:1) = \begin{vmatrix} 2a & b & 0 \\ b & 2c & d-1 \\ 0 & d-1 & 0 \end{vmatrix} = -2a(d-1)^2.$$

On the other hand, we have $I_P(C,L) = \mathrm{Ord}_{P,L}(F(x,y)) = \deg_x(ax^2 + F_3(x,0) + \cdots)$. So,

$$H_{\tilde{F}}(0:0:1) \neq 0 \;\Leftrightarrow\; a \neq 0 \;\Leftrightarrow\; I_P(C,L) = 2,$$

or also (by virtue of Exercise 386),

$$H_{\tilde{F}}(0:0:1) = 0 \;\Leftrightarrow\; I_P(C,L) \geq 3.$$

The fact that $N \in [\![1, 3d(d-2)]\!]$ follows from Strong Bézout's Theorem 340 (and we have $N = 3d(d-2)$ if we count multiplicities).

2.d) For $F = X^3 - Y^2 Z - YZ^2$, we have $H_F = -24X(Y^2 + YZ + Z^2)$. Clearly we have exactly two points of inflexion at $(0:0:1)$ and $(0:-1:1)$ in the real affine plane (the only ones we can see in Fig. 5.24), an inflexion point at $(0:1:0)$ (at ∞), and 6 other inflexion points $\xi_1 = (e^{\frac{2i\pi}{9}}, e^{\frac{2i\pi}{3}}, 1)$, $\xi_2 = (e^{\frac{8i\pi}{9}}, e^{\frac{2i\pi}{3}}, 1)$, $\xi_3 = (e^{\frac{-4i\pi}{9}}, e^{\frac{2i\pi}{3}}, 1)$, $\bar{\xi}_1, \bar{\xi}_2, \bar{\xi}_3$, over $\mathbb{C}$. Of course, all these 9 inflexion points are simple by virtue of Strong Bézout's Theorem 340.

2.e) By Strong Bézout's Theorem 340, a smooth cubic C has nine points of inflexion counting multiplicities. At each point of inflexion P with tangent line L, we have $I_P(C,L) \geq 3$ by definition and $I_P(C,L) \leq 3$ by Strong Bézout's Theorem 340. So, $I_P(C,L) = 3$ and P is simple.

Exercise 388:

First notice the following immediate fact:

$$P \in C \cap D \;\Leftrightarrow\; I_P(C,D) \geq 1.$$

By a coordinates change, we can work with affine coordinates (x,y) and suppose that $P = (0,0)$. Now write $C: F(x,y) = F_0(x,y) + F_1(x,y) + F_2(x,y) + \cdots = 0$, $D: G(x,y) = G_0(x,y) + G_1(x,y) + G_2(x,y) + \cdots = 0$, where $F_i, G_i \in \mathbf{K}[X,Y]$ are homogeneous of degree i. Since $P \in C \cap D$, we can suppose that $F_0 = G_0 = 0$, that is, both F and G are in the maximal ideal $\mathfrak{m} := \langle X, Y \rangle$ of the local ring $\mathbf{K}[X,Y]_{\langle X,Y \rangle}$. Now:

$I_P(C,D) = \dim_{\mathbf{K}}(\mathbf{K}[X,Y]_{\mathfrak{m}} / \langle F, G \rangle) = 1$

$\Leftrightarrow \dim_{\mathbf{K}}(\mathbf{K}[X,Y]_{\mathfrak{m}} / \langle F, G \rangle) = \dim_{\mathbf{K}}(\mathbf{K}[X,Y]_{\mathfrak{m}} / \mathfrak{m}) \;\Leftrightarrow\; \langle F, G \rangle = \mathfrak{m}$

$\Leftrightarrow \langle \bar{F}, \bar{G} \rangle = \mathfrak{m}/\mathfrak{m}^2$ (by Nakayama's lemma; classes being modulo $\mathfrak{m}^2$)

$\Leftrightarrow \mathbf{K}F_1 + \mathbf{K}G_1 = \mathbf{K}X + \mathbf{K}Y$

$\Leftrightarrow F_1$ and G_1 are not proportional

$\Leftrightarrow C$ and D intersect transversally at P.

Exercise 389:

1) If $\langle F, G \rangle \subseteq \langle F, H \rangle$ then obviously $\dim_{\mathbf{K}}(\mathcal{O}_P / \langle F, G \rangle) \geq (\dim_{\mathbf{K}} \mathcal{O}_P / \langle F, H \rangle)$.

2) By Remark 343, we know that $I_P(C, L) = 1$ and, more generally, $I_P(C, V(\ell^n) = n \ \forall \, n \in \mathbb{N}$. Moreover, by virtue of Exercise 388, we know that $\langle F, \ell \rangle = \mathfrak{m}_P$.

We have $\langle F, G \rangle \subseteq \mathfrak{m}_P = \langle F, \ell \rangle$ and $\langle F, G \rangle \not\subseteq \langle F, \ell^{I_P(C,D)+1} \rangle$ (by virtue of 1)). Denote by $N = \max(\mathcal{N})$ where $\mathcal{N} = \{n \in \mathbb{N}^* \mid \langle F, G \rangle \not\subseteq \langle F, \ell^n \rangle\}$. Let $U, V \in \mathcal{O}_P$ such that $G = UF + V\ell^N$. Since $\mathfrak{m}_P = \langle F, \ell \rangle$ and $N + 1 \notin \mathcal{N}$, we infer that $V \in \mathcal{O}_P^\times$, and, thus, $\langle F, G \rangle = \langle F, \ell^N \rangle$ and $N = I_P(C, D)$ as desired.

3) This is an immediate consequence of 1) and 2).

4) If C is singular at P then 3) is no longer true. For example, take $F = Y^2 - X^3, G = Y, H = X \in \mathbb{R}[X, Y], C = V(F), D = V(G), E = V(H)$.

The curve C is singular at $P = (0,0)$. We have $I_p(C, D) = \dim_{\mathbb{R}}(\mathbb{R}[X]/\langle X^3 \rangle) = 3 > I_p(C, E) = \dim_{\mathbb{R}}(\mathbb{R}[Y]/\langle Y^2 \rangle) = 2$, yet $\langle F, G \rangle = \langle X^3, Y \rangle \not\subseteq \langle F, H \rangle = \langle X, Y^2 \rangle$.

Exercise 390:

The irreducible projective plane curve $C = V(H)$, where $H(X, Y, Z) = X^2 Z^2 - Y^3 Z + Y^4 \in \mathbb{R}[X, Y, Z]$, has two singular points at $P = (0 : 0 : 1)$ (nonordinary double point) and $R = (1 : 0 : 0)$ (nonordinary double point). Thus, $g^*(C) = 1$ and C does not fulfill the hypotheses of Proposition 354.

If we send (X, Y, Z) to $(X - Y, X + Y, X + Z)$, the equation becomes $G(x, y, z) = 0$ with

$$G(X, Y, Z) = X^4 - X^3 Y + X^3 Z + 4X^2 Y^2 - 7X^2 YZ + X^2 Z^2 + 3XY^3 - XY^2 Z$$
$$- 2XYZ^2 + Y^4 - Y^3 Z + Y^2 Z^2.$$

Let us check that G fulfills the hypotheses of Proposition 354:

For $z = 0$, the equation becomes $x^4 - x^3 y + 4x^2 y^2 + 3xy^3 + y^4 = 0$, with disc$(t^4 - t^3 + 4t^2 + 3t + 1) = 5744 \neq 0$ (4 simple points distinct from $(0 : 1 : 0)$ and $(1 : 0 : 0)$).

For $y = 0$, the equation becomes $x^2(x^2 + xz + z^2) = 0$. So, we have besides P two simple points distinct from $(1 : 0 : 0)$) (disc$(t^2 + t + 1) \neq 0$).

For $x = 0$, the equation becomes $y^2(y^2 - yz + z^2) = 0$. So, we have besides P two simple points distinct from $(0 : 1 : 0)$ (disc$(t^2 - t + 1) \neq 0$).

We find

$$\bar{G} = \frac{G(YZ, XZ, XY)}{Z^2}$$
$$= X^4 Y^2 - X^4 YZ + X^4 Z^2 - 2X^3 Y^3 - X^3 Y^2 Z + 3X^3 YZ^2 + X^2 Y^4$$
$$- 7X^2 Y^3 Z + 4X^2 Y^2 Z^2 + XY^4 Z - XY^3 Z^2 + Y^4 Z^2.$$

At $z = 0$, the equation of $V(\bar{G})$ becomes $x^2 y^2 (x - y)^2 = 0$. We see that besides $Q = (0 : 1 : 0)$ and $R = (1 : 0 : 0)$, $V(\bar{G})$ intersects $V(Z)$ at the point $M = (1 : 1 : 0)$ with intersection multiplicity 2. This is a nonsingular point of $V(\bar{G})$ as the component of lowest degree one of $\bar{G}(1, Y + 1, Z)$ is $-8Z$.

As a conclusion, $V(\bar{G})$ is birationally equivalent to C and has ordinary singularities at P, Q, R with multiplicities $4, 2, 2$, respectively, and a nonordinary double point at $N = (1 : -1 : -1)$ (inherited from C). Note that $g^*(V(\bar{G})) = \binom{5}{2} - \binom{4}{2} - 3\binom{2}{2} = 1 = g^*(C)$ but $V(\bar{G})$ has less nonordinary singular points than C. Now, we have to continue with the curve $V(\bar{G})$. After one or two quadratic transformations eventually we find an irreducible projective curve birationally equivalent to C without nonordinary singularities and with genus zero, and, thus, the genus of C is zero.

Chapter 6

Elliptic curves

An elliptic curve is a particular case of an algebraic curve equipped, among other properties, with a geometric addition of its points. Formally, an elliptic curve is a projective smooth algebraic curve of genus 1 over a field with a rational point. Elliptic curves have numerous applications in widely different fields of mathematics. For example, they were used in the proof, by Andrew Wiles, of Fermat's Last Theorem. They also find applications in elliptic curve cryptography and integer factorization (the elliptic-curve factorization method, see Remark 403). One of the main reasons for interest in cryptosystems based on elliptic curves is that these curves are a source of a considerable number of finite Abelian groups providing the same security with smaller key size (compared to the multiplicative group of a finite field).

The main references of this chapter are [9, 15, 40, 61].

6.1 Elliptic curves over a field

In this section, all fields are supposed to be perfect.

Definition 391. An elliptic curve defined over a field $\mathbf{K}$ is a smooth projective plane curve $C = V(H)$ (H being an irreducible homogeneous polynomial in $\mathbf{K}[X,Y,Z]$) of genus 1 and with a rational point $O \in C(\mathbf{K}) := \{(\alpha : \beta : \gamma) \in \mathbb{P}^2(\mathbf{K}) \mid H(\alpha,\beta,\gamma) = 0\}$.

For example, the curve $C = V(X^3 + Z^3 - Y^2Z)$ over a field $\mathbf{K}$ (char($\mathbf{K}$) $\neq$ 2, 3) is elliptic with rational point $O = (0 : 1 : 0)$. Its genus is $\frac{(3-1)(3-2)}{2} = 1$ by Theorem 295.

The genus is a "topological invariant" which is a nonnegative integer. Two curves that have different genus cannot be transformed into each other in a continuous way without introducing singularities. Recall that algebraic curves of

221

genus 0 are exactly those that admit a parametrization (Corollary 304), and, so, elliptic curves do not admit a parametrization. Over $\mathbb{C}$, one can see that the genus of an elliptic curve is 1. First, one has to understand that an elliptic curve can be given as $\mathbb{C}$ modulo a lattice Λ ($\Lambda = \mathbb{Z}\omega_1 + \mathbb{Z}\omega_2$ with $\frac{\omega_2}{\omega_1} \notin i\mathbb{R}$) through the so-called Weierstrass functions. But $\mathbb{C}$ modulo a lattice is a complex torus[1] and it is well known that the genus counts the number of "holes" or "handles" in a compact Riemann surface.[2]

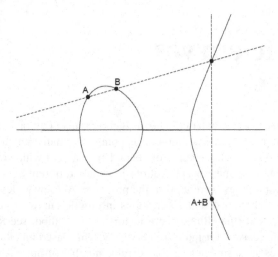

Figure 6.1: Addition of points on an elliptic curve

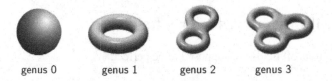

Figure 6.2: Gena 0, 1, 2, 3

The following important result gives a concrete characterization of elliptic curves as projective completions of smooth affine plane curves given by a Weierstrass equation.

---

[1]The source of Fig. 6.3 is `https://perso.univ-rennes1.fr/christophe.ritzenthaler/cours/elliptic-curve-course.pdf`.

[2]The source of Fig. 6.2 is
`https://en.wikipedia.org/wiki/Genus_(mathematics)`.

Figure 6.3: A complex torus

Definition 392. Let $\mathbf{K}$ be a field. A *Weierstrass equation* over $\mathbf{K}$ is an equation of the form:

$$C : y^2 + a_1 xy + a_3 y = x^3 + a_2 x^2 + a_4 x + a_6 \quad (a_i \in \mathbf{K}). \tag{6.1}$$

Denoting by $H(x) = a_1 x + a_3$, if we replace y with $-y - H(x)$, Equation (6.1) remains unchanged, and so, C has an involution $\mathfrak{s}_C : (x,y) \mapsto (x, -y - H(x))$.

If $\mathrm{char}(\mathbf{K}) \neq 2, 3$, then by the coordinates change $y' = y + 2^{-1} H(x)$ and then $x' = x + 3^{-1} a_2$, a Weierstrass equation becomes *reduced*, i.e., of the form $y^2 = x^3 + ax + b$ with $a, b \in \mathbf{K}$. In that case, $\mathfrak{s}_C(x,y) = (x, -y)$.

Theorem 393. *Let C be a smooth affine plane curve defined over a field $\mathbf{K}$ and given by a Weierstrass equation (6.1). Then, the projective completion of C is an elliptic curve over $\mathbf{K}$. Conversely, every elliptic curve over $\mathbf{K}$ is isomorphic to the projective completion of a curve given by (6.1).*

Proof. The projective completion $\tilde{C}$ of C is given by the equation

$$\tilde{C} : y^2 z + a_1 xyz + a_3 yz^2 = x^3 + a_2 x^2 z + a_4 xz^2 + a_6 z^3.$$

The only point at ∞ is $O = (0 : 1 : 0) \in \tilde{C}(\mathbf{K})$. In the affine chart $y = 1$, the curve becomes $V(G)$ with $G = Z + a_1 XZ + a_3 Z^2 - X^3 - a_2 X^2 Z - a_4 XZ^2 - a_6 Z^3 \in \mathbf{K}[X,Z]$. As $\frac{\partial G}{\partial Z}(0,0) = 1$, we infer that O is a nonsingular point of $\tilde{C}$. On the other hand, the genus of $\tilde{C}$ is $\frac{(3-1)(3-2)}{2} = 1$ by Theorem 295. We conclude that $\tilde{C}$ is an elliptic curve over $\mathbf{K}$.

Conversely, consider an elliptic curve E of genus $g = 1$ defined over a field $\mathbf{K}$ with a rational point $O \in E(\mathbf{K})$. Then, by Corollary 302 of Riemann-Roch Theorem 301, considering a canonical divisor K_E of E, we have $\deg(K_E) = 2g - 2 = 0$, and, for all $D \in \mathbb{Z}[E]$, if $\deg(D) \geq 0$ then $\ell(D) = \deg(D)$. In particular, $\ell(n[O]) = n$ for all $n \geq 1$. We admit that the $\mathbf{K}$-vector space $\mathscr{L}(n[O]) \cap \mathbf{K}(E)$ has dimension $\ell(n[O])$ (here we need the hypothesis that $\mathbf{K}$ is perfect). As $1 \in \mathscr{L}(2[O])$ and $\ell(2[O]) = 2$, there exists $x \in \mathbf{K}(E)$ such that $(1, x)$ is a basis of $\mathscr{L}(2[O])$. In particular, O is the unique pole of x (it is a double pole because $x \notin \mathscr{L}([O])$). Since $\ell(3[O]) = 3$, this latter basis can be completed into a basis $(1, x, y)$ of $\mathscr{L}(3[O])$. In particular, O is the unique pole of y (it is a triple pole).

It follows that O is the unique pole of each of the 7 functions $1, x, y, x^2, xy, y^2, x^3$ with order ≤ 6. Since $\ell(6[O]) = 6$, there exists $(b_1, \ldots, b_7) \in \mathbf{K}^7 \setminus \{0\}$ such that $b_1 + b_2 x + b_3 y + b_4 x^2 + b_5 xy + b_6 y^2 + b_7 x^3 = 0$. Considering the order of O as a pole, we can deduce that $b_6 \neq 0$ and $b_7 \neq 0$. Replacing x with $-b_6 b_7 x$ and y with $b_6 b_7^2 x$, we can suppose that $b_6 = b_7$, and, thus, x and y satisfy an equation of the form (6.1). Denote by D the affine plane curve defined by this equation and by $\tilde{D}$ its projective completion. We will prove that the rational map $\phi = (x : y : 1) :$ $E \dashrightarrow \tilde{D}$ is an isomorphism. Since E is smooth, ϕ is regular (Proposition 309) and we can check that $\phi(O)$ is the point at ∞ of D. We will first prove that $\tilde{D}$ is nonsingular, and, in particular, it is irreducible (by Theorem 295). We already saw that D is nonsingular at ∞. By way of contradiction, suppose that D has a singular point (x_0, y_0). By coordinates change, we can suppose that $(x_0, y_0) = (0, 0)$. Thus, $a_6 = 0$, and as the partial derivatives vanish at $(0, 0)$, we obtain $a_3 = a_4 = 0$. The equation of D becomes $y^2 + a_1 xy = x^3 + a_2 x^2$. The rational function $t = \frac{y}{x}$ on E has a simple pole at O. It is the unique pole of t since, from the latter equation, ensues the relation $t(t + a_1) = x + a_2$. Since $t \in \mathscr{L}([O])$ and $\ell([O]) = 1$, we obtain that $t \in \mathbf{K}^\times$, contradicting the fact that x and y are linearly independent over $\mathbf{K}$.

As in Definition 314, we obtain a one-to-one morphism between fields $\phi^* :$ $\bar{\mathbf{K}}(\tilde{D}) \to \bar{\mathbf{K}}(E)$. Since $\deg(x) = 2$, we have $[\bar{\mathbf{K}}(E) : \bar{\mathbf{K}}(x^2)] = 2$ (O is the unique pole of x and it is of order 2; then use Theorem 286). The facts that $[\bar{\mathbf{K}}(E) : \bar{\mathbf{K}}(x^2)] = 2$ and $y \notin \bar{\mathbf{K}}(x^2)$ (because its order at O is not even) yield that $\bar{\mathbf{K}}(E) = \bar{\mathbf{K}}(x^2, y)$, and, thus, ϕ^* is surjective. The isomorphism $(\phi^*)^{-1}$ allows to define a rational map $\psi : \tilde{D} \dashrightarrow E$ (which is regular as $\tilde{D}$ is smooth) reciprocal to ϕ.

The following proposition gives a necessary and sufficient condition for a curve given by a reduced Weierstrass equation to be elliptic.

Proposition 394. *Let $\mathbf{K}$ be a field with $\mathrm{char}(\mathbf{K}) \neq 2, 3$, and $a, b \in \mathbf{K}$. Then the affine plane curve $E : y^2 = x^3 + ax + b$ is nonsingular if and only if $\Delta :=$ $\mathrm{disc}(X^3 + aX + b) = 4a^3 + 27b^2 \neq 0$. In such case, it defines an elliptic curve.*

Proof. $P = (\alpha, \beta)$ is a singular point of E if and only if $\beta = 0$ and α is a double root of $X^3 + aX + b$.

In particular, over the reals, an affine elliptic curve is given by a reduced Weierstrass equation

$$E : y^2 = x^3 + ax + b, \quad a, b \in \mathbb{R},$$

with $\Delta = 4a^3 + 27b^2 \neq 0$. Here, it is worth pointing out that, in case $\Delta > 0$ (see, e.g., the first curve in Fig. 6.4, the elliptic curve has two connected components, while the case $\Delta < 0$ corresponds to a single connected component (see, e.g.,

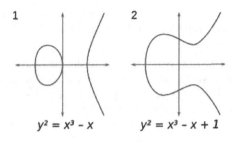

$$y^2 = x^3 - x \qquad y^2 = x^3 - x + 1$$

Figure 6.4:

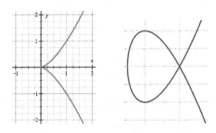

Figure 6.5:   Figure 6.6:

the second curve in Fig. 6.4[3]). The singular case $\Delta = 0$ corresponds to a cusp (for example, $y^2 = x^3$ has a cusp at $(0,0)$, see Fig. 6.5) or a self-intersection (for example, $y^2 = x^3 - 3x + 2 = (x-1)^2(x+2)$ with a self intersection at $(1,0)$, see Fig. 6.6).

The addition of two points on an elliptic curve E over the reals is made possible by the following property which is a particular case of Strong Bézout's Theorem 340. In order to define the sum of two distinct points P, Q on the curve, we consider the line (PQ) passing through P and Q. This line intersects the curve at a third point R (distinct or not). Intuitively, this point R which is well-defined from the points P, Q is a good candidate to be their sum. But such choice don't give the properties that we expect from a "good" addition: for example, we cannot define the zero element. We make the construction a bit more complicated by taking as $P + Q$ the symmetric of R with respect to the x-axis. This is a point on the curve by virtue of the form of a reduced Weierstrass equation (see Fig. 6.7.1). Some particular cases arise:

• The line (PQ) is vertical. In that case the point R is the point at ∞ and it is its own symmetric with respect to the x-axis. So, $P + Q$ is the point at ∞ (see Fig. 6.7.3).

---

[3]The source of Fig. 6.4 is https://commons.wikimedia.org/wiki/File:ECClines-3.svg.

• The line (PQ) is tangent to the curve at Q. Then the point R is nothing but Q (Q being a double intersection point). So, $P + Q$ is the symmetric of Q with respect to the x-axis (see Fig. 6.7.2).

• $P = Q$. Then the line (PQ) is the tangent to the curve at P. It intersects the curve at a third point R which can be equal to P (when P is a point of inflexion, see Fig. 6.8) or to another point (for example, the point at ∞, see Fig. 6.7.4). So, $P + Q$ is the symmetric of R with respect to the x-axis.

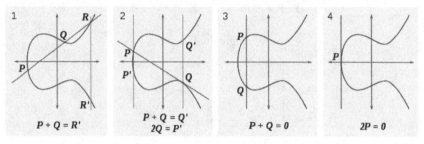

Figure 6.7: The sum of two points on an elliptic curve[4]

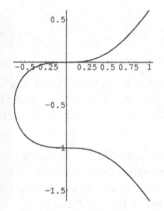

Figure 6.8: The elliptic curve $y^2 + y = x^3$ has two points of inflexion at $(0,0)$ and $(0,-1)$ in the real affine plane

Now, we will give a rigorous explanation on where the group structure on the points of an elliptic curve over a field $\mathbf{K}$ stems from. Let E be an elliptic curve with a point $O \in E(\mathbf{K})$. Recall that the Picard group $\mathrm{Pic}(E)$ is the quotient of the free group $\mathrm{Div}(E) = \mathbb{Z}[E]$ by the group $\mathrm{Princ}(E) = \mathrm{div}(\bar{\mathbf{K}}(E)^{\times})$ of principal divisors. We define also the Abelian group $\mathrm{Pic}^0(E)$ as the quotient of the group $\mathrm{Div}^0(E)$ of degree-zero divisors by $\mathrm{Princ}(E)$. So far, we have three groups built from the points of E:

---

[4]The source of Fig. 6.7 is https://en.wikipedia.org/wiki/File:ECClines.svg.

$$\mathrm{Princ}(E) \subsetneq \mathrm{Div}^0(E) \subsetneq \mathrm{Div}(E),$$

with quotient groups $\mathrm{Pic}(E) = \mathrm{Div}(E)/\mathrm{Princ}(E)$, $\mathrm{Pic}^0(E) = \mathrm{Div}^0(E)/\mathrm{Princ}(E)$, and $\mathrm{Div}(E)/\mathrm{Div}^0(E) \cong \mathrm{Pic}(E)/\mathrm{Pic}^0(E) \cong \mathbb{Z}$.

The map $D \mapsto (D - deg(D)[O], deg(D))$ defines an isomorphism between $\mathrm{Div}(E)$ and $\mathrm{Div}^0(E) \times \mathbb{Z}$. So, $\mathrm{Div}^0(E)$ is as interesting as $\mathrm{Div}(E)$. A much more interesting group is $\mathrm{Pic}^0(E)$ which describes the difference between $\mathrm{Div}^0(E)$ and $\mathrm{Princ}(E)$. The trick is to see degree-zero divisors as "parts of functions" as the following example suggests.

Example 395. (Example 290 continued)

Consider the elliptic curve $C : y^2 = x^3 + 1$ over $\mathbb{F}_{13}$. We saw that at the point $P = (0 : 1 : 1)$ (with affine coordinates $(0,1)$) the function x is a uniformizing parameter. The same thing is valid for the point $P' = (0 : -1 : 1)$.

Consider the divisors $D_1 = [(0 : 1 : 1)] - [O]$ and $D_2 = [(0 : -1 : 1)] - [O]$, where $O = (0 : 1 : 0)$ is the point of C at ∞. Both D_1 and D_2 are in $\mathrm{Div}^0(E)$, but neither is in $\mathrm{Princ}(E)$ (this is because a rational function should have the same order at the points P and P' since they have the same uniformizing parmeter x). However, $D_1 + D_2 = \mathrm{div}(x) \in \mathrm{Princ}(E)$. So, we could view D_1 and D_2 as being "pieces" of x.

Let E be an elliptic curve with a point $O \in E(\mathbf{K})$. The following important result puts the Abelian group $\mathrm{Pic}^0(E)$ in bijection with E. By structure transportation, this allows to equip E with an Abelian group structure.

Theorem 396. *Let E be an elliptic curve with a point $O \in E(\mathbf{K})$. Then the map*

$$\varphi : \quad \begin{array}{ccc} E & \longrightarrow & \mathrm{Pic}^0(E) \\ P & \longmapsto & [P] - [O] \end{array}$$

is bijective.

Proof. Let us first prove that φ is one-to-one. By way of contradiction, suppose that there are two points $P \neq Q$ on E with $\varphi(P) = \varphi(Q)$. Therefore, $[P] - [Q] = \mathrm{div}(f)$ for some $f \in \bar{\mathbf{K}}(E)^\times$, that is, f has only a simple zero at P and a simple pole at Q. Necessarily, f is nonconstant and belongs to $\mathscr{L}([Q])$, and, thus, $\ell([Q]) \geq 2$. This contradicts Corollary 302 of Riemann-Roch Theorem 301 stipulating that $\ell([Q]) = 1$.

Now, let us prove that φ is surjective. Let $D \in \mathrm{Div}^0(E)$. As $\deg(D + [O]) = 1$, then, by Corollary 302 of Riemann-Roch Theorem 301, we have $\ell(D + [O]) = 1$. Thus, there exists $f \in \bar{\mathbf{K}}(E)^\times$ such that $\mathrm{div}(f) \geq -D - [O]$, that is,

$$\mathrm{div}(f) = -D - [O] + \sum_{\substack{P_j \in E, n_j \geq 0 \\ 1 \leq j \leq \ell}} n_j[P_j].$$

But, since $\deg(\operatorname{div}(f)) = 0$ (by Corollary 287), we have $\displaystyle\sum_{1 \leq j \leq \ell} n_j = 1$, and, thus, all the n_j's are zero except one, say n_1, which is equal to 1. We conclude that $\varphi(P_1) = D$.

As a consequence of Theorem 396, for every $P_1, P_2 \in E$, there exists a unique $P_3 \in E$ and a function $f \in \bar{\mathbf{K}}(E)^\times$ (unique up to a constant function) such that

$$[P_1] - [O] + [P_2] - [O] = [P_3] - [O] + \operatorname{div}(f).$$

But given P_1 and P_2, how to compute the point P_3 and the function f?

Let $g = \alpha x + \beta y + \gamma$ ($\alpha, \beta, \gamma \in \mathbf{K}$) be the polynomial function defining the line through P_1 and P_2. It has zeroes at P_1, P_2 and some other point $R = (r_1, r_2)$ (by Strong Bézout's Theorem 340), and a triple pole at ∞. The polynomial function $h = x - r_1$ defines a "vertical" line through R and O: it has zeroes at R and $\mathfrak{s}_E(R)$, and a double pole at ∞. It follows that

$$\operatorname{div}\left(\frac{g}{h}\right) = ([P_1] + [P_2] + [R] - 3[O]) - ([R] + [\mathfrak{s}_E(R)] - 2[O]),$$

and so

$$[P_1] - [O] + [P_2] - [O] = [\mathfrak{s}_E(R)] - [O] + \operatorname{div}\left(\frac{g}{h}\right).$$

Thus, $P_3 = \mathfrak{s}_E(R)$ and $f = \frac{g}{h}$.

Comments 397.

- The bijection between E and $\operatorname{Pic}^0(E)$ allows to define an Abelian group structure to E: $(P_1, P_2) \mapsto P_1 + P_2 := P_3 = \mathfrak{s}_E(R)$.

- Note that $P_1 + P_2$ means the point on E representing the sum $[P_1] - [O] + [P_2] - [O]$, as opposed to the divisor $[P_1] + [P_2]$.

- The zero element of the group E is O since it corresponds to $[O] - [O] = 0$.

- For $P = (c_1, c_2) \in E$, we have $\operatorname{div}(x - c_1) = [P] + [\mathfrak{s}_E(P)] - 2[O]$, and, thus, $[P] - [O] + [\mathfrak{s}_E(P)] - [O] = 0$, showing that $-P = \mathfrak{s}_E(P)$.

- In the following, we summarize all the possible situations for the intersection between a projective line and an elliptic curve E. We know by Strong Bézout's Theorem 340 that they intersect at $1 \times 3 = 3$ points counted with their intersection multiplicities.

  (1) If L meets E at three distinct points P, Q, R, then $P + Q + R = O$ (see Fig. 6.7.1 and Fig. 6.7.3).

  (2) If L is tangent to E at P and meets E at a points Q distinct from P, then $2P + Q = O$ (see Fig. 6.7.2 and Fig. 6.7.4).

(3) If L is tangent to E at P and does not meet E at another point (we say that P is a *point of inflexion* of E, see Exercise 387), then $3P = O$ (see Fig. 6.8).

Now we are in position to give explicit formulas for the addition of two points on an elliptic curve. To simplify, we suppose that the base field is $\mathbb{R}$.

Explicit formulas for addition on an elliptic curve 398. Let E be an elliptic curve given by a reduced Weierstrass equation $E : y^2 = x^3 + ax + b$ with $a, b \in \mathbb{R}$, and consider two points $P = (x_P, y_P)$, $Q = (x_Q, y_Q) \in E$. Four cases may arise:

- Case 1: $x_P \neq x_Q$ (see Fig. 6.7.1 and Fig. 6.7.2). The line passing through P et Q is $(PQ) : y = sx + t$, with $s = \frac{y_P - y_Q}{x_P - x_Q}$ (the slope) and $t = \frac{y_Q x_P - x_Q y_P}{x_P - x_Q}$.
We have $(PQ) \cap E = \{P, Q, R = (x_R, y_R)\}$, with $x_R = s^2 - x_P - x_Q$ and $y_R = s x_R + t$. Finally, the coordinates of $P + Q$ (the symmetric of R with respect to the x-axis) are:

$$x_{P+Q} = s^2 - x_P - x_Q; \; y_{P+Q} = -s(s^2 - x_P - x_Q) - t.$$

- Case 2: $x_P = x_Q$ and $y_P \neq y_Q$ (see Fig. 6.7.3). We have necessarily $y_P = -y_Q$, and $P + Q$ is the point at ∞ (we consider that the point at ∞ is its own symmetric with respect to the x-axis).

- Case 3: $x_P = x_Q$ and $y_P = y_Q \neq 0$ (see Fig. 6.9). In other terms, $P = Q$ and this point is not on the x-axis. The equation of the tangent line to E at P is $y = sx + t$, with $s = \frac{3x_P^2 + a}{2y_P}$ and $t = y_P - \frac{(3x_P^2 + a)x_P}{2y_P}$. The coordinates of the point $P + P$, denoted by $2P$, are:

$$x_{2P} = s^2 - 2x_P; \; y_{2P} = -y_P + s(x_P - x_{2P}).$$

- Case 4: $x_P = x_Q$ and $y_P = y_Q = 0$ (see Fig. 6.7.4). In other terms, $P = Q$ and this point is on the x-axis. In that case $2P$ is the point at ∞.

Example 399. (Generating rational points, from
`http://www.geometer.org/mathcircles/ecc.pdf`)

Let E be the elliptic curve given by the reduced Weierstrass equation $E : y^2 = x^3 - 2x + 5$ (see Fig. 6.10).

We see that the point $P = (2, 3) \in E(\mathbb{Q})$. Starting from P and drawing the tangent line to E at P, it intersects E at another point Q (by Strong Bézout's Theorem 340). Because Q is in on the curve, $-Q$ (the symmetric of Q with respect to the x-axis, this is in fact the point $2P$) is also on the curve. Again, by Strong Bézout's Theorem 340, the line passing through P and $-Q$ intersects E at another point R. From there, we can find $-R$ or intersect the line (QR) with the

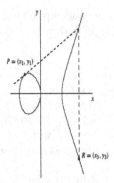

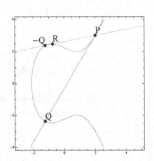

Figure 6.9: Doubling of a point
P on an elliptic curve

Figure 6.10: Generating rational points
on the elliptic curve $y^2 = x^3 - 2x + 5$

curve to obtain more and more points. This process can be extended forever, and since the original coordinates at the point P are rational, all the other will be too. In the example, the coordinates of the first generated rational points are:

$$P = (2,3), \; Q = \left(-\frac{11}{9}, -\frac{64}{27}\right), \; -Q = \left(-\frac{11}{9}, \frac{64}{27}\right), \; R = \left(-\frac{622}{841}, -\frac{60111}{24389}\right).$$

The coordinates are rational, but the rational numbers get messier and messier as we go on.

Now we give explicit formulas for the addition of two points on an elliptic curve given by a general Weierstrass equation.

Explicit formulas for addition on an elliptic curve 400. Let E be an elliptic curve defined over a field $\mathbf{K}$ and given by a Weierstrass equation:

$$y^2 + a_1xy + a_3y = x^3 + a_2x^2 + a_4x + a_6 \; (a_i \in \mathbf{K}).$$

Let $P_1 = (x_1, y_1)$ and $P_2 = (x_2, y_2)$ be two points of E. The point $P_3 = P_1 + P_2$ is given by:

- if $x_1 = x_2$ and $y_1 + y_2 + a_1x_1 + a_3 = 0$, then $P_3 = O$, and, thus, $P_2 = -P_1$;

- if $x_1 \neq x_2$ then $P_3 = (x_3, y_3) = (\lambda^2 + a_1\lambda - a_2 - x_1 - x_2, -(\lambda + a_1)x_3 - v - a_3)$ with

$$(\lambda, v) = \begin{cases} \left(\frac{y_2-y_1}{x_2-x_1}, \frac{y_1x_2-y_2x_1}{x_2-x_1}\right) & \text{if} \quad x_1 \neq x_2 \\ \left(\frac{3x_1^2+2a_2x_1+a_4-a_1y_1}{2y_1+a_1x_1+a_3}, \frac{-x_1^3+a_4x_1+2a_6-a_3y_1}{2y_1+a_1x_1+a_3}\right) & \text{if} \quad x_1 = x_2. \end{cases}$$

Definition 401.

- Let E be an elliptic curve with a point $O \in E(\mathbf{K})$. For $m \in \mathbb{Z}$, we denote by $[m] : E \to E$ the map $P \mapsto mP$ defined with the group structure in E, that is, $mP = \underbrace{P + \cdots + P}_{m \text{ times}}$ if $m > 0$, $mP = -(-m)P$ if $m < 0$, and $0P = O$. As already pointed out, the coordinates of mP are rational functions of the co-ordinates of P. Thus, by virtue of Proposition 309, $[m]$ is regular. Moreover, we can prove that $[m]$ is surjective for $m \neq 0$ (by virtue of Proposition 317, it suffices to prove that $[m]$ is nonconstant).

- Let E, E' be two elliptic curves over a field $\mathbf{K}$. An *isogeny* $\phi : E \to E'$ is a regular map such that $\phi(O) = O$. If $E = E'$, an isogeny from E to E is simply called *an endomorphism* of E. An example is $[m]$. It is worth pointing out that an isogeny between elliptic curves is always a group homomorphism.

Example 402. Let $E : y^2 = x^3 + ax^2 + bx$ and $E' : y^2 = x^3 - 2ax^2 + (a^2 - 4b)x$ be two elliptic curves over a field $\mathbf{K}$. Consider

$$\phi : \quad E \longrightarrow E'$$
$$(x, y) \longmapsto (\tfrac{y^2}{x^2}, y\tfrac{b-x^2}{x^2}).$$

The map ϕ is rational and, thus, regular (by virtue of Proposition 309). To show that ϕ is an isogeny, we have to check that $\phi(O) = O$. We projectivize ϕ as $\tilde{\phi}(x : y : z) = (y^2 z : byz^2 - x^2 y : x^2 z)$ which is not defined at $(0 : 1 : 0)$. But we can show that $\tilde{\phi}$ can be defined on all points of E (using the equation $y^2 z = x^3 + ax^2 z + bxz$ of the projective completion of E) as:

$$\tilde{\phi}(x : y : z) = (y^2 z : byz^2 - x^2 y : x^2 z)$$
$$= \left(xy^2 : bxyz - \frac{x^3 y}{z} : x^3 \right) = (xy^2 : 2bxyz - y^3 + ax^2 y : x^3).$$

Now we see that $\tilde{\phi}(0 : 1 : 0) = (0 : 1 : 0)$. Note that the degree of ϕ is 2 since this is the cardinality of a generic fibre.

Remark 403.

(1) Let us say a few words on the dynamical method in algebra. As a typical example, let us consider the classical theorem "any polynomial P in $\mathbf{K}[X]$ is a product of irreducible polynomials ($\mathbf{K}$ a field)". This leads to an interesting problem: it seems like no general algorithm that produces the irreducible factors. What, then, is the constructive content of this theorem? A possible answer is as follows: when performing computations with P, proceed as if its decomposition into irreducible polynomials were known (at the beginning, proceed as if P were irreducible). When something strange happens (e.g., when the gcd of P and

another polynomial Q is a strict divisor of P), use this fact to improve the decomposition of P. This trick was invented in Computer Algebra as the D5-philosophy [16], and later taken up in the form of the dynamical proof method in algebra [13]. It indeed enables one to carry out computations inside the algebraic closure $\widetilde{\mathbf{K}}$ of $\mathbf{K}$ even if it is not possible to effectively construct $\widetilde{\mathbf{K}}$, for in general this would require transfinite methods as Zorn's Lemma. The foregoing has been referred to as "dynamical evaluation" of the algebraic closure. Dynamical Gröbner bases [32, 64, 66] are an example of use of this philosophy (see Subsection 1.3).

(2) Another important example of use of the dynamical method is the elliptic-curve factorization method (in short, ECM) to find a nontrivial factor of a given composite natural number n. It is a fast method for integer factorization due to Lenstra [45] and it works as follows: pick random elements x_0, y_0, $a \in \mathbb{Z}/n\mathbb{Z}$ and consider the "elliptic curve" $E : y^2 = x^3 + ax + b$ over $\mathbb{Z}/n\mathbb{Z}$ with $b = y_0^2 - x_0^3 - ax_0$. Then we can form repeated multiples mP of the point $P = (x_0, y_0) \in E$ using the Explicit formulas 400 for addition on an elliptic curve as if $\mathbb{Z}/n\mathbb{Z}$ were a field, that is, as if n were a prime number. When something strange happens (when computing a certain slope $\frac{u}{v}$, we find that v is a zero-divisor in $\mathbb{Z}/n\mathbb{Z}$, that is, $\gcd(v, n) \in [\![2, n-1]\!]$), this shows that our elliptic curve is not a group and $\gcd(v, n)$ is a nontrivial factor of n. If we were able to finish all the calculations above without encountering non-invertible elements mod n, then we need to try again with some other curve and starting point. If at some stage we found $mP = O$ (the point at infinity on the "elliptic curve"), we should start over with a new curve and starting point, since this point is the group identity element, so is unchanged under any further addition operations.

6.2   Elliptic curves over a finite field

Let E be an elliptic curve defined over a finite field $\mathbb{F}_q$ with q elements and denote by O its point at ∞. The set $E(\mathbb{F}_q)$ of $\mathbb{F}_q$-rational points is a finite Abelian group whose group structure is given by the following theorem.

Theorem 404. *Let E be an elliptic curve defined over a finite field $\mathbb{F}_q$. Then the group $E(\mathbb{F}_q)$ is either cyclic ($\simeq \mathbb{Z}/d\mathbb{Z}$, where $d = |E(\mathbb{F}_q)|$), or the product of two cyclic groups, more precisely, it is isomorphic to $\mathbb{Z}/d_1\mathbb{Z} \times \mathbb{Z}/d_2\mathbb{Z}$, where $d_1 \mid d_2$, $d_1 \mid q-1$, and $d_1 d_2 = |E(\mathbb{F}_q)|$.*

Definition 405. Let E be an elliptic curve defined over a field $\mathbf{K}$ and $n \in \mathbb{N}^*$. A point $P \in E(\bar{\mathbf{K}})$ is said to be of *n-torsion* if $nP = O$. The subgroup of $E(\bar{\mathbf{K}})$ of *n*-torsion points is denoted by $E[n]$.

The following theorem describes the group structure of $E[n]$.

Theorem and Definition 406. *Let E be an elliptic curve defined over a field $\mathbf{K}$, and $n \in \mathbb{N}^*$. Then*

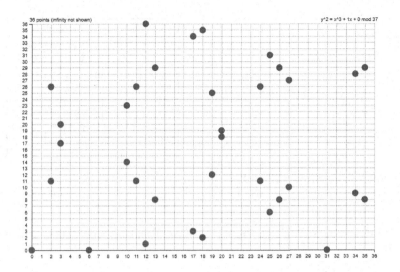

Figure 6.11: The elliptic curve $y^2 = x^3 + x$ over $\mathbb{F}_{37}$

- *if* char($\mathbf{K}$) $= 0$ *or* gcd(n, char($\mathbf{K}$)) $= 1$, *then* $E[n] \simeq \mathbb{Z}/n\mathbb{Z} \times \mathbb{Z}/n\mathbb{Z}$;

- *if* char($\mathbf{K}$) $= p \neq 0$, *then either*

  $E[p] = \{O\}$ *(in such case, E is called* supersingular *(see Exercise 432), and we have $E[p^r] = \{O\}$ for all $r \geq 1$), or*

  $E[p] \simeq \mathbb{Z}/p\mathbb{Z}$ *(in such case, E is called* ordinary, *and we have $E[p^r] \simeq \mathbb{Z}/p^r\mathbb{Z}$ for all $r \geq 1$);*

- *more generally, if* char($\mathbf{K}$) $= p \neq 0$ *and* $n = p^r n'$ *with* $p \nmid n'$, *then*
  $E[n] \simeq \mathbb{Z}/n'\mathbb{Z} \times \mathbb{Z}/n'\mathbb{Z}$ *or* $\mathbb{Z}/n\mathbb{Z} \times \mathbb{Z}/n'\mathbb{Z}$.

In particular, for all $n \geq 2$, the group $E[n]$ is finite of order at most n^2.

Corollary 407. *Let E be an elliptic curve defined over a field $\mathbf{K}$ and ℓ a prime number $\neq$ char($\mathbf{K}$). Then $E[\ell]$ is an $\mathbb{F}_\ell$-vector space of dimension 2.*

Example 408. Consider the elliptic curve $E : y^2 = x^3 + 4x + 1$ defined over the field $\mathbb{F}_5$ (see Fig. 6.12).[5]

We have $E(\mathbb{F}_5) = \{O, (0,1), (0,4), (1,1), (1,4), (3,0), (4,1), (4,4)\}$. We can check that the point $(0, 1)$ has order 8 and so $E(\mathbb{F}_5) \simeq \mathbb{Z}/8\mathbb{Z}$. The group $E(\mathbb{F}_5)$ does not contain a nontrivial 5-torsion point while, over the field $\mathbb{F}_{5^8} = \mathbb{F}_5[t]/\langle t^8 + 2 \rangle$, the point $(2t^4 + 1, 2t^6 + t^2)$ is a 5-torsion point (it cannot be defined in a

---

[5]Figures 6.12, 6.11 and 6.13 are obtained via https://graui.de/code/elliptic2.

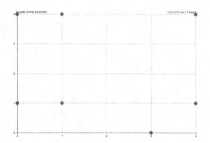

Figure 6.12: The elliptic curve $y^2 = x^3 + 4x + 1$ over $\mathbb{F}_5$

+	∞	(0,1)	(0,4)	(1,1)	(1,4)	(3,0)	(4,1)	(4,4)
∞	∞	(0,1)	(0,4)	(1,1)	(1,4)	(3,0)	(4,1)	(4,4)
(0,1)	(0,1)	(4,1)	∞	(4,4)	(3,0)	(1,1)	(1,4)	(0,4)
(0,4)	(0,4)	∞	(4,4)	(3,0)	(4,1)	(1,4)	(0,1)	(1,1)
(1,1)	(1,1)	(4,4)	(3,0)	(4,1)	∞	(0,1)	(0,4)	(1,4)
(1,4)	(1,4)	(3,0)	(4,1)	∞	(4,4)	(0,4)	(1,1)	(0,1)
(3,0)	(3,0)	(1,1)	(1,4)	(0,1)	(0,4)	∞	(4,4)	(4,1)
(4,1)	(4,1)	(1,4)	(0,1)	(0,4)	(1,1)	(4,4)	(3,0)	∞
(4,4)	(4,4)	(0,4)	(1,1)	(1,4)	(0,1)	(4,1)	∞	(3,0)

Figure 6.13: Table of point additions

smaller extension field of $\mathbb{F}_5$). Thus, $E[5] \simeq \mathbb{Z}/5\mathbb{Z}$ and the curve is not supersingular.

The use of elliptic curves for the discrete logarithm problem requires that we know the order $|E(\mathbb{F}_q)|$ of the group $E(\mathbb{F}_q)$ (or at least the order of the base point). An estimation of this order is given by the following theorem.

Theorem 409. (Hasse-Weil)

Let E be an elliptic curve defined over $\mathbb{F}_q$. We have:

$$\big||E(\mathbb{F}_q)| - q - 1\big| \leq 2\sqrt{q}.$$

Deuring has proved that if q is prime, Hasse's bound above is optimal: for $t \in \mathbb{Z}$ with $|t| \leq 2\sqrt{q}$, there exists an elliptic curve E such that $|E(\mathbb{F}_q)| = q + 1 - t$. As application of Theorem 409, if there exists a point $P \in E(\mathbb{F}_q)$ of order $> \frac{q+1}{2} + \sqrt{q}$ then necessarily $E(\mathbb{F}_q)$ is cyclic generated by P. But, in practice, we rather first compute $|E(\mathbb{F}_q)|$ and then try to deduce the order of P. For example, if the curve is given by a Weierstrass equation $y^2 = x^3 + ax^2 + bx + c$ over $\mathbb{F}_p$ ($p \neq 2$), then we have the formula:

$$|E(\mathbb{F}_p)| = p + 1 + \sum_{x \in \mathbb{F}_p} \left(\frac{x^3 + ax^2 + bx + c}{p} \right),$$

where $\left(\frac{z}{p}\right)$ denotes the Legendre Symbol (see definition and proposition below). This allows to compute $|E(\mathbb{F}_p)|$ with complexity $\varnothing(p^{1+\varepsilon})$.

Definition 410.

(1) Fix a prime number p. An integer a not divisible by p is a *quadratic residue* modulo p if a is a square modulo p; otherwise, a is a *quadratic nonresidue*. For example, the squares modulo 5 are $1^2 = 1$, $2^2 = 4$, $3^2 = 4$, $4^2 = 1$. So, 1 and 4 are both quadratic residues and 2 and 3 are quadratic nonresidues.

(2) Let p be an odd prime and let a be an integer. Set

$$\left(\frac{a}{p}\right) := \begin{cases} 0 & \text{if } p \mid a, \\ +1 & \text{if } a \text{ is a quadratic residue modulo } p, \text{ and} \\ -1 & \text{if } a \text{ is a quadratic nonresidue modulo } p. \end{cases}$$

We call this symbol the *Legendre Symbol*. Note that, by Exercise 373, we have $\left(\frac{a}{p}\right) \equiv a^{\frac{p-1}{2}} \bmod p$ (*Euler's criterion*). As a consequence, the map $a \mapsto \left(\frac{a}{p}\right)$ is multiplicative.

For example, we have

$$\left(\frac{1}{5}\right) = 1, \left(\frac{2}{5}\right) = -1, \left(\frac{3}{5}\right) = -1, \left(\frac{4}{5}\right) = 1, \left(\frac{5}{5}\right) = 0.$$

Proposition 411. *Let E be an elliptic curve given by a Weierstrass equation $y^2 = x^3 + ax^2 + bx + c$ over $\mathbb{F}_p$ ($p \neq 2$), then*

$$|E(\mathbb{F}_p)| = p + 1 + \sum_{x \in \mathbb{F}_p} \left(\frac{x^3 + ax^2 + bx + c}{p}\right).$$

Proof. For $x \in \mathbb{F}_p$, there are exactly two points of abscissa x in $E(\mathbb{F}_p)$ if $\left(\frac{x^3+ax^2+bx+c}{p}\right) = 1$, none if $\left(\frac{x^3+ax^2+bx+c}{p}\right) = -1$, and only one if $\left(\frac{x^3+ax^2+bx+c}{p}\right) = 0$. Thus, counting the point at ∞, we have $|E(\mathbb{F}_p)| = \sum_{x \in \mathbb{F}_p} (1 + (\frac{x^3 + ax^2 + bx + c}{p}))$.

Example 412. Let E be the elliptic curve $E : y^2 = x^3 + x + 5$ defined over $\mathbb{F}_{11}$. As $\left(\frac{3}{11}\right) = \left(\frac{4}{11}\right) = \left(\frac{5}{11}\right) = 1$ and $\left(\frac{2}{11}\right) = \left(\frac{6}{11}\right) = \left(\frac{7}{11}\right) = \left(\frac{8}{11}\right) = -1$, then

$$|E(\mathbb{F}_{11})| = 12 + \left(\frac{5}{11}\right) + \left(\frac{7}{11}\right) + \left(\frac{4}{11}\right) + \left(\frac{2}{11}\right) + \left(\frac{7}{11}\right)$$

$$+ \left(\frac{3}{11}\right) + \left(\frac{7}{11}\right) + \left(\frac{3}{11}\right) + \left(\frac{8}{11}\right) + \left(\frac{6}{11}\right) + \left(\frac{3}{11}\right) = 11.$$

Definition and Proposition 413. Let E be an elliptic curve defined over $\mathbb{F}_q$, and write $|E(\mathbb{F}_q)| = q + 1 - a$ with $|a| \leq 2\sqrt{q}$. The integer a is called the *trace of Frobenius*.

The curve E is supersingular if and only if $\text{char}(\mathbb{F}_q) \mid a$ (see Exercise 432).

The characteristic polynomial of E is

$$P_E(T) := T^2 - aT + q \in \mathbb{Z}[T].$$

Note that $|E(\mathbb{F}_q)| = P_E(1)$. The polynomial $P_E(T)$ being the characteristic polynomial of the Frobenius endomorphism (see Exercise 420), we have $P_E(\sigma_q) = 0$, that is, for all $P = (x, y) \in E(\bar{\mathbb{F}}_q)$, we have

$$(x^{q^2}, y^{q^2}) - a \cdot (x^q, y^q) + q \cdot (x, y) = O.$$

The complex roots α and $\beta = \bar{\alpha}$ of $P_E(T)$ are called the *characteristic roots of Frobenius*.

The following theorem enables to compute $|E(\mathbb{F}_{q^n})|$ as soon as we know $|E(\mathbb{F}_q)|$. See also the recursive formula given in Exercise 431 for computing $\alpha^n + \beta^n$.

Theorem 414. *Keeping the notation of Definition and Proposition 413, we have:*

$$|E(\mathbb{F}_{q^n})| = q^n + 1 - (\alpha^n + \beta^n).$$

Example 415. Consider the elliptic curve $E : y^2 + xy = x^3 + x^2 + 1$ defined over the field $\mathbb{F}_2$. We have $E(\mathbb{F}_2) = \{O, (0,1)\}$, and so the trace of Frobenius is $a = 1$, the characteristic polynomial of E is $P_E(T) = T^2 - T + 2$ with characteristic roots of Frobenius $\frac{1}{2}(1 \pm i\sqrt{7})$. It follows that

$$|E(\mathbb{F}_{2^{100}})| = 2^{100} + 1 - \left(\frac{1 + i\sqrt{7}}{2}\right)^{100} - \left(\frac{1 - i\sqrt{7}}{2}\right)^{100}$$

$$= 1267650600228229382588845215376.$$

Example 416. (Example 408 continued)

Consider the elliptic curve $E : y^2 = x^3 + 4x + 1$ defined over the field $\mathbb{F}_5$. We saw that $|E(\mathbb{F}_5)| = 8$, and so the trace of Frobenius is $a = -2$, the characteristic polynomial of E is $P_E(T) = T^2 + 2T + 5$ with characteristic roots of Frobenius $-1 \pm 2i$. It follows that

$$|E(\mathbb{F}_{25})| = 25 + 1 - (-1 + 2i)^2 - (-1 - 2i)^2 = 32.$$

As $a = -2 \not\equiv 0 \bmod 5$, the curve E is ordinary (as already seen), and, thus, there exists an extension field $\mathbb{F}_{5^n}$ of $\mathbb{F}_5$ such that $E(\mathbb{F}_{5^n})$ contains a 5-torsion point, or equivalently, such that 5 divides $|E(\mathbb{F}_{5^n})|$. Computing repeatedly $|E(\mathbb{F}_{5^n})| = 5^n + 1 - (-1 + 2i)^n - (-1 - 2i)^n$ for $n = 1, 2, 3, \ldots$, we obtain for $n = 8$:

$$|E(\mathbb{F}_{5^8})| = 391680.$$

6.3   Exercises

Exercise 417. Let $E = E(\bar{\mathbf{K}})$ be an elliptic curve defined over a field $\mathbf{K}$, and $n \in \mathbb{N}^*$. Show that $|E[n]| = |E/nE|$. Are these groups isomorphic?

Exercise 418. Let E be an elliptic curve given by a reduced Weierstrass equation $E : y^2 = x^3 + ax + b$ with $a, b \in \mathbf{K}$ ($\mathbf{K}$ a field). Show that a uniformizing parameter at the point O at ∞ is $\frac{x}{y}$.

Exercise 419. Show that a divisor $D = \sum_{i=1}^{r} n_i.[P_i] \in \mathbb{Z}[E]$ on an elliptic curve E with a point O at ∞ is principal if and only if $\deg(D) = 0$ and $\sum_{i=1}^{r} n_i P_i = O$ (sum of points on E).

Exercise 420. Let E be an elliptic curve defined over a finite field $\mathbf{K} = \mathbb{F}_q$. Show that the map $\phi : E \to E$ with $\phi(x,y) = (x^q, y^q)$ and $\phi(O) = O$ is an isogeny (it is called the *Frobenius endomorphism* of E). Show that it is bijective but not an isomorphism.

Exercise 421. Let E be an elliptic curve defined over a field $\mathbf{K}$ given by a Weierstrass equation

$$E : y^2 + a_1 xy + a_3 y = x^3 + a_2 x^2 + a_4 x + a_6 \ \ (a_i \in \mathbf{K}). \qquad (6.2)$$

We consider the differential form $\omega := \frac{dx}{2y + a_1 x + a_3}$.

1) Show that $\omega = \frac{dy}{3x^2 + 2a_2 x + a_4 - a_1 y}$.

2) Show that $\operatorname{div}(\omega) = 0$.

Exercise 422. Consider the points $P = (-3,9)$ and $Q = (-2,8)$ on the elliptic curve $y^2 = x^3 - 36x$ over $\mathbb{Q}$. Calculate $P + Q$.

Exercise 423. Consider the point $P = (2,3)$ on the elliptic curve $E : y^2 = x^3 + 1$ over $\mathbb{Q}$. Calculate $2P$ and $4P$. Deduce the order of P.

Exercise 424. Show that $E : y^2 = x^3 + a_2 x^2 + a_4 x + a_6 \ (a_i \in \mathbf{K})$ is not an elliptic curve when $\operatorname{char}(\mathbf{K}) = 2$.

Exercise 425. Let E be an elliptic curve over a field of characteristic $\neq 2, 3$, and $P \in E$.

1) Show that $3P = O$ if and only if P is a point of inflexion (see Exercise 387).

2) Suppose that $P \neq O$ is a point of inflexion of E with tangent line $L : ax + by + c = 0$. What is $\operatorname{div}(ax + by + c)$?

Exercise 426.

1) Let p be an odd prime number. Show that -1 is a quadratic residue modulo p if and only if $p \equiv 1 \bmod 4$.

2) Let a be an odd integer and consider the elliptic curve E over $\mathbb{Q}$ given by the equation $y^2 = x^3 + (2a)^3 - 1$. Show that E does not contain any point of the form (x,y) with $x,y \in \mathbb{Z}$.

Exercise 427. We admit that if $P = (x_1, y_1)$, $Q = (x_2, y_2)$ and $P + Q = (x_3, y_3)$ are points in the affine plane on a supersingular elliptic curve $E : y^2 + a_3 y = x^3 + a_4 x + a_6 \ (a_i \in \mathbf{K}, a_3 \neq 0, \operatorname{char}(\mathbf{K}) = 2)$, then:

$x_3 = (\frac{y_1+y_2}{x_1+x_2})^2 + x_1 + x_2$, $y_3 = (\frac{y_1+y_2}{x_1+x_2})(x_1+x_3) + y_1 + a_3$ if $P \neq Q$ and

$x_3 = \frac{x_1^4+a_4^2}{a_3^2}$, $y_3 = (\frac{x_1^2+a_4}{a_3})(x_1+x_3) + y_1 + a_3$ if $P = Q$.

Let E be the supersingular elliptic curve $E : y^2 + y = x^3$ defined over $\mathbb{F}_2$.

1) Let $P = (x,y) \in E(\mathbb{F}_{16})$. Give the coordinates of $-P$ and $2P$ as functions of x and y.

2) Show that every $P \in E(\mathbb{F}_{16})$ distinct from the point at infinity has order 3.

3) Show that every point in $E(\mathbb{F}_{16})$ is in fact in $E(\mathbb{F}_4)$.

4) Deduce $|E(\mathbb{F}_{16})|$ using Hasse's Theorem 409. What is the group structure of $E(\mathbb{F}_{16})$?

Exercise 428.

1) Let E be the elliptic curve $E : y^2 = x^3 + 7x + 1$ defined over $\mathbb{F}_{101}$. Knowing that $(0,1)$ is a point of $E(\mathbb{F}_{101})$ of order 116, deduce $|E(\mathbb{F}_{101})|$.

2) Let C be the elliptic curve $C : y^2 = x^3 - 10x + 21$ defined over $\mathbb{F}_{557}$. Knowing that $(2,3)$ is a point of $C(\mathbb{F}_{557})$ of order 189, deduce $|C(\mathbb{F}_{557})|$.

Exercise 429. Let E be the elliptic curve $E : y^2 = x^3 + 2$ defined over $\mathbb{F}_7$.

1) Check that the points $P = (-1,1)$ and $Q = (0,4)$ have order 3.

2) Deduce that $E(\mathbb{F}_7) = E[3] \cong (\mathbb{Z}/3\mathbb{Z}) \times (\mathbb{Z}/3\mathbb{Z})$.

Exercise 430. Consider an elliptic curve $E : y^2 + xy = x^3 + a_2x^2 + a_6$ defined over $\mathbb{F}_{2^n}$. Let $P = (x,y) \in E(\mathbb{F}_{2^n})$. Give the coordinates of $-P$ as functions of x and y.

Exercise 431. Let E be an elliptic curve defined over $\mathbb{F}_q$, and write $|E(\mathbb{F}_q)| = q + 1 - a$, where the integer a is the trace of Frobenius. Recall that the characteristic polynomial of E is $P_E(T) = T^2 - aT + q \in \mathbb{Z}[T]$. Recall also that, denoting the characteristic roots of Frobenius by α and $\beta = \bar{\alpha}$ (the complex roots of $P_E(T)$), we have $|E(\mathbb{F}_{q^n})| = q^n + 1 - (\alpha^n + \beta^n)$.

For $k \in \mathbb{N}$, we set $s_k = \alpha^k + \beta^k$. We have $s_0 = 2$ and $s_1 = a$. Show that, for $k \geq 1$,

$$s_{k+1} = a s_k - q s_{k-1}$$

(this recursive formula enables to compute $|E(\mathbb{F}_{q^n})|$ as soon as we know $|E(\mathbb{F}_q)|$).

Exercise 432. Let E be an elliptic curve defined over a finite field $\mathbb{F}_q$ of characteristic p and Frobnenius trace a. Show that:

$$E \text{ is supersingular} \iff |E(\mathbb{F}_q)| \equiv 1 \bmod p \iff p \mid a.$$

Exercise 433. Let E be an elliptic curve defined over $\mathbb{F}_p$, where p is a prime number > 3. Show that E is supersingular if and only if $|E(\mathbb{F}_p)| = p + 1$.

Exercise 434. Let $E : y^2 = x^3 + ax + b$ be an elliptic curve defined over a field $\mathbf{K}$ of characteristic ≥ 5 with $\Delta = 4a^3 + 27b^2 \neq 0$, and denote by $O = (0 : 1 : 0)$ its point at ∞. We denote by α, β, γ the three roots of $X^3 + aX + b$ in $\bar{\mathbf{K}}$.

1) Show that $E[2] = \{O, (\alpha : 0 : 1), (\beta : 0 : 1), (\gamma : 0 : 1)\}$ and check that $E[2] \cong (\mathbb{Z}/2\mathbb{Z}) \times (\mathbb{Z}/2\mathbb{Z})$ (the Klein group).

2) Let $G = 3X^4 + 6aX^2 + 12bX - a^2 \in \mathbf{K}[X]$.

2.a) Show that G has 4 distinct roots in $\bar{\mathbf{K}}$.

2.b) Let $P = (x, y)$ be a point in $E(\bar{\mathbf{K}})$ distinct from O. Show that:

$$P \in E[3] \quad \Leftrightarrow \quad G(x) = 0.$$

2.c) Deduce that $E[3] \cong (\mathbb{Z}/3\mathbb{Z}) \times (\mathbb{Z}/3\mathbb{Z})$.

3) Compute $E[2]$ and $E[3]$ when $E : y^2 = x^3 - 2$ over $\mathbb{Q}$. Give a basis $\mathscr{B}_\ell$ for $E[\ell]$ ($\ell = 2, 3$) as an $\mathbb{F}_\ell$-vector space.

4) Compute $E[2]$ and $E[3]$ when $E : y^2 = x^3 + x + 3$ over $\mathbb{F}_{11}$. Give a basis $\mathscr{B}_\ell$ for $E[\ell]$ ($\ell = 2, 3$) as an $\mathbb{F}_\ell$-vector space.

5) Suppose that $\mathbf{K} = \mathbb{R}$. Let $\tau : E(\mathbb{C}) \to E(\mathbb{C}); P = (x, y) \mapsto (\bar{x}, \bar{y})$ and $\tau(O) = O$, where $\bar{z}$ denotes the conjugate of $z \in \mathbb{C}$.

5.a) Show that τ is a group automorphism.

5.b) As τ induces (by restriction) an endomorphism of the rank 2 free $\mathbb{Z}/n\mathbb{Z}$-module $E[n]$, using the basis found in 3), write the matrix representing the action of τ on $E[2]$.

6) More generally, for $\sigma \in \mathrm{Gal}(\bar{\mathbf{K}}/\mathbf{K})$, let

$$\rho_\sigma : \begin{array}{ccc} E(\bar{\mathbf{K}}) & \longrightarrow & E(\bar{\mathbf{K}}) \\ P = (x, y) & \longmapsto & (\sigma(x), \sigma(y)), \end{array}$$

and $\sigma(O) = O$. Show that ρ_σ is a group automorphism.

Suppose that $\mathrm{char}(\mathbf{K}) \nmid n$. By restriction to $E[n]$, ρ_σ induces an automorphism $\rho_{n,\sigma}$ of the rank 2 free $\mathbb{Z}/n\mathbb{Z}$-module $E[n]$ (see Theorem and Definition 406). Show that

$$\rho_n : \begin{array}{ccc} \mathrm{Gal}(\bar{\mathbf{K}}/\mathbf{K}) & \longrightarrow & \mathrm{GL}_2(\mathbb{Z}/n\mathbb{Z}) \\ \sigma & \longmapsto & \rho_{n,\sigma} \end{array}$$

is a group homomorphism (we identify $\mathrm{Aut}(E[n])$ with $\mathrm{GL}_2(\mathbb{Z}/n\mathbb{Z})$).

Exercise 435.

1) Show that if p is a prime number then

$$|\mathrm{GL}_2(\mathbb{F}_p)| = (p^2 - 1)(p^2 - p).$$

2) Show that if p is a prime number and $\alpha \geq 1$ then

$$|GL_2(\mathbb{Z}/p^\alpha\mathbb{Z})| = p^{4\alpha} \left(1 - \frac{1}{p}\right)\left(1 - \frac{1}{p^2}\right).$$

3) Show that for $n \geq 2$, we have

$$|GL_2(\mathbb{Z}/n\mathbb{Z})| = n^4 \prod_{p|n;\, p\,\text{prime}} \left(1 - \frac{1}{p}\right)\left(1 - \frac{1}{p^2}\right).$$

Exercise 436. Let E be the elliptic curve $E : y^2 = x^3 + x + 3$ defined over $\mathbb{F}_{11}$.

1) Compute $|E(\mathbb{F}_{11})|$.

2) Compute $|E(\mathbb{F}_{11^3})|$.

Exercise 437. Let E be the elliptic curve $E : y^2 = x^3 + 2x$ defined over $\mathbb{F}_{13}$.

1) Compute the points of $E(\mathbb{F}_{13})$. What is the group structure of $E(\mathbb{F}_{13})$?

2) Determine the points of $E(\mathbb{F}_{13})$ of order 2.

3) Let $P = (1,4) \in E(\mathbb{F}_{13})$. Compute $2P$ and $4P$. Give a point in $E(\mathbb{F}_{13})$ of order 5.

4) We consider the extension field $\mathbb{F}_{13^2} = \mathbb{F}_{13}(\theta)$ with $\theta^2 = -2$. What are the points of $E(\mathbb{F}_{13^2})$ of order 2? Is the group $E(\mathbb{F}_{13^2})$ cyclic?

5) Compute $|E(\mathbb{F}_{13^2})|$.

Exercise 438. Let E be the elliptic curve $E : y^2 = x^3 - ux$ defined over $\mathbb{F}_p$, where $p \geq 5$ is a prime number $\equiv 3 \bmod 4$ and $u \in \mathbb{F}_p^\times$.

Show that $|E(\mathbb{F}_p)| = p + 1$. Deduce that E is supersingular.

Exercise 439. Let E be an elliptic curve defined over a finite field $\mathbf{K} = \mathbb{F}_q$. We saw in Exercise 420 that the map $\phi_q : E \to E$ with $\phi_q(x,y) = (x^q, y^q)$ and $\phi_q(O) = O$ is an isogeny (it is called the *Frobenius endomorphism* of E). It is bijective but not an isomorphism.

Let $n \geq 2$ and suppose that $\text{char}(\mathbf{K}) \nmid n$. By restriction to $E[n]$, ϕ_q induces an endomorphism $\phi_{q,n}$ of the rank 2 free $\mathbb{Z}/n\mathbb{Z}$-module $E[n]$ (see Theorem and Definition 406).

Considering a basis for $E[n]$, we can represent $\phi_{q,n}$ by a 2×2 matrix with entries in $\mathbb{Z}/n\mathbb{Z}$. Then, a general result, that we admit here, says that its trace is $\text{Tr}(\phi_{q,n}) = a + n\mathbb{Z}$ and its determinant is $\det(\phi_{q,n}) = q + n\mathbb{Z}$, where a is the trace of Frobenius of E.

Check this result on $E[3]$ computed in Exercise 434.4.

Exercise 440. Let $E : y^2 = x^3 + b$ be an elliptic curve defined over $\mathbb{F}_p$ (p a prime number) with $b \in \mathbb{F}_p$ and $p \equiv 2 \bmod 3$.

1) Show that $b \neq 0$ and $p > 3$.

2) Give the group structure of $E[2]$.

3) Show that application $\varphi : \mathbb{F}_p^\times \to \mathbb{F}_p^\times$; $x \mapsto x^3$, is bijective.

4) Show that $E[2] \nsubseteq E(\mathbb{F}_p)$ and that $E[2] \subseteq E(\mathbb{F}_{p^2})$.

5) Show that $|E(\mathbb{F}_p)| = p + 1$. Deduce that the curve is supersingular.

6) Give $|E(\mathbb{F}_{p^n})|$ for $n \neq 1$.

7) Show that $E(\mathbb{F}_p) \cong \mathbb{Z}/(p+1)\mathbb{Z}$.

Exercise 441.

1) Show that $\mathbf{K} = \mathbb{F}_3[X]/\langle X^2 + X + 2 \rangle$ is a field. Give the characteristic and cardinality of $\mathbf{K}$.

2) Let ω be the image of X in $\mathbf{K}$. Show that ω is a generator of the multiplicative group $\mathbf{K}^\times$.

3) Give the elements of $\mathbf{K}$ which are squares in $\mathbf{K}$.

4) We consider the cubic E defined over $\mathbf{K}$ by the equation

$$y^2 = x^3 + x + \omega.$$

Check that E is an elliptic curve. Give the set of points in $\mathbb{P}^2(\mathbf{K})$ satisfying the equation of the curve E. Show that the Abelian group constructed from E has order 7.

5) What is the nature of the group constructed from E?

Exercise 442. Let E be an elliptic curve over a finite field $\mathbb{F}_q$. Show that if $q - 1$ is prime and $q \geq 5$ then $E(\mathbb{F}_q)$ is a cyclic group.

Exercise 443. We consider the affine plane curve $E : y^2 = x^3 + \theta x^2 + \theta$ defined over the field $\mathbb{F}_9 = \mathbb{F}_3(\theta)$ with $\theta^2 = -1$.

1) Show that E is an elliptic curve.

2) Compute the points of $E(\mathbb{F}_9)$.

3) Deduce the value of the Frobnenius trace a of E. Is E supersingular?

4) Is the group $E(\mathbb{F}_9)$ cyclic?

5) What is the smallest extension field $\mathbf{K}$ of $\mathbb{F}_9$ such that $E(\mathbf{K})$ has a point of order 2.

6) Compute $|E(\mathbb{F}_{9^3})|$.

Exercise 444. (Brunault)

Let p be a prime number. We consider the affine plane curve $E_0 : y^2 + y = x^3$ defined over $\mathbb{F}_p$.

1) Show that E_0 is irreducible.

2) What is the projective completion E of E_0?

3) For which p the curve E is an elliptic curve?

4) Compute $|E(\mathbb{F}_p)|$ for $p = 2, 3, 5$.

5) Compute $|E(\mathbb{F}_{16})|$.

6) We suppose that E is an elliptic curve. Let $P = (x, y) \in E \setminus \{O\}$. Give $-P$ and $2P$ as functions of x and y.

7) Deduce that for $P = (x, y) \in E \setminus \{O\}$, $3P = O$ if and only if $x = 0$ or $x^3 = -1$.

8) We suppose that $p \equiv 2 \bmod 3$. Show that $x \mapsto x^3$ is a bijection of $\mathbb{F}_p$. Deduce that $|E(\mathbb{F}_p)| = p + 1$.

9) We suppose that $p \equiv 1 \bmod 3$.

9.a) Show that $\mathbb{F}_p^\times$ has an element j of order 3 and that $u : (x, y) \mapsto (jx, y)$ is an automorphism of E.

9.b) Show that u commutes with the Frobenius endomorphism φ of E.

9.c) What are the fixed points of u?

9.d) Deduce that $|E(\mathbb{F}_p)|$ is divisible by 3.

9.e) Show that $u^2 + u + 1 = 0$ in $\text{End}(E)$, where $1 = \text{id}_E$.

9.f) Determine the points of E of order 3.

9.g) Deduce that $|E(\mathbb{F}_p)|$ is divisible by 9.

9.h) Show that the groups $E[3]$ and $E(\mathbb{F}_p)$ are not cyclic (without using Theorem and Definition 406).

10) Determine the group structure of $E(\mathbb{F}_7)$ and $E(\mathbb{F}_{11})$.

Exercise 445. (Reduction modulo p of an elliptic curve over the rationals [38, 61])

Let $E : y^2 = x^3 + ax^2 + bx + c$ be an elliptic curve defined over $\mathbb{Q}$, with $a, b, c \in \mathbb{Z}$ and discriminant $\Delta = -4a^3c + a^2b^2 + 18abc - 4b^3 - 27c^2 \neq 0$.

Let p be a prime number $\neq 2$ and non dividing Δ (in short, $p \nmid 2\Delta$). Then $\bar{E} : y^2 = x^3 + \bar{a}x^2 + \bar{b}x + \bar{c}$ defined over $\mathbb{F}_p$ is again an elliptic curve called *the reduction modulo p of E*.

For $P = (x : y : z) \in E(\mathbb{Q})$ with $x, y, z \in \mathbb{Z}$ and $\gcd(x, y, z) = 1$, the point $\bar{P} = (\bar{x} : \bar{y} : \bar{z})$ is well-defined as a point of $\mathbb{P}^2(\mathbb{F}_p)$ and it is clearly in $\bar{E}(\mathbb{F}_p)$. So the map

$$\mathrm{Red}_p : \begin{array}{ccc} E(\mathbb{Q}) & \longrightarrow & \bar{E}(\mathbb{F}_p) \\ P & \longmapsto & \bar{P}, \end{array}$$

is well-defined. It can be proven that Red_p is a group homomorphism. We denote its kernel by N_p, that is, $N_p = \{P \in E(\mathbb{Q}) \mid \mathrm{Red}_p(P) = O\}$.

As in Example 35, $\mathbb{Z}_{p\mathbb{Z}}$ denotes the discrete valuation domain $\mathbb{Z}_{p\mathbb{Z}} = \{\frac{a}{b} \in \mathbb{Q} \mid a \in \mathbb{Z} \text{ and } b \in \mathbb{Z} \setminus p\mathbb{Z}\} = \{x \in \mathbb{Q} \mid v_p(x) \geq 0\}$, where for $x \in \mathbb{Q}$, $v_p(x)$ denotes the valuation of x at p.

1) Show that

$$N_p = \{(x, y) \in E(\mathbb{Q}) \mid v_p(x) < 0\} \cup \{O\} = \{(x, y) \in E(\mathbb{Q}) \mid v_p(y) < 0\} \cup \{O\}.$$

2) Let $E(\mathbb{Q})_{\mathrm{tors}}$ be the subgroup of $E(\mathbb{Q})$ consisting of all rational points of finite order. It can be proven that $N_p \cap E(\mathbb{Q})_{\mathrm{tors}} = \{O\}$, and, thus, $E(\mathbb{Q})_{\mathrm{tors}}$ can be seen as a subgroup of $\bar{E}(\mathbb{F}_p)$, $E(\mathbb{Q})_{\mathrm{tors}}$ finite, and $|E(\mathbb{Q})_{\mathrm{tors}}|$ divides $|\bar{E}(\mathbb{F}_p)|$.

2.a) Consider the elliptic curve $E : y^2 = x^3 + 80$ defined over $\mathbb{Q}$. Using the fact that $|\bar{E}(\mathbb{F}_7)| = 13$ and $|\bar{E}(\mathbb{F}_{11})| = 12$, give $E(\mathbb{Q})_{\mathrm{tors}}$.

2.b) Consider the elliptic curve $E : y^2 = x^3 + x$ defined over $\mathbb{Q}$. Using the fact that $\bar{E}(\mathbb{F}_3) = \{O, (0,0), (2,1), (2,2)\} \cong \mathbb{Z}/4\mathbb{Z}$ and $\bar{E}(\mathbb{F}_5) = \{O, (0,0), (2,0), (3,0)\} \cong (\mathbb{Z}/2\mathbb{Z}) \times (\mathbb{Z}/2\mathbb{Z})$, give $E(\mathbb{Q})_{\mathrm{tors}}$.

Exercise 446. Let $E : y^2 = x^3 + bx$ be an elliptic curve defined over $\mathbb{Q}$, with $b \in \mathbb{Z}^*$. Let $E(\mathbb{Q})_{\mathrm{tors}}$ be the subgroup of $E(\mathbb{Q})$ consisting of all rational points of finite order.

Show that $|E(\mathbb{Q})_{\mathrm{tors}}| = 2$ or 4.

Hint: use Exercises 438 and 445, and *Dirichlet's theorem on arithmetic progressions* stating that for any two positive coprime integers a and d, there are infinitely many primes of the form $a + nd$, where $n \in \mathbb{N}$.

Exercise 447. (Brunault)

We consider the affine plane curve $y^2 + y = x^3 - x^2$ defined over $\mathbb{Q}$.

1) Give the equation of its projective completion E.

2) Check that E is an elliptic curve over $\mathbb{Q}$.

3) Show that x is a uniformizing parameter at $P_0 = (0, 0)$.

4) What is $\mathrm{Ord}_{P_0}(y)$?

5) Determine the divisor of the rational function y, x^2, and $g = y + x^2$.

6) Interpret geometrically the positive part of $\text{div}(g)$.

7) Show that P_0 has order 5 in $E(\mathbb{Q})$.

8) For which prime numbers p, E has a good reduction modulo p, that is, the reduction $\bar{E}$ of E modulo p is an elliptic curve over $\mathbb{F}_p$?

9) Show that $\bar{\mathbb{P}}_0 \in \bar{E}(\mathbb{F}_p)$ is a nonsingular point of $\bar{E}$.

10) Show that if E has a good reduction modulo p, then 5 divides $|\bar{E}(\mathbb{F}_p)|$.

11) Show that $E(\mathbb{Q})_{\text{tors}}$ is generated by P_0.

Exercise 448. Let p be a prime number ≥ 5, and consider an elliptic curve E over $\mathbb{F}_p$. Suppose that $m := |E(\mathbb{F}_p)|$ has a prime divisor ℓ such that

$$\ell > (\sqrt[4]{p}+1)^2.$$

Show that there exists a point $P \in E(\mathbb{F}_p)$ such that $mP = O$ and $\left(\frac{m}{\ell}\right)P \neq O$.

Exercise 449. (Elliptic curves over $\mathbb{Z}/n\mathbb{Z}$)

Let n be an integer ≥ 2. For $x,y,z \in \mathbb{Z}/n\mathbb{Z}$, we set $\gcd(x,y,z,n) := \gcd(\tilde{x},\tilde{y},\tilde{z},n)$, where $\tilde{x},\tilde{y},\tilde{z}$ are representatives of x,y,z in $\mathbb{Z}$, respectively. This definition makes sense since it does not depend on the chosen representatives. Set

$$T := \{(x,y,z) \in (\mathbb{Z}/n\mathbb{Z})^3 \mid \gcd(x,y,z,n) = 1\}.$$

Note that: $(x,y,z) \in T \Leftrightarrow \langle x,y,z \rangle = \mathbb{Z}/n\mathbb{Z}$. We define $\mathbb{P}^2(\mathbb{Z}/n\mathbb{Z}) := T/\sim$, where $\sim$ is the equivalence relation

$$(x,y,z) \sim (x',y',z') \Leftrightarrow \exists \lambda \in (\mathbb{Z}/n\mathbb{Z})^\times \mid (x',y',z') = \lambda(x,y,z).$$

The equivalence class of (x,y,z) will be denoted by $(x:y:z)$.

1) Show that $\sharp(\mathbb{P}^2(\mathbb{Z}/p^r\mathbb{Z})) = p^{2r}(1 + \frac{1}{p} + \frac{1}{p^2})$, where p is a prime number and $r \in \mathbb{N}^*$.

2) Let n be an integer ≥ 2 such that $n \wedge 6 = 1$. An elliptic curve E over $\mathbb{Z}/n\mathbb{Z}$ is a projective curve contained in $\mathbb{P}^2(\mathbb{Z}/n\mathbb{Z})$ whose equation is of the from

$$y^2z = x^3 + axz^2 + bz^3 \tag{6.3}$$

for some $a,b \in \mathbb{Z}/n\mathbb{Z}$ such that $4a^3 + 27b^2 \in (\mathbb{Z}/n\mathbb{Z})^\times$. The curve $E(\mathbb{Z}/n\mathbb{Z})$ contains the "affine" part

$$E_{\text{aff}}(\mathbb{Z}/n\mathbb{Z}) := \{(x:y:1) \in \mathbb{P}^2(\mathbb{Z}/n\mathbb{Z}) \mid y^2 = x^3 + ax + b\}$$

as well as the point $O = (0:1:0)$. We denote by (x,y) the point $(x:y:1)$ and set

$$V_{n,E} := E_{\text{aff}}(\mathbb{Z}/n\mathbb{Z}) \cup \{O\}.$$

Note that when n is prime, we have $V_{n,E} = E(\mathbb{F}_n)$. Compute $V_{n,E}$ and $E(\mathbb{F}_n)$ when E is the cubic over $\mathbb{Z}/25\mathbb{Z}$ given by the equation $y^2z = x^3 + z^3$.

Exercise 450. (ECPP, Elliptic Curve Primality Proving, Goldwasser and Killian, 1986)

We keep the notation of Exercise 449. Let n be an integer ≥ 2 such that $n \wedge 6 = 1$, and consider an elliptic curve E over $\mathbb{Z}/n\mathbb{Z}$ whose equation is $y^2 z = x^3 + axz^2 + bz^3$ for some $a, b \in \mathbb{Z}/n\mathbb{Z}$ such that $4a^3 + 27b^2 \in (\mathbb{Z}/n\mathbb{Z})^\times$.

For every prime divisor p of n, by reducing the coordinates of the points of E modulo p via the canonical morphism $\varphi_p : \mathbb{Z}/n\mathbb{Z} \to (\mathbb{Z}/n\mathbb{Z})/(p\mathbb{Z}/n\mathbb{Z}) = \mathbb{Z}/p\mathbb{Z}$, we obtain an elliptic curve over $\mathbb{F}_p$ again called E. So, we have a reduction modulo p map:

$$\begin{aligned} E(\mathbb{Z}/n\mathbb{Z}) &\longrightarrow E(\mathbb{Z}/p\mathbb{Z}) \\ (x : y : z) &\longmapsto (\varphi_p(x) : \varphi_p(y) : \varphi_p(z)). \end{aligned}$$

For every point $P \in V_{n,E}$, we denote by P_p its image in $E(\mathbb{Z}/p\mathbb{Z})$.

The following algorithm is called *pseudo-addition algorithm in $V_{n,E}$*. By this algorithm, given $P, Q \in V_{n,E}$, either one finds a nontrivial divisor of n, or constructs a point $R \in V_{n,E}$ such that $R_p = P_p + Q_p$ for every prime divisor p of n.

1. If $P = O$ (resp. $Q = O$), then set $R = Q$ (resp. $R = P$).

   Suppose now that $P \neq O$ and $Q \neq O$. Set $P = (x_1, y_1)$, $Q = (x_2, y_2)$, and $d = \gcd(x_2 - x_1, n)$.

2. If $d \neq 1$ and $d \neq n$, then d is a nontrivial divisor of n.

3. If $d = 1$, then, using the fact that $x_2 - x_1 \in (\mathbb{Z}/n\mathbb{Z})^\times$, as in the field case (Explicit formulas for addition on an elliptic curve 400), set:
   $R = (\lambda^2 - x_1 - x_2, -\lambda(\lambda^2 - x_1 - x_2) - v)$, where $\lambda = (y_2 - y_1)(x_2 - x_1)^{-1}$, and $v = (y_1 x_2 - y_2 x_1)(x_2 - x_1)^{-1}$.

4. Suppose that $d = n$, that is, $x_1 = x_2$. Set $d = \gcd(y_1 + y_2, n)$.

   (a) If $d' \neq 1$ and $d' \neq n$, then d' is a nontrivial divisor of n.

   (b) If $d' = n$, then $y_1 = -y_2$, and set $R = O$.

   (c) If $d' = 1$, then from the equalities $y_1^2 = x_1^3 + ax_1 + b$, $y_2^2 = x_2^3 + ax_2 + b$, and $x_1 = x_2$, we infer that $(y_1 - y_2)(y_1 + y_2) = 0$, and hence $y_1 = y_2$ since $y_1 + y_2 \in (\mathbb{Z}/n\mathbb{Z})^\times$. Thus, $P = Q$. Using that $2y_1 \in (\mathbb{Z}/n\mathbb{Z})^\times$, as in the field case (Explicit formulas for addition on an elliptic curve 400), set:
   $R = (\lambda^2 - 2x_1, -\lambda(\lambda^2 - 2x_1) - v)$, where $\lambda = (3x_1^2 + a)(2y_1)^{-1}$, and $v = (-x_1^3 + ax_1 + 2b)(2y_1)^{-1}$.

A) Let E be the elliptic curve over $\mathbb{Z}/77\mathbb{Z}$ whose equation is $y^2 = x^3 + x + 1$, and consider $P = (0,1) \in V_{77,E}$. Compute $3P$, $6P = 3P + 3P$, and then $9P = 3P + 6P$. Compute $4P$, $8P = 4P + 4P$, and then $9P = P + 8P$. What do you find out?

B) Let N be a large number that is very likely prime (for example, using the Miller-Rabin primality test which is a probabilistic test of type Monte Carlo). The goal is to give a proof that n is actually prime. This can be done by the following primality criterion that we owe to Goldwasser and Killian, and which is the elliptic analogue of Pocklington criterion ($N - 1$ being replaced with m).

Let N be an integer ≥ 2 such that $N \wedge 6 = 1$. Let E be an elliptic curve over $\mathbb{Z}/N\mathbb{Z}$, $m \in \mathbb{N}$, and $P \in V_{N,E}$ satisfying the following conditions:

1. The pseudo-addition algorithm works for the point mP and we have $mP = O$.

2. There exists a prime divisor ℓ of m such that:

   (a) $\ell > (\sqrt[4]{N} + 1)^2$,

   (b) and the pseudo-addition algorithm works for the point $(\frac{m}{\ell})P$ and we have $(\frac{m}{\ell})P \neq O$.

Show that N is prime.

C) Using B) and Exercise 448, deduce an algorithm giving a primality certificate (in case the primality is confirmed) for a large prime number N that is very likely prime.

Exercise 451. (Elliptic curves, finite fields, and Gröbner bases)

Let $\mathbb{F}_q$ be a finite field with characteristic p and cardinality q. For each subfield $\mathbf{K}$ of $\bar{\mathbb{F}}_q$ and ideal I of $\mathbf{K}[X_1, \ldots, X_n]$, we denote by $V_{\mathbf{K}}(I)$ the algebraic subvariety of $\mathbf{K}^n$ consisting of the zeroes of I in $\mathbf{K}^n$.

In the same way, for each subset E of $\bar{\mathbb{F}}_q^n$ and subfield $\mathbf{K}$ of $\bar{\mathbb{F}}_q$, we denote by $\mathscr{I}_{\mathbf{K}}(E)$ the ideal of $\mathbf{K}[X_1, \ldots, X_n]$ consisting of all the polynomials $f \in \mathbf{K}[X_1, \ldots, X_n]$ such that $\tilde{f}|_E = 0$, where $\tilde{f} \colon \bar{\mathbb{F}}_q^n \to \bar{\mathbb{F}}_q$ is the polynomial map associated to f.

We denote by $\mathscr{N}$ the ideal of $\mathbb{F}_q[X_1, \ldots, X_n]$ generated by $X_1^q - X_1, \ldots, X_n^q - X_n$. For any ideal $I = \langle f_1, \ldots, f_s \rangle$ of $\mathbb{F}_q[X_1, \ldots, X_n]$, we denote by $\bar{I}$ the ideal generated by $f_1, \ldots, f_s$ in $\bar{\mathbb{F}}_q[X_1, \ldots, X_n]$.

1) Show that $\bar{\mathbb{F}}_q^n$ is an algebraic subvariety of $\bar{\mathbb{F}}_q^n$.

2) Show that $G = \{X_1^q - X_1, \ldots, X_n^q - X_n\}$ is a universal Gröbner basis for the ideal $\mathscr{N}$ of $\mathbb{F}_q[X_1, \ldots, X_n]$ (i.e., it is a Gröbner basis with respect to any monomial order).

3) Show that the ideals $\mathscr{N}$ and $\bar{\mathscr{N}}$ are zero-dimensional radical ideals.

4) Show that $\mathcal{N} = \mathscr{I}_{\mathbb{F}_q}(\mathbb{F}_q^n)$.

5) Give the dimension as $\mathbb{F}_q$-vector space of the $\mathbb{F}_q$-algebra $\mathbb{F}_q[X_1, \ldots, X_n]/\mathcal{N}$.

6) We denote by $\mathscr{F}(\mathbb{F}_q^n, \mathbb{F}_q)$ the $\mathbb{F}_q$-algebra of maps from $\mathbb{F}_q^n$ to $\mathbb{F}_q$. Show that the canonical homomorphism

$$\varphi: \quad \mathbb{F}_q[X_1, \ldots, X_n] \quad \longrightarrow \quad \mathscr{F}(\mathbb{F}_q^n, \mathbb{F}_q)$$
$$f \quad \longmapsto \quad \tilde{f}|_{\mathbb{F}_q^n}$$

induces an $\mathbb{F}_q$-algebras isomorphism $\mathbb{F}_q[X_1, \ldots, X_n]/\mathcal{N} \cong \mathscr{F}(\mathbb{F}_q^n, \mathbb{F}_q)$.

7) Show that every subset E of $\mathbb{F}_q^n$ is an algebraic subvariety of $\bar{\mathbb{F}}_q^n$.

8) Show that for every ideal I of $\mathbb{F}_q[X_1, \ldots, X_n]$, we have $\mathscr{I}_{\mathbb{F}_q}(V_{\mathbb{F}_q}(I)) = I + \mathcal{N}$.

9) Show that the map $I \mapsto V_{\bar{\mathbb{F}}_q}(\bar{I})$ is a bijection between the set of ideals of $\mathbb{F}_q[X_1, \ldots, X_n]$ containing $\mathcal{N}$ and the set of subsets of $\mathbb{F}_q^n$.

10) Show that for every ideal I of $\mathbb{F}_q[X_1, \ldots, X_n]$, the cardinality $\sharp(V_{\mathbb{F}_q}(I))$ is equal to the dimension as $\mathbb{F}_q$-vector space of the $\mathbb{F}_q$-algebra $\mathbb{F}_q[X_1, \ldots, X_n]/(I + \mathcal{N})$.

11) Take $q = 7$, $n = 2$, and consider the elliptic curve C over $\mathbb{F}_7$:

$$C: \quad y^2 = x^3 + 2x + 4.$$

Determine, using Gröbner bases, the number of rational points of C, i.e., $\sharp(C \cap \mathbb{F}_7^2)$.

6.4 Solutions to the exercises

Exercise 417:

This is due to the exact sequence:

$$0 \to E[n] \to E \to nE \to 0.$$

In general, the groups $E[n]$ and E/nE are not isomorphic, for example when E has an n^2-torsion point.

Exercise 418:

The projective completion of E is $\bar{E}: y^2 z = x^3 + axz^2 + bz^3$. The point at ∞ of E is $O = (0:1:0)$. In the affine chart $(y \neq 0)$, $\bar{E}$ becomes $\hat{E}: z = x^3 + axz^2 + bz^3$ and O becomes $(0,0)$. Setting $h(x,z) = z - x^3 - axz^2 - bz^3$, since $\frac{\partial h}{\partial z}(0,0) = 1 \neq 0$, x is a uniformizing parameter at $(0,0)$. Going back to the projective coordinates and homogenizing with y (see Remark 288), we infer that $\frac{x}{y}$ is a uniformizing parameter at $O = (0:1:0)$.

Exercise 419:

Let $D = \sum_{i=1}^{r} n_i.[P_i] \in \mathbb{Z}[E]$ with $\deg(D) = \sum_{i=1}^{r} n_i = 0$ (this is a necessary condition to be principal). Then:

$$D \text{ is principal} \quad \Leftrightarrow \quad D = 0 \text{ in } \mathrm{Pic}^0(E)$$

$$\Leftrightarrow \quad \sum_{i=1}^{r} n_i.([P_i] - [O]) + (\sum_{i=1}^{r} n_i)[O] = 0 \text{ in } \mathrm{Pic}^0(E)$$

$$\Leftrightarrow \quad \sum_{i=1}^{r} n_i.([P_i] - [O]) = [O] - [O] \text{ in } \mathrm{Pic}^0(E)$$

$$\Leftrightarrow \quad \sum_{i=1}^{r} n_i P_i = O \text{ (as a sum of points, by Theorem 396).}$$

Exercise 420:

The map $\phi : E \to E$ with $\phi(x,y) = (x^q, y^q)$ and $\phi(O) = O$ is rational and, thus, regular (by virtue of Proposition 309). As seen in Chapter 3, it is a group homomorphism. Moreover, it is bijective since $x \mapsto x^q$ is an $\mathbb{F}_q$-automorphism of $\bar{\mathbf{K}}$ (see Exercise 237). Nevertheless, ϕ is not an isomorphism since $\deg(\phi) = [\bar{\mathbf{K}}(E) : \bar{\mathbf{K}}(E)^q] = [\bar{\mathbf{K}}(x) : \bar{\mathbf{K}}(x)^q] = q$ (this is because $[\bar{\mathbf{K}}(E) : \bar{\mathbf{K}}(x)] = 2$).

Exercise 421:

1) Just differentiate Equation 6.2.

2) First, since E is smooth, the partial derivatives of Equation 6.2 don't vanish simultaneously over E, and, thus, using the two expressions of ω, we deduce that ω is regular over $E \setminus \{O\}$. Let us compute $\mathrm{Ord}_O(\omega)$.

Let $\tilde{E} : y^2 z + a_1 xyz + a_3 yz^2 = x^3 + a_2 x^2 z + a_4 xz^2 + a_6 z^3$ be the projective completion of E. The point at ∞ of E is $O = (0 : 1 : 0)$. The homogenization of ω is $\tilde{\omega} := \frac{z}{2y + a_1 x + a_3 z} d(\frac{x}{z})$.

In the affine chart $(y \neq 0)$, $\tilde{E}$ becomes $\hat{E} : z + a_1 xz + a_3 z^2 = x^3 + a_2 x^2 z + a_4 xz^2 + a_6 z^3$, O becomes $(0,0)$, and $\tilde{\omega}$ becomes $\hat{\omega} := \frac{z}{2 + a_1 x + a_3 z} d(\frac{x}{z})$.

Setting $h(x,z) = z + a_1 xz + a_3 z^2 - x^3 - a_2 x^2 z - a_4 xz^2 - a_6 z^3$, since $\frac{\partial h}{\partial z}(0,0) = 1 \neq 0$, x is a uniformizing parameter at $(0,0)$.

From the equation of $\hat{E}$, we have $z = \frac{x^3}{1 + g(x,z)}$ with $g(0,0) = 0$, and, thus, $\mathrm{Ord}_{(0,0)}(z) = 3$.

Case 1: $\mathrm{char}(\mathbf{K}) \neq 2$. Then $\mathrm{Ord}_{(0,0)}(z) = 3 \Rightarrow \mathrm{Ord}_{(0,0)}(\frac{x}{z}) = -2 \Rightarrow \mathrm{Ord}_{(0,0)}(d(\frac{x}{z})) = -3 \Rightarrow \mathrm{Ord}_{(0,0)}(\hat{\omega}) = 0$ (because $2 \neq 0$) $\Rightarrow \mathrm{Ord}_O(\omega) = 0$.

Case 2: $\mathrm{char}(\mathbf{K}) = 2$. From 1), we have $\omega = \frac{dy}{x^2 + a_4 + a_1 y}$, and, thus, $\tilde{\omega} = \frac{z^2}{x^2 + a_4 z^2 + a_1 yz} d(\frac{y}{z})$, and $\hat{\omega} = \frac{z^2}{x^2 + a_4 z^2 + a_1 z} d(\frac{1}{z}) = \frac{1}{x^2 + a_4 z^2 + a_1 z} dz$. Hence, $\mathrm{Ord}_O(\omega) = \mathrm{Ord}_{(0,0)}(\hat{\omega}) = -2 + 2 = 0$.

Since $\operatorname{div}(\omega) \geq 0$ and $\deg(\operatorname{div}(\omega)) = \deg(K_E) = 2g - 2 = 0$ (Corollary 302), we conclude that $\operatorname{div}(\omega) = 0$ (implying that $K_E = 0$).

Exercise 422:

$P + Q = (6,0)$.

Exercise 423:

The tangent line to the curve E at P is $L : y = 2x - 1$. In the affine plane, we have $E \cap L = \{P, (0,-1)\}$. So, $2P = P + P = (0,1)$. After computation, we find $4P = 2P + 2P = (0,1) + (0,1) = (0,-1)$. As $4P = -2P$, we have $6P = O$, and, thus, the order of P divides 6. Since $2P \neq 0$ and $3P \neq O$ (because $2P \neq -P$), we deduce that the order of P is 6.

Exercise 424:

This is because the discriminant is zero.

Exercise 425:

1) This is immediate by virtue of Exercise 387 and the definition of addition of points on an elliptic curve.

2) $\operatorname{div}(ax + by + c) = 3[P] - 3[O]$.

Exercise 426:

1) Recall that an element $x \in \mathbb{F}_p^{\times}$ is called a perfect square if there exists $y \in \mathbb{F}_p^{\times}$ such that $x = y^2$. The set H of prefect squares in $\mathbb{F}_p$ is a subgroup of $\mathbb{F}_p^{\times}$ of cardinality $\frac{p-1}{2}$ (see Exercise 242). Let $x \in \mathbb{F}_p^{\times}$ such that $x = y^2$ for some $y \in \mathbb{F}_p^{\times}$. By Fermat's little theorem (or by Lagrange's theorem), we have

$$x^{\frac{p-1}{2}} = y^{p-1} = 1,$$

and hence x is a root of the polynomial $P = X^{\frac{p-1}{2}} - 1$. As P as at most $\frac{p-1}{2}$ roots (counted with multiplicities), we deduce that all the roots of P are simple and $H = \{\text{roots of } P\}$. Now

$$-1 \in H \Leftrightarrow (-1)^{\frac{p-1}{2}} = 1 \Leftrightarrow \frac{p-1}{2} \equiv 0 \bmod 2 \Leftrightarrow p \equiv 1 \bmod 4.$$

2) Let $x, y \in \mathbb{Z}$ such that $y^2 = x^3 + (2a)^3 - 1$. As $y^2 \equiv 0$ or $1 \bmod 4$, necessarily x is odd. We have the equality

$$y^2 + 1 = (x + 2a)((x - a)^2 + 3a^2).$$

As $x - a$ is even and $a^2 \equiv 1 \bmod 4$, we infer that $(x - a)^2 + 3a^2 \equiv 3 \bmod 4$, and, thus, $(x - a)^2 + 3a^2$ is divisible by a prime number $p \equiv 3 \bmod 4$. The fact that p divides $y^2 + 1$ contradicts 1).

Exercise 427:

1) Let $P = (x, y) \in E(\mathbb{F}_{16})$ and denote $-P = (x, y')$. As $y'^2 + y' = x^3 = y^2 + y$, we have $(y' + y)^2 = y' + y$, and, thus, $y' + y \in \mathbb{F}_2$, that is, $y' + y = 0$ or $y' + y = 1$. We deduce that $-P = (x, y + 1)$. After computation, we find $2P = (x^4, y^4 + 1)$.

2) If $P = (x, y) \in E(\mathbb{F}_{16})$, we have $4P = 2(2P) = (x^{16}, y^{16}) = (x, y) = P$. So, $3P = O$. If P distinct from the point at infinity then its order is 3.

3) Let $P = (x, y) \in E(\mathbb{F}_{16})$. Since $3P = O$, we have $2P = -P$, and, thus, $x^4 = x$, $y^4 = y$, and $P \in E(\mathbb{F}_4)$.

4) Since $E(\mathbb{F}_{16}) = E(\mathbb{F}_4)$, denoting $\omega = |E(\mathbb{F}_{16})| = |E(\mathbb{F}_4)|$ and applying twice Hasse's Theorem 409, we have $|\omega - 17| \leq 8$ and $|\omega - 5| \leq 4$, and, thus, $\omega = 9$. Taking account of 3), we infer that $E(\mathbb{F}_{16}) \cong \mathbb{Z}/3\mathbb{Z} \times \mathbb{Z}/3\mathbb{Z}$.

Exercise 428:

1) By Hasse's Theorem 409, we have $82 \leq |E(\mathbb{F}_{101})| \leq 122$. As $|E(\mathbb{F}_{101})|$ is a multiple of 116, it is necessarily equal to 116.

2) By Hasse's Theorem 409, we have $511 \leq |C(\mathbb{F}_{557})| \leq 605$. As $|C(\mathbb{F}_{557})|$ is a multiple of 189, it is necessarily equal to 567.

Exercise 429:

1) We find that $3P = 3Q = O$ with $2P \neq O$, $2Q \neq O$, $2P \neq Q$.

2) We know by Theorem and Definition 406 that $E[3] \cong (\mathbb{Z}/3\mathbb{Z}) \times (\mathbb{Z}/3\mathbb{Z})$. By Hasse's Theorem 409, we have $3 \leq |E(\mathbb{F}_7)| \leq 13$. As both P and Q have order 3 and Q is not on the $\mathbb{F}_3$-line $\{O, P, 2P\}$ generated by P, we infer that $|E(\mathbb{F}_7)| = 9$, and, thus, $E(\mathbb{F}_7) = E[3] \cong (\mathbb{Z}/3\mathbb{Z}) \times (\mathbb{Z}/3\mathbb{Z})$.

Exercise 430:

Let $P = (x, y) \in E(\mathbb{F}_{2^n})$ and set $-P = (x, y')$. We have $y'^2 + xy' = y^2 + xy$ and, thus, $(y' + y)(y' + y + x) = 0$. We conclude that $-P = (x, y + x)$.

Exercise 431:

This follows immediately by adding the equalities $\alpha^{k+1} - a\alpha^k + q\alpha^{k-1} = 0$ and $\beta^{k+1} - a\beta^k + q\beta^{k-1} = 0$.

Exercise 432:

Since $|E(\mathbb{F}_q)| = q + 1 - a$ with q a power of p, then $|E(\mathbb{F}_q)| \equiv 1 \bmod p \Leftrightarrow p \mid a$.

Set $X^2 - aX + q = (X - \alpha)(X - \beta)$, where $\alpha, \beta \in \mathbb{C}$. For $n \in \mathbb{N}$, setting $s_n = \alpha^n + \beta^n$, we have (by Exercise 431), $s_0 = 2$, $s_1 = a$, and $s_{n+1} = as_n - qs_{n-1}$.

Suppose that $|E(\mathbb{F}_q)| \equiv 1 \bmod p$ (or equivalently, $p \mid a$) and let us prove that $E(\overline{\mathbb{F}_q})$ has no torsion point of order p, or equivalently, $E(\mathbb{F}_{q^n})$ has no torsion point of order p for $n \geq 1$. We have $|E(\mathbb{F}_{q^n})| = q^n + 1 - s_n \equiv 1 \bmod p$ and the desired result clearly follows.

Conversely, suppose that $p \nmid a$. By Fermat's little theorem, we have $a^{p-1} \equiv 1 \bmod p$. Moreover, for $n \geq 1$, we have $s_{n+1} \equiv a s_n \bmod p \equiv a^{n-1} s_1 \bmod p \equiv a^n \bmod p$. It follows that

$$|E(\mathbb{F}_{q^{p-1}})| = q^{p-1} + 1 - s_{p-1} \equiv 1 - a^{p-1} \equiv 0 \bmod p.$$

We infer that $E(\mathbb{F}_{q^{p-1}})$ contains an element of order p.

Exercise 433:

Denoting by a the trace of Frobenius, by Hasse's Theorem 409, we have $|a| \leq 2\sqrt{p} < p$. So, $p \mid a \Leftrightarrow a = 0 \Leftrightarrow |E(\mathbb{F}_p)| = p + 1$. The desired result follows from Exercise 432.

Exercise 434:

1) Let $P = (x, y) \in E$ with $P \neq O$. Then: $P \in E[2] \Leftrightarrow P = -P \Leftrightarrow (x, y) = (x, -y) \Leftrightarrow y = 0 \Leftrightarrow x^3 + ax + b = 0$.

2.a) The discriminant of G is $-2^8 \times 3^3 (4a^3 + 27b^2)^2 \neq 0$.

2.b) Let $P = (x, y) \in E[3]$ distinct from O. As $P \neq O$, $2P \neq O$ and, thus, $y \neq 0$ and

$$2P = (\lambda^2 - 2x, -\lambda(\lambda^2 - 2x) - v), \tag{6.4}$$

where $\lambda = (2y)^{-1}(3x^2 + a)$ and $v = (2y)^{-1}(-x^3 + ax + 2b)$. Using that $-2P = P$ and $y^2 = x^3 + ax + b$, we deduce that $G(x) = 0$.

Conversely, suppose that $G(x) = 0$. As $(3X^2 + 4a)G(X) - (X^3 + aX + b)(3X^3 + 21aX + 27b) = -(4a^3 + 27b^2) \neq 0$, we deduce that $x^3 + ax + b \neq 0$, or also, $y \neq 0$. From the expression (6.4) of $2P$, we infer that $3P = O$.

2.c) Each root of G in $\bar{\mathbf{K}}$ is the x-coordinate of two points in E. Then, the group $E[3]$ has order 9 and is isomorphic to $(\mathbb{Z}/3\mathbb{Z}) \times (\mathbb{Z}/3\mathbb{Z})$.

3) Let α be the real cubic root of 2 and ξ a primitive cubic root of unity. Then $E[2] = \{O, (\alpha, 0), (\xi\alpha, 0), (\xi^2\alpha, 0)\}$ and $\mathscr{B}_2 = ((\alpha, 0), (\xi\alpha, 0))$ is a basis for $E[2]$ as an $\mathbb{F}_2$-vector space.

We have $G(X) = 3X^4 - 24X = 3X(X - 2)(X^2 + 2X + 4)$. Thus, the x-coordinates of order 3 points in E are $0, 2, -1 + \sqrt{-3}, -1 - \sqrt{-3}$. We deduce that

$$E[3] = \{O, (0, \pm\sqrt{-2}), (-1 + \sqrt{-3}, \pm\sqrt{-6}), (-1 - \sqrt{-3}, \pm\sqrt{6}), (2, \pm\sqrt{6})\}.$$

$\mathscr{B}_3 = ((0, \sqrt{-2}), (-1 + \sqrt{-3}, \sqrt{-6}))$ is a basis for $E[3]$ as an $\mathbb{F}_3$-vector space.

4) In $\mathbb{F}_{11}[X]$, we have $X^3 + X + 3 = (X - 3)(X^2 + 3X - 1)$ with $X^2 + 3X - 1$ irreducible.

We find $E[2] = \{O, (3, 0), (\alpha, 0), (-\alpha + 8, 0)\}$ with basis $\mathscr{B}_2 = ((3, 0), (\alpha, 0))$ as an $\mathbb{F}_2$-vector space, where $\alpha^2 + 3\alpha - 1 = 0$.

We have $G(X) = 3X^4 + 6X^2 + 36X - 1 = 3X(X-1)(X-2)(X^2+3X+9) \in$ $\mathbb{F}_{11}[X]$ with $X^2 + 3X + 9$ irreducible over $\mathbb{F}_{11}$. The points $(1, \pm 4)$ are of order 3 in $E(\mathbb{F}_{11})$. Let β be a square root of 2 in $\overline{\mathbb{F}_{11}}$. The points $(2, \pm \beta)$ are in $E[3]$. The roots of $X^2 + 3X + 9$ are $4 \pm 3\beta$. Then

$$E[3] = \{O, (1, \pm 4), (2, \pm \beta), (4 + 3\beta, \pm(2 + 9\beta)), (4 - 3\beta, \pm(2 + 2\beta))\}.$$

$\mathcal{B}_3 = ((1,4), (2, \beta))$ is a basis for $E[3]$ as an $\mathbb{F}_3$-vector space.

5.a) See 6).

5.b) $\mathrm{Mat}(\tau_{|E[2]}, \mathcal{B}_2) = \begin{pmatrix} 1 & 1 \\ 0 & 1 \end{pmatrix}$.

6) The point $\rho_\sigma(P) = (\sigma(x), \sigma(y))$ is on E since both a and b are fixed by σ. By virtue of the Explicit formulas 400 for addition on an elliptic curve, we infer that ρ_σ is a group endomorphism of E. In addition, it is bijective since σ is bijective. In 5.a), the complex conjugation $\mathbf{c} : z \mapsto \bar{z}$ is the nontrivial element in $\mathrm{Gal}(\mathbb{C}/\mathbb{R}) = \{\mathbf{id}_\mathbb{C}, \mathbf{c}\}$.

ρ_n is a group homomorphism since $\forall \sigma, \tau \in \mathrm{Gal}(\bar{\mathbf{K}}/\mathbf{K})$, $\rho_n(\sigma \circ \tau) = \rho_n(\sigma) \cdot \rho_n(\tau)$.

Exercise 435:

1) $|\mathrm{GL}_2(\mathbb{F}_p)|$ is the number of bases in the $\mathbb{F}_p$-vector space $\mathbb{F}_p^2$. We start by choosing a nonzero vector u_1 ($p^2 - 1$ possibilities) that we complete with a vector which is not in $\mathbb{F}_p \cdot u_1$ ($p^2 - p$ possibilities).

2) First note that $\mathbb{Z}/p\mathbb{Z} \cong (\mathbb{Z}/p^\alpha \mathbb{Z})/(p\mathbb{Z}/p^\alpha \mathbb{Z})$. We consider the surjective group homomorphism

$$\varphi : \qquad \mathrm{GL}_2(\mathbb{Z}/p^\alpha \mathbb{Z}) \quad \longrightarrow \quad \mathrm{GL}_2(\mathbb{Z}/p\mathbb{Z})$$

$$\begin{pmatrix} a + p^\alpha \mathbb{Z} & b + p^\alpha \mathbb{Z} \\ c + p^\alpha \mathbb{Z} & d + p^\alpha \mathbb{Z} \end{pmatrix} \quad \longmapsto \quad \begin{pmatrix} a + p\mathbb{Z} & b + p\mathbb{Z} \\ c + p\mathbb{Z} & d + p\mathbb{Z} \end{pmatrix}$$

Its kernel is made up of matrices

$$\begin{pmatrix} a + p^\alpha \mathbb{Z} & b + p^\alpha \mathbb{Z} \\ c + p^\alpha \mathbb{Z} & d + p^\alpha \mathbb{Z} \end{pmatrix}$$

with $a, b, c, d \in [\![1, p^\alpha]\!]$, $a, d \equiv 1 \bmod p$, and $b, c \equiv 0 \bmod p$. There are $p^{\alpha-1}$ multiples of p and $p^{\alpha-1}$ elements in the form $1 + kp$ in $[\![1, p^\alpha]\!]$. So, $|\mathrm{Ker}(\varphi)| = p^{4(\alpha-1)}$. The desired result follows from 1).

3) Write $n = p_1^{\alpha_1} \cdots p_s^{\alpha_s}$, where the p_i's are pairwise distinct prime numbers. By the Chinese theorem we know that $\mathbb{Z}/n\mathbb{Z} \cong \prod_{i=1}^{s} (\mathbb{Z}/p_i^{\alpha_i} \mathbb{Z})$, and, thus

$$\mathrm{GL}_2(\mathbb{Z}/n\mathbb{Z}) \cong \prod_{i=1}^{s} \mathrm{GL}_2(\mathbb{Z}/p_i^{\alpha_i} \mathbb{Z}).$$

The desired result follows from 2).

Exercise 436:

1) By Proposition 411, as $(\frac{3}{11}) = (\frac{4}{11}) = (\frac{5}{11}) = 1$ and $(\frac{2}{11}) = (\frac{6}{11}) = (\frac{7}{11}) = (\frac{8}{11}) = -1$, we obtain:

$$|E(\mathbb{F}_{11})| = 12 + \left(\frac{3}{11}\right) + \left(\frac{5}{11}\right) + \left(\frac{2}{11}\right) + \left(\frac{0}{11}\right) + \left(\frac{5}{11}\right)$$
$$+ \left(\frac{1}{11}\right) + \left(\frac{5}{11}\right) + \left(\frac{1}{11}\right) + \left(\frac{6}{11}\right) + \left(\frac{4}{11}\right) + \left(\frac{1}{11}\right) = 18.$$

2) Using the notation of Exercise 431, by virtue of 1), the trace of Frobenius is $a = 12 - 18 = -6$. We have: $s_0 = 2$, $s_1 = a = -6$, $s_2 = a s_1 - p s_0 = 6 \times 6 - 11 \times 2 = 14$, $s_3 = a s_2 - p s_1 = -6 \times 14 + 11 \times 6 = -18$. It follows that $|E(\mathbb{F}_{11^3})| = 11^3 + 1 - s_3 = 1350$.

Exercise 437:

1)

x	0	± 1	± 2	± 3	± 4	± 5	± 6
x^2	0	1	4	-4	3	-1	-3
x^3	0	± 1	± 5	± 1	± 1	± 5	± 5
$x^3 + 2x$	0	± 3	± 1	± 6	± 6	± 5	± 6

Thus, $E(\mathbb{F}_{13}) = \{O, (0,0), (1,4), (1,-4), (-1,6), (-1,-6), (2,5), (2,-5), (-2,1), (-2,-1)\}$, and $|E(\mathbb{F}_{13})| = 10$.

2) Let $P = (x,y) \in E(\mathbb{F}_{13})$ with $P \neq O$. Then: $P \in E[2] \Leftrightarrow P = -P \Leftrightarrow (x,y) = (x,-y) \Leftrightarrow y = 0 \Leftrightarrow x^3 + 2x = x(x^2 + 2) = 0 \Leftrightarrow x = 0$. So $(0,0)$ is the only point in $E(\mathbb{F}_{13})$ of order 2.

3) $2P = (1,-6)$, $4P = (1,-4) = -P$. Thus, $5P = O$ and P has order 5.

4) By 1), we have $E[2] = \{O, (0,0), (\theta,0), (-\theta,0)\} \subseteq E(\mathbb{F}_{13^2})$. As $E[2] \cong (\mathbb{Z}/2\mathbb{Z}) \times (\mathbb{Z}/2\mathbb{Z})$, the group $E(\mathbb{F}_{13^2})$ is not cyclic.

5) Trace of Frobenius is $a = 13 + 1 - |E(\mathbb{F}_{13})| = 4$. Using the notation of Exercise 431, we have $|E(\mathbb{F}_{13^n})| = 13^n + 1 - s_n$, with $s_0 = 2$, $s_1 = a = 4$, $s_{n+1} = a s_n - q s_{n-1} = 4 s_n - 13 s_{n-1}$. So, $s_2 = 4 s_1 - 13 s_0 = -10$, and $|E(\mathbb{F}_{13^2})| = 13^2 + 1 + 10 = 180$.

Exercise 438:

For $x \in \mathbb{F}_p$, we have $(\frac{(-x)^3 - u(-x)}{p}) = (\frac{-1}{p})(\frac{x^3 - ux}{p})$. But $(\frac{-1}{p}) = (-1)^{\frac{p-1}{2}} = -1$ since $p \equiv 3 \bmod 4$ (see Exercise 373). Hence

$$|E(\mathbb{F}_p)| = p + 1 + \sum_{x \in \mathbb{F}_p} \left(\frac{x^3 - ux}{p}\right) = p + 1.$$

The fact that E is supersingular follows from Exercise 433.

Exercise 439:

Let E be the elliptic curve $E : y^2 = x^3 + x + 3$ defined over $\mathbb{F}_{11}$. Let β be a square root of 2 in $\overline{\mathbb{F}_{11}}$. We saw in Exercise 434.4 that $\mathscr{B}_3 = (P_1 = (1,4), P_2 = (2,\beta))$ is a basis for $E[3]$ as an $\mathbb{F}_3$-vector space. We have $\phi_{11,3}(P_1) = P_1$ since $P_1 \in E(\mathbb{F}_{11})$. On the other hand, $\phi_{11,3}(P_2) = (2^{11}, \beta^{11}) = (2, 13^5\beta) = (2, -\beta) = -P_2$. Thus, the matrix of $\phi_{11,3}$ in $\mathscr{B}_3$ is

$$\begin{pmatrix} 1 & 0 \\ 0 & -1 \end{pmatrix}.$$

We saw in Exercise 436 that the trace of Frobenius of E is $a = -6$. We have $\mathrm{Tr}(\phi_{11,3}) = 0 \equiv -6 \bmod 3$ and $\det(\phi_{11,3}) = -1 \equiv 11 \bmod 3$.

Exercise 440:

1) If $p = 2$ or $p = 3$ or $b = 0$ then the curve is singular.

2) $E[2] \cong \mathbb{Z}/2\mathbb{Z} \times \mathbb{Z}/2\mathbb{Z}$ since $\gcd(p,2) = 1$.

3) As $p \equiv 2 \bmod 3$, $p - 2 = 3k$ for some $k \in \mathbb{N}$. Let $x, x' \in \mathbb{F}_p^\times$ such that $\varphi(x) = \varphi(x')$. Since $x^{p-1} = x'^{p-1}$ (by Lagrange's theorem) and $x^{p-2} = x'^{p-2}$, we infer that $x = x'$. It follows that φ is injective, and, thus, bijective.

4) The 2-torsion points in $E(\bar{\mathbb{F}}_p)$ are of the form $(x,0)$ with $x^3 = -b$. This equation has a unique solution $x_1 \in \mathbb{F}_p$ and the two other solutions are in $\mathbb{F}_{p^2} = \mathbb{F}_p[x]/\langle x^2 + \alpha x + \beta \rangle$, where $x^3 + b = (x - x_1)(x^2 + \alpha x + \beta)$ and $x^2 + \alpha x + \beta$ is an irreducible polynomial in $\mathbb{F}_p[x]$.

5) As φ is bijective, when x varies over the elements in $\mathbb{F}_p$, $x^3 + b$ varies over all of the elements of $\mathbb{F}_p$. If $x^3 + b$ is a nonzero square in $\mathbb{F}_p$ then there will be two points of the form (x,y) on E; if $x^3 + b = 0$ then there is one corresponding point $(x,0)$ on E; and if $x^3 + b$ is not a square then there are no points on E with that x-coordinate. Now since p is odd, exactly $1/2$ of the elements of $\mathbb{F}_p^\times$ are squares (see Exercise 242). So we get $2(1/2)(p-1) + 1 = p$ points on the curve in the affine plane. Counting the point at infinity O, we get $p + 1$ points on E.

6) As the trace of Frobenius is $a = p + 1 - |E(\mathbb{F}_p)| = 0$, the characteristic roots of Frobenius are $\pm i\sqrt{p}$, and by Theorem 414, we have

$$|E(\mathbb{F}_{p^n})| = \begin{cases} p^n + 1 & \text{if } n \text{ is odd} \\ p^n + 1 - 2(-p)^{\frac{n}{2}} & \text{if } n \text{ is even.} \end{cases}$$

Or, also, using the notation of Exercise 431, we have $|E(\mathbb{F}_{p^n})| = p^n + 1 - s_n$, with $s_0 = 2$, $s_1 = a = 0$, $s_{n+1} = a s_n - p s_{n-1} = -p s_{n-1}$. So,

$$s_n = \begin{cases} 0 & \text{if } n \text{ is odd} \\ 2(-p)^{\frac{n}{2}} & \text{if } n \text{ is even.} \end{cases}$$

7) By Theorem 404, $E(\mathbb{F}_p)$ is isomorphic to $\mathbb{Z}/d_1\mathbb{Z} \times \mathbb{Z}/d_2\mathbb{Z}$, where $d_1 \mid d_2$, $d_1 \mid p - 1$, and $d_1 d_2 = |E(\mathbb{F}_p)| = p + 1$. Since $d_1 \mid p - 1$ and $d_1 \mid p + 1$, then $d_1 \mid 2$,

and so $d_1 = 1$ or 2. We know, by 4), that $\mathbb{Z}/2\mathbb{Z} \times \mathbb{Z}/2\mathbb{Z} \cong E[2] \nsubseteq E(\mathbb{F}_p)$. But the group $E(\mathbb{F}_p)$ contains a subgroup isomorphic to $\mathbb{Z}/d_1\mathbb{Z} \times \mathbb{Z}/d_1\mathbb{Z}$. We deduce that $d_1 \neq 2$, and so $d_1 = 1$ and $E(\mathbb{F}_p) \cong \mathbb{Z}/(p+1)\mathbb{Z}$.

Exercise 441:

1) Denoting $P = X^2 + X + 2$, we have $P(0) = P(2) = 2 \neq 0$ and $P(1) = 1 \neq 0$. It follows that P is irreducible over $\mathbb{F}_3$, and, thus, $\mathbf{K}$ is a field (of characteristic 3 and cardinality of 9).

2) $\omega^2 = 2\omega + 1$, $\omega^4 = 2 \neq 1$ $\Rightarrow$ ω is a generator of $\mathbf{K}^\times$.

3)

y	0	1	2	ω	$\omega+1$	2ω	$2\omega+1$	$2\omega+2$
y^2	0	1	1	$2\omega+1$	2	$2\omega+1$	2	$\omega+2$

4) $\Delta = 2 \neq 0$.

$E(\mathbf{K}) = \{O, (1, \omega+1), (1, 2\omega+2), (\omega, \omega+1), (\omega, 2\omega+2), (2\omega, \omega), (2\omega, 2\omega)\}$.

5) $E \cong \mathbb{Z}/7\mathbb{Z}$ as 7 is prime.

Exercise 442:

$E(\mathbb{F}_q) \cong (\mathbb{Z}/d_1\mathbb{Z}) \times (\mathbb{Z}/d_2\mathbb{Z})$ with $d_1 \mid d_2$ and $d_1 \mid q-1$. If $d_1 = q-1$ we would have $d_2 \geq q-1$, and, thus, $|E(\mathbb{F}_q)| = d_1 d_2 \geq (q-1)^2 > q+1+2\sqrt{q}$, in contradiction with Hasse's Theorem 409. We deduce that $d_1 = 1$, that is, $E(\mathbb{F}_q)$ is cyclic.

Exercise 443:

1) By Theorem 393, it suffices to show that E is smooth. This is obvious by examining the partial derivatives of $F(x,y) = y^2 - x^3 - \theta x^2 - \theta$.

2)

x	0	1	-1	θ	$-\theta$	$\theta+1$	$-\theta-1$	$\theta-1$	$1-\theta$
x^2	0	1	1	-1	-1	$-\theta$	$-\theta$	θ	θ
x^3	0	1	-1	$-\theta$	θ	$1-\theta$	$\theta-1$	$-1-\theta$	$1+\theta$
$x^3+\theta x^2+\theta$	θ	$1-\theta$	$-1-\theta$	$-\theta$	θ	-1	$-\theta$	1	$-\theta$

Thus, $E(\mathbb{F}_9) = \{O, (0, \theta-1), (0, 1-\theta), (-\theta, \theta-1), (-\theta, 1-\theta), (\theta, \theta+1), (\theta, -\theta-1), (-\theta-1, \theta+1), (-\theta-1, -\theta-1), (1-\theta, \theta+1), (1-\theta, -\theta-1), (\theta-1, 1), (\theta-1, -1), (\theta+1, \theta), (\theta+1, -\theta)\}$, and $|E(\mathbb{F}_9)| = 15$.

3) $a = 9 + 1 - 15 = -5$. As $3 \nmid -5$, E is not supersingular.

4) Since $15 = 3 \times 5$ is square-free, the group $E(\mathbb{F}_9)$ is cyclic.

5) Let $P = (x,y) \in E$ with $P \neq O$. Then: $P \in E[2] \Leftrightarrow P = -P \Leftrightarrow (x,y) = (x,-y) \Leftrightarrow y = 0 \Leftrightarrow x^3 + \theta x^2 + \theta = 0$. The polynomial $X^3 + \theta X^2 + \theta$ has no roots in $\mathbb{F}_9$ and, hence, is irreducible over $\mathbb{F}_9$. It follows that the smallest extension field $\mathbf{K}$ of $\mathbb{F}_9$ such that $E(\mathbf{K})$ has a point of order 2 is $\mathbf{K} = \mathbb{F}_{9^3} = \mathbb{F}_9[X]/\langle X^3 + \theta X^2 + \theta \rangle$.

6) Using the notation of Exercise 431, we have $|E(\mathbb{F}_{9^n})| = 9^n + 1 - s_n$, with $s_0 = 2$, $s_1 = a = -5$, $s_{n+1} = a s_n - q s_{n-1} = -5 s_n - 9 s_{n-1}$. So, $s_2 = -5 s_1 - 9 s_0 = 7$, $s_3 = -5 s_2 - 9 s_1 = 10$, and $|E(\mathbb{F}_{9^3})| = 9^3 + 1 - 10 = 720$.

Exercise 444:

1) Let us prove that $F = Y^2 + Y - X^3$ is irreducible in $\bar{\mathbb{F}}_p[X, Y]$. By way of contradiction, suppose that $F = (Y + G_1(X))(Y + G_2(X))$, with $G_i \in \bar{\mathbb{F}}_p[X]$. This would imply that $G_1 + G_2 = 1$ and $G_1 G_2 = X^3$, a contradiction by degree consideration.

2) The projective completion of E_0 is $E = V(Y^2 Z + Y Z^2 - X^3)$.

3) If (x, y) is a singular point of E_0 the $3x^2 = 2y + 1 = 0$. If $p \neq 3$, then $x = 0$ and $y = 1$ or -1, which is impossible. If $p = 3$ then $(-1, 1)$ is a singular point of E_0. We conclude that E is an elliptic curve if and only if $p \neq 3$.

4) By an exhaustive search, $E(\mathbb{F}_2) = \{(0,0), (0,1), O\}$, $E(\mathbb{F}_3) = \{(0,0), (-1,1), (0,-1), O\}$, and $E(\mathbb{F}_5) = \{(0,0), (-2,1), (0,-1), (1,2), (-2,-2), O\}$.

5) Using the notation of Exercise 431, by virtue of 4), when $p = 2$, the trace of Frobenius is $a = 2 + 1 - 3 = 0$. We have: $s_0 = 2$, $s_1 = a = 0$, $s_2 = a s_1 - p s_0 = -4$, $s_3 = a s_2 - p s_1 = 0$, and $s_4 = a s_3 - p s_2 = 8$. It follows that $|E(\mathbb{F}_{2^4})| = 2^4 + 1 - s_4 = 9$.

6) We suppose that $p \neq 3$. Let $P = (x, y) \in E \setminus \{O\}$. By the Explicit formulas 400 for addition on an elliptic curve, we have $-P = (x, -y - 1)$. If $2y + 1 = 0$ then $2P = O = (0 : 1 : 0)$. If $2y + 1 \neq 0$, then setting $\lambda = \frac{3x^2}{2y+1}$ and $v = -\frac{x^3 + y}{2y+1}$, and using the relation $(2y + 1)^2 = 4(y^2 + y) + 1 = 4x^3 + 1$, we obtain

$$2P = \begin{cases} (\frac{x^4 - 2x}{4x^3 + 1}, -\lambda(\lambda^2 - 2x) - v - 1) & \text{if} \quad 2y + 1 \neq 0 \\ O & \text{if} \quad 2y + 1 = 0. \end{cases}$$

7) Let $P = (x, y) \in E \setminus \{O\}$. We have:

$$3P = O \iff 2P = -P \overset{\text{as } 2P \neq P}{\iff} x_{2P} = x_P \iff 4x^3 + 1 \neq 0 \text{ and } \frac{x^4 - 2x}{4x^3 + 1} = x$$

$$\iff x = 0 \text{ or } x^3 = -1.$$

8) By Exercise 440.3, we know that $x \mapsto x^3$ is a bijection of $\mathbb{F}_p$. For each $y \in \mathbb{F}_p$, there is exactly one $x \in \mathbb{F}_p$ such that $y^2 + y = x^3$. It follows that $|E(\mathbb{F}_p)| = p + 1$ (counting the point at ∞).

9.a) As 3 divides $|\mathbb{F}_p^\times|$, there exists $j \in \mathbb{F}_p^\times$ of order 3. The map $u : (x, y) \mapsto (jx, y)$ is well-defined because $(jx)^3 = j^3 x^3 = x^3$. Since u is rational, it is regular (see Proposition 309). Moreover, as $u^3 = \text{id}_E$, u is an automorphism of E. Note that $u(P + Q) = u(P) + u(Q)$ for all $P, Q \in E$; this follows, for example, from the fact that u is an affine transformation, and, thus, preserves collinearity.

9.b) The Frobenius endomorphism of E is $\varphi : (x,y) \mapsto (x^p, y^p)$. For $(x,y) \in E$, we have:

$$\varphi \circ u(x,y) = (j^p x^p, y^p) \overset{\text{as } p \equiv 1 \bmod 3}{=} (jx^p, y^p) = u \circ \varphi(x,y).$$

9.c) We have $u(O) = O$ and if $u(x,y) = (x,y)$ then $x = 0$ and $y \in \{0, -1\}$. It follows that the fixed points of u are O, $(0,0)$, and $(0,-1)$.

9.d) Because u acts on $E(\mathbb{F}_p)$ and u has order 3, the orbits of u have cardinalities 1 or 3. As u has 3 fixed points, we deduce that $|E(\mathbb{F}_p)|$ is divisible by 3.

9.e) Let $P = (x_0, y_0) \in E$. The line $D : y = y_0$ intersects E at the points (x_0, y_0), (jx_0, y_0), and $(j^2 x_0, y_0)$. Hence $D \cap E = \{P, u(P), u^2(P)\}$ and $P + u(P) + u^2(P) = O$. This being true for any $P \in E$, we conclude that $u^2 + u + 1 = 0$.

9.f) By virtue of 7), $P \in E \setminus \{O\}$ has order 3 if and only if $x = 0$ (and, thus, $y = 0$ or $y = -1$), or $x^3 = -1$, that is, $x \in \{-1, -j, -j^2\}$. In such case, $y^2 + y + 1 = 0$, and, thus, $y \in \{j, j^2\}$. The 8 points of order 3 are then

$$\{(0,0), (0,-1), (-1,j), (-1,j^2), (-j,j), (-j,j^2), (-j^2,j), (-j^2,j^2)\}.$$

9.g) In 8), we saw that $E[3]$ has order 9 and $E[3] \subseteq E(\mathbb{F}_p)$. So, $|E(\mathbb{F}_p)|$ is divisible by 9 by Lagrange's theorem.

9.h) By the fundamental theorem of finite Abelian groups, we know that $E[3] \cong \mathbb{Z}/9\mathbb{Z}$ or $(\mathbb{Z}/3\mathbb{Z}) \times (\mathbb{Z}/3\mathbb{Z})$. But the elements of $E[3]$ are all of order 1 or 3. We infer that $E[3] \cong (\mathbb{Z}/3\mathbb{Z}) \times (\mathbb{Z}/3\mathbb{Z})$, and, thus, it is not cyclic. Since $E[3]$ is a subgroup of $E(\mathbb{F}_p)$, this latter also is not cyclic.

10) By 9.g), we know that 9 divides $|E(\mathbb{F}_7)|$. Since to each $x \in E(\mathbb{F}_7)$ correspond at most two points in $E(\mathbb{F}_7)$, we have $|E(\mathbb{F}_7)| \leq 2 \times 7 + 1 < 18$. We conclude that $E(\mathbb{F}_7) = E[3] \cong (\mathbb{Z}/3\mathbb{Z}) \times (\mathbb{Z}/3\mathbb{Z})$.

By 8), we know that $|E(\mathbb{F}_{11})| = 12$. So, by the fundamental theorem of finite Abelian groups, we know that $E(\mathbb{F}_{11}) \cong \mathbb{Z}/12\mathbb{Z}$ or $(\mathbb{Z}/6\mathbb{Z}) \times (\mathbb{Z}/2\mathbb{Z})$. To reach a conclusion, it suffices to determine the points $E(\mathbb{F}_{11})$ of order 2 (note that, in $\mathbb{Z}/12\mathbb{Z}$, there is only one element of order 2, while, in $(\mathbb{Z}/6\mathbb{Z}) \times (\mathbb{Z}/2\mathbb{Z})$, there are exactly 3 elements of order 2). For this, if $P = (x,y) \in E(\mathbb{F}_{11})$ has order 2 then $2y + 1 = 0$, and, thus, $y = 5$ and $x = 2$. We conclude that $E(\mathbb{F}_{11}) \cong \mathbb{Z}/12\mathbb{Z}$.

Exercise 445:

1) Clearly $O \in N_p$ since $\bar{O} = O$. Let $P = (x : y : 1) \in E(\mathbb{Q})$. If $x = \frac{a}{b} \in \mathbb{Z}_{p\mathbb{Z}}$ then $y = \frac{c}{d} \in \mathbb{Z}_{p\mathbb{Z}}$, with $a, c \in \mathbb{Z}$ and $b, d \in \mathbb{Z} \setminus p\mathbb{Z}$. Then $P = (ad : cb : bd)$ and $\bar{P} \neq O = (0 : 1 : 0)$ since $bd \in \mathbb{Z} \setminus p\mathbb{Z}$.

Now suppose that $v_p(x) < 0$ or $v_p(y) < 0$. Then there exists $k \in \mathbb{N}^*$ such that $v_p(x) = -2k$ and $v_p(y) = -3k$. Writing $x = p^{-2k}\frac{a}{b}$, $y = p^{-3k}\frac{c}{d}$, with $a,b,c,d \in \mathbb{Z} \setminus p\mathbb{Z}$, we have $P = (adp^k : cb : bdp^{3k})$ and $\bar{P} = O = (0 : 1 : 0)$ since $cb \in \mathbb{Z} \setminus p\mathbb{Z}$.

2.a) $|E(\mathbb{Q})_{\text{tors}}|$ divides $\gcd(12,13) = 1$ and so $E(\mathbb{Q})_{\text{tors}} = \{O\}$.

2.b) Since $E(\mathbb{Q})_{\text{tors}}$ embeds into both $\mathbb{Z}/4\mathbb{Z}$ and $(\mathbb{Z}/2\mathbb{Z}) \times (\mathbb{Z}/2\mathbb{Z})$, we infer that $E(\mathbb{Q})_{\text{tors}}$ is cyclic and has exponent a divisor of 2. As $E(\mathbb{Q})_{\text{tors}}$ contains the subgroup $\{O,(0,0)\}$ of order 2, we deduce that $E(\mathbb{Q})_{\text{tors}} = \{O,(0,0)\}$.

Exercise 446:

By Exercises 438 and 445, $N = |E(\mathbb{Q})_{\text{tors}}|$ divides $p+1$ for all primes $p \geq 5$ such that $p \equiv 3 \bmod 4$ and $p \nmid b$. So, every prime greater than b and 5 which is congruent to 3 mod 4 is also congruent to $-1 \bmod N$.

By Dirichlet's theorem on arithmetic progressions, the sequence $(7+12r)_{r \in \mathbb{N}} = (3 + 4(1 + 3r))_{r \in \mathbb{N}}$ contains infinitely many primes. These primes are congruent to 3 mod 4 but congruent to $1 \neq -1 \bmod 3$. We infer that $3 \nmid N$.

Again, by Dirichlet's theorem on arithmetic progressions, the sequence $(4Nn + 3)_{n \in \mathbb{N}}$ contains infinitely many primes. These primes are congruent to 3 mod 4 but congruent to $3 \neq -1 \bmod N$ if $N \geq 5$. It follows that $N \leq 4$. As $E(\mathbb{Q})_{\text{tors}}$ contains the point $(0,0)$ (of order 2), we deduce that $N = 2$ or 4.

Exercise 447:

1) $E = V(Y^2 Z + Y Z^2 - X^3 + X^2 Z)$.

2) If $(x_0, y_0) \in E$ satisfies $2y_0 + 1 = 3x_0^2 - 2x_0 = 0$ then $y_0 = -\frac{1}{2}$ and $x_0 \in \{0, \frac{2}{3}\}$, which is impossible.

3) Let us denote by $C = V(F) \subseteq \mathbb{A}^2(\bar{\mathbb{Q}})$, where $F(X,Y) = Y^2 + Y - X^3 + X^2$. By Proposition 283 (keeping its notation), we know that the ideal $\mathfrak{M}_{C,P_0}$ is principal generated by any $t \in \mathfrak{M}_{P_0} \setminus \mathfrak{M}_{P_0}^2$. We know also that as $\bar{\mathbb{Q}}$-vector space, $\mathfrak{M}_{P_0}/\mathfrak{M}_{P_0}^2$ is generated by $\bar{x}$ and $\bar{y}$, where the classes are modulo $\mathfrak{M}_{P_0}^2$. But, as $y(y+1) = x^2(x-1) \in \mathfrak{M}_{P_0}^2$ and $y + 1 \notin \mathfrak{M}_{P_0}$, we conclude that $y \in \mathfrak{M}_{C,P_0}^2$. As, a generator for $\mathfrak{M}_{C,P_0}$ is either x or y and this latter is zero modulo $\mathfrak{M}_{C,P_0}^2$, we infer that $\mathfrak{M}_{C,P_0}$ is generated by x, that is, x is a uniformizing parameter at $P_0 = (0,0) \in C$.

Or, alternatively, since $\frac{\partial F}{\partial y}(0,0) = 1 \neq 0$, we deduce that x is a uniformizing parameter at $P_0 = (0,0)$.

4) As $y = x^2 \cdot \frac{x-1}{x+1}$, we deduce that $\text{Ord}_{P_0}(y) = 2$.

5) The rational function y has two zeroes $P_0 = (0,0)$ and $P_1 = (1,0)$. As $\frac{\partial F}{\partial x}(P_1) \neq 0$, y is a uniformizing parameter at P_1. Since $\deg(\text{div}(y)) = 0$, we infer that

$$\text{div}(y) = 2[P_0] + [P_1] - 3[O].$$

The rational function x has two zeroes $P_0 = (0,0)$ and $P_2 = (0,-1)$. As $\frac{\partial F}{\partial y}(P_2) \neq 0$, x is a uniformizing parameter at P_2. It follows that $\text{div}(x) = [P_0] + [P_2] - 2[O]$, and, thus,

$$\text{div}(x^2) = 2\,\text{div}(x) = 2[P_0] + 2[P_2] - 4[O].$$

The rational function $g = y + x^2 = x^3 - y^2$ has two zeroes P_0 and $P_3 = (1, -1)$. As $\mathrm{Ord}_{P_0}(x^3) = 3 \neq \mathrm{Ord}_{P_0}(y^2) = 4$, we have $\mathrm{Ord}_{P_0}(g) = \min\{3, 4\} = 3$. Similarly, as $\mathrm{Ord}_O(y) = -3 \neq \mathrm{Ord}_O(x^2) = -4$, we have that $\mathrm{Ord}_O(g) = -4$. We conclude that

$$\mathrm{div}(g) = 3[P_0] + [P_3] - 4[O].$$

6) In the affine plane, the parabola with equation $y = -x^2$ intersects E at the points $P_0 = (0, 0)$ and $P_3 = (1, -1)$ with multiplicities 3 and 1, respectively. We also deduce, by virtue of Strong Bézout's Theorem 340, that, in the projective plane, the parabola with equation $yz = -x^2$ intersects E at the points $(0 : 0 : 1)$, $(1 : -1 : 1)$, and $O = (0 : 1 : 0)$ with multiplicities 3, 1, and 2, respectively.

7) The tangent line to E at P_0 is the line $y = 0$. Plugging $y = 0$ in the equation, we find $x \in \{0, 1\}$, and, therefore, $-2P_0 = (1, 0)$. On the other hand, the divisor of g being principal, we have $3P_0 = -P_3 = (1, 0) = 2P_0$, and, thus, P_0 has order 5. Explicitly, $2P_0 = (1, -1)$, $3P_0 = (1, 0)$, and $4P_0 = (0, -1)$.

8) The discriminant of the equation of E is $\Delta = -11$. So, E has a good reduction modulo p for $p \neq 11$.

9) This follows from the fact that $\frac{\partial F}{\partial y}(0, 0) = 1 \not\equiv 0 \bmod p$.

10) If E has a good reduction modulo p, we have a surjective group homomorphism

$$\mathrm{Red}_p : \quad \begin{aligned} E(\mathbb{Q}) &\longrightarrow \bar{E}(\mathbb{F}_p) \\ P &\longmapsto \bar{P}. \end{aligned}$$

As $\mathrm{Red}_p(P_0) = \bar{P}_0 \neq O$, the point $\bar{P}_0$ has order 5, and, thus, 5 divides $|\bar{E}(\mathbb{F}_p)|$ by Lagrange's theorem.

11) The order of the group $\bar{E}(\mathbb{F}_2)$ is both $\leq 2 \times 2 + 1 = 5$ and divisible by 5. So, $\bar{E}(\mathbb{F}_2) \cong \mathbb{Z}/5\mathbb{Z}$. But, as stated in Exercise 445, $E(\mathbb{Q})_{\mathrm{tors}}$ can be seen as a subgroup of $\bar{E}(\mathbb{F}_p)$ for $p \neq 11$, in particular, as a subgroup of $\bar{E}(\mathbb{F}_2)$. As the point P_0 has order 5 in $E(\mathbb{Q})_{\mathrm{tors}}$, we deduce that $E(\mathbb{Q})_{\mathrm{tors}} = \langle P_0 \rangle \cong \mathbb{Z}/5\mathbb{Z}$.

Exercise 448:

As $m = |E(\mathbb{F}_p)|$, we have $mP = O$ for any $P \in E(\mathbb{F}_p)$. By way of contradiction, suppose that for any $P \in E(\mathbb{F}_p)$, we have $(\frac{m}{\ell})P = O$. In that case, the order of every $P \in E(\mathbb{F}_p)$ divides $\frac{m}{\ell}$. In particular, the exponent of the group $E(\mathbb{F}_p)$ (that is, the LCM of the orders of its elements) divides $\frac{m}{\ell}$. Furthermore, we know that $E(\mathbb{F}_p)$ is isomorphic to $\mathbb{Z}/d_1\mathbb{Z} \times \mathbb{Z}/d_2\mathbb{Z}$, where $d_1 \mid d_2$, and $d_1 d_2 = m$. Since the exponent of $E(\mathbb{F}_p)$ is d_2, we have $m = d_1 d_2 \leq d_2^2 \leq (\frac{m}{\ell})$, and, thus, $\ell^2 \leq m$. On the other hand, by Hasse's Theorem 409, we have $m \leq (\sqrt{p}+1)^2$. Combining the last two inequalities with the inequality $\ell > (\sqrt[4]{p}+1)^2$, we get $\sqrt{p}+1 > (\sqrt[4]{p}+1)^2 = \sqrt{p}+1+2\sqrt[4]{p}$, a contradiction.

Exercise 449:

1) As the ring $\mathbb{Z}/p^r\mathbb{Z}$ is local with maximal ideal $\mathfrak{M}_p := p\mathbb{Z}/p^r\mathbb{Z}$, we know that $(x : y : z) \in T$ if and only if at least one among x, y, z is not divisible by p. It follows that $\mathbb{P}^2(\mathbb{Z}/p^r\mathbb{Z})$ is the disjoint union of $A = \{(a : b : 1) \mid a, b \in \mathbb{Z}/p^r\mathbb{Z}\}$, $B = \{(1 : a : pb) \mid a, b \in \mathbb{Z}/p^r\mathbb{Z}\}$, and $C = \{(pa : 1 : pb) \mid a, b \in \mathbb{Z}/p^r\mathbb{Z}\}$. We have $\sharp A = p^{2r}$, $\sharp B = p^r p^{r-1}$, and $\sharp C = (p^{r-1})^2$. So, $\sharp(\mathbb{P}^2(\mathbb{Z}/p^r\mathbb{Z})) = p^{2r} + p^{2r-1} + p^{2r-2}$.

2) As $4a^3 + 27b^2 = 27 \in (\mathbb{Z}/25\mathbb{Z})^\times$, E is an elliptic curve over $\mathbb{Z}/25\mathbb{Z}$. We can check that $V_{25,E}$ has cardinality is made up of O and the 25 following points:

$$(0,\pm1), \; (2,\pm3), \; (5,\pm1), \; (7,\pm12), \; (10,\pm1), \; (12,\pm2), \; (15,\pm1), \; (17,\pm8),$$
$$(20,\pm1), \; (22,\pm7), \; (24,0), \; (24,\pm5), \; (24,\pm10).$$

Furthermore, let $(x : y : z) \in E(\mathbb{Z}/25\mathbb{Z})$ with z non invertible (that is, $z \in \mathfrak{M}_5$). Then x is not invertible (since $x^3 = y^2 z - z^3 \in \mathfrak{M}_5$), and, a fortiori, y is invertible. We can suppose that $y = 1$. Since $5^3 = 0$ in $\mathbb{Z}/25\mathbb{Z}$, we have $x^3 = z^3 = 0$, and hence $z = 0$. Thus, we obtain the following 4 points:

$$(5 : 1 : 0), \; (10 : 1 : 0), \; (15 : 1 : 0), \; (20 : 1 : 0).$$

Note that $\sharp E(\mathbb{Z}/25\mathbb{Z}) = 30$ while $\sharp V_{25,E} = 26$.

Exercise 450:

A) $3P = (72,72)$, $6P = (0,43)$, and $3P + 6P = 9P = (14,69)$. Furthermore, $4P = (28,27)$, $4P + 4P = 8P = (44,23)$, but we cannot compute $P + 8P$ as $\gcd(44,77) = 11$. Also, we cannot compute $P + 4P$ as $\gcd(28,77) = 7$. We conclude that a chain computation for mP can work while another may fail.

B) Let p be a prime divisor of N. The image of O in $E(\mathbb{F}_p)$ will also be denoted by O. As $mP_p = O$, the order d of P_p in $E(\mathbb{F}_p)$ divides m. Write $(\frac{m}{\ell})P = (x : y : 1) \in V_{N,E}$ for some $x, y \in \mathbb{Z}/N\mathbb{Z}$. Thus, $(\frac{m}{\ell})P_p \neq O$, and d does not divides $\frac{m}{\ell}$. Since ℓ is prime, we deduce that ℓ divides d, and, thus, $\ell \leq |E(\mathbb{F}_p)|$. On the other hand, by Hasse's Theorem 409, we have $|E(\mathbb{F}_p)| \leq (\sqrt{p}+1)^2$. So, $\ell \leq (\sqrt{p}+1)^2$. Now, by way of contradiction, suppose that N is not prime and let p be its smallest prime divisor. We have $p \leq \sqrt{N}$ and so $\ell \leq (\sqrt[4]{N}+1)^2$, a contradiction.

C) Let N be a large number that is very likely prime.

1. We randomly choose $a, b \in \mathbb{N}$ so that $\gcd(4a^3 + 27b^2, N) = 1$, and we consider the elliptic curve over $\mathbb{Z}/N\mathbb{Z}$ whose equation is $y^2 = x^3 + ax + b$. We compute $|E(\mathbb{Z}/N\mathbb{Z})|$ as if N were prime. Then, Exercise 448 suggests to choose $m = |E(\mathbb{Z}/N\mathbb{Z})|$.

2. We try to factorize m hoping to find a prime divisor $\ell > (\sqrt[4]{N} + 1)^2$. If we do not succeed in this, we choose a new elliptic curve E.

3. If m has such prime divisor ℓ, then we determine a point $P \in V_{N,E}$. For this, we randomly choose $x \in \mathbb{Z}/N\mathbb{Z}$ so that $(\frac{x^3+ax+b}{N}) = 1$ (the Legendre Symbol). We then use one of the available algorithms for computing square roots modulo N to find $y \in \mathbb{Z}/N\mathbb{Z}$ such that $y^2 = x^3 + ax + b$. If N is actually prime, we automatically have $mP = O$ and often we find in practice that $(\frac{m}{\ell})P \neq O$ in which case the hypotheses in B) are fulfilled. Thus, we obtain a proof that N is prime. The quadruplet (E, m, ℓ, P) can be seen as a primality certificate for N.

Exercise 451:

1) Let $x = (x_1, \ldots, x_n) \in \bar{\mathbb{F}}_q^n$. We know that, for $1 \leq i \leq n$, $x_i^q = x_i$ if and only if $x_i \in \mathbb{F}_q$. So, $\mathbb{F}_q^n = V_{\mathbb{F}_q}(\mathcal{N}) = V_{\bar{\mathbb{F}}_q}(\bar{\mathcal{N}})$.

2) We have $\mathrm{LM}(X_i^q - X_i) = X_i^q$. So, the leading monomials of the $X_i^q - X_i$ are pairwise relatively prime, and, thus, $G = \{X_1^q - X_1, \ldots, X_n^q - X_n\}$ is a Gröbner basis for the ideal $\mathcal{N} = \langle X_1^q - X_1, \ldots, X_n^q - X_n \rangle$ of $\mathbb{F}_q[X_1, \ldots, X_n]$ (by virtue of Exercise 111).

3) The ideal $\mathcal{N}$ (resp. $\bar{\mathcal{N}}$) is zero-dimensional since, for each $1 \leq i \leq n$, it contains a nonzero univariate polynomial at X_i.

Denote by $P_i = X_i^q - X_i$. We have $P_i' = -1$ and, thus, $P_i \wedge P_i' = 1$. It follows, by virtue of Seidenberg's lemma (see Exercise 204), that $\mathcal{N}$ and $\bar{\mathcal{N}}$ are radical.

4) $\mathscr{I}_{\mathbb{F}_q}(\mathbb{F}_q^n) = \mathscr{I}_{\bar{\mathbb{F}}_q}(\mathbb{F}_q^n) \cap \mathbb{F}_q[X_1, \ldots, X_n] \overset{\text{by 1)}}{=} \mathscr{I}_{\bar{\mathbb{F}}_q}(V_{\bar{\mathbb{F}}_q}(\bar{\mathcal{N}})) \cap \mathbb{F}_q[X_1, \ldots, X_n]$

$\overset{\text{by Hilbert's Nullstellensatz, Theorem 134}}{=} \sqrt{\bar{\mathcal{N}}} \cap \mathbb{F}_q[X_1, \ldots, X_n] \overset{\text{by 3)}}{=} \bar{\mathcal{N}} \cap \mathbb{F}_q[X_1, \ldots, X_n] = \mathcal{N}$.

5) By virtue of 2), $(X_1^{\alpha_1} \cdots X_n^{\alpha_n}, 0 \leq \alpha_i \leq q - 1)$ is a basis of $\mathbb{F}_q[X_1, \ldots, X_n]/\mathcal{N}$ as $\mathbb{F}_q$-vector space, and so its dimension is q^n.

6) By virtue of 4), $\mathcal{N} = \mathscr{I}_{\mathbb{F}_q}(\mathbb{F}_q^n) = \mathrm{Ker}(\varphi)$, and so the homomorphism

$$\bar{\varphi}: \quad \mathbb{F}_q[X_1, \ldots, X_n]/\mathcal{N} \quad \longrightarrow \quad \mathscr{F}(\mathbb{F}_q^n, \mathbb{F}_q)$$
$$\bar{f} \quad \longmapsto \quad \tilde{f}|_{\mathbb{F}_q^n}$$

induced by φ is injective. As $\sharp(\mathbb{F}_q[X_1,\ldots,X_n]/\mathcal{N}) \overset{\text{by 5)}}{=} q^{q^n} = \sharp(\mathscr{F}(\mathbb{F}_q^n,\mathbb{F}_q))$, the desired result follows.

7) Let $E \subseteq \mathbb{F}_q^n$. By virtue of 6), there is a polynomial $f \in \mathbb{F}_q[X_1,\ldots,X_n]$ such that $\tilde{f}|_{\mathbb{F}_q^n}$ is the characteristic function of E. Then, $E = V_{\bar{\mathbb{F}}_q}(\langle f \rangle + \mathcal{N})$.

8) Note that the ideals $I + \mathcal{N}$ and $\bar{I} + \mathcal{N}$ are radical by virtue of Seidenberg's lemma (see Exercise 204). Moreover, we have:

$$V_{\bar{\mathbb{F}}_q}(\bar{I}+\mathcal{N}) = V_{\bar{\mathbb{F}}_q}(\bar{I}) \cap V_{\bar{\mathbb{F}}_q}(\mathcal{N}) = V_{\bar{\mathbb{F}}_q}(\bar{I}) \cap \mathbb{F}_q^n = V_{\mathbb{F}_q}(I),$$

$$\mathscr{I}_{\mathbb{F}_q}(V_{\mathbb{F}_q}(I)) = \mathscr{I}_{\mathbb{F}_q}(V_{\bar{\mathbb{F}}_q}(\bar{I}+\mathcal{N})) = \mathscr{I}_{\bar{\mathbb{F}}_q}(V_{\bar{\mathbb{F}}_q}(\bar{I}+\mathcal{N})) \cap \mathbb{F}_q[X_1,\ldots,X_n]$$

$$\overset{\text{by Hilbert's Nullstellensatz, Theorem 134}}{=} \sqrt{(\bar{I}+\mathcal{N})} \cap \mathbb{F}_q[X_1,\ldots,X_n]$$

$$= (\bar{I}+\mathcal{N}) \cap \mathbb{F}_q[X_1,\ldots,X_n] = I + \mathcal{N}.$$

9) This map is surjective as, for $E \subseteq \mathbb{F}_q^n$, keeping the notation of 7), $E = V_{\bar{\mathbb{F}}_q}(\langle f \rangle + \mathcal{N})$. Injectivity follows from the fact that for an ideal I of $\mathbb{F}_q[X_1,\ldots,X_n]$ containing $\mathcal{N}$, we have $V_{\mathbb{F}_q}(I) = V_{\bar{\mathbb{F}}_q}(\bar{I})$ and $\mathscr{I}_{\mathbb{F}_q}(V_{\mathbb{F}_q}(I)) = I$ (by 8)).

10) Since the ideal $I + \mathcal{N}$ is zero-dimensional and radical, then

$$\sharp(V_{\mathbb{F}_q}(I)) = \dim_{\mathbb{F}_q}(\mathbb{F}_q[X_1,\ldots,X_n]/(I+\mathcal{N})).$$

11) $\sharp(C \cap \mathbb{F}_7^2) = \dim_{\mathbb{F}_7}(\mathbb{F}_7[X,Y]/J)$, with $J = \langle Y^2 - X^3 - 2X - 4, X^7 - X, Y^7 - Y \rangle$. We calculate the reduced Gröbner basis G of J with respect to the lexicographic order with $X > Y$ using the computing software MAPLE. We obtain $G = \{Y^7 + 6Y, 2Y^4 + XY^2 + 2Y^2 + 5X + 2, 3Y^6 + 2Y^4 + X^2 + 6Y^2 + 2X + 4\}$, and, thus, $\text{LT}(J) = \langle Y^7, XY^2, X^2 \rangle$. The staircase of the Gröbner basis G is $(1, Y,\ldots,Y^6, X, XY)$, and, thus, $\sharp(C \cap \mathbb{F}_7^2) = 9$, that is, C has exactly 9 rational points.

Notation list

$\gcd(a,b) = a \wedge b$ the greatest common divisor of a and b.

$\operatorname{lcm}(a,b) = \operatorname{LCM}(a,b) = a \vee b$ the least common multiple of a and b.

$\mathbb{A}^n(\mathbf{K})$ the affine space of dimension.

$V(f_1, \ldots, f_s)$ the algebraic variety defined by $f_1 = \cdots = f_s = 0$.

$\mathscr{I}(V)$ the ideal formed by the polynomials vanishing at the V.

$\bar{S}$ the Zariski closure of a subset S, i.e., the smallest algebraic variety containing S.

$\operatorname{char}(\mathbf{K})$ the characteristic of the field $\mathbf{K}$.

$\mathbf{K}_0$ the prime subfield of $\mathbf{K}$.

$[\mathbf{Ł} : \mathbf{K}]$ the degree of the field extension $\mathbf{Ł}/\mathbf{K}$.

$\varphi(n)$ the Euler function at n.

$\mathbb{P}^n(\mathbf{K})$ the projective space over $\mathbf{K}$ of dimension n.

$\bar{\mathbf{K}}[C]$ the ring of regular functions over C.

$\bar{\mathbf{K}}(C)$ the field of rational functions over the irreducible curve C; the quotient field of $\bar{\mathbf{K}}[C]$.

$\operatorname{div}(f)$ the divisor of the function f.

$\operatorname{Div}(E)$ the free group of divisors on E.

$\operatorname{Div}^0(E)$ the subgroup of $\operatorname{Div}(E)$ of degree-zero divisors.

$\operatorname{Princ}(E)$ the subgroup of $\operatorname{Div}(E)$ of principal divisors.

$\operatorname{Pic}(E)$ the Picard group $\operatorname{Div}(E)/\operatorname{Princ}(E)$ of E.

$\operatorname{Pic}^0(E) = \operatorname{Div}^0(E)/\operatorname{Princ}(E)$.

$E(\mathbf{K})$ the points of E with coordinates in $\mathbf{K}$.

$[m]$ the map $P \mapsto mP$.

O the point at infinity of an elliptic curve.

$\left(\frac{a}{p}\right)$ the Legendre Symbol.

$\mathscr{L}(D)$ the Riemann space of the divisor D.

$\ell(D)$ the dimension of $\mathscr{L}(D)$ as $\bar{\mathbf{K}}$-vector space.

$\operatorname{Aut}(\mathbf{K})$ the automorphisms group of $\mathbf{K}$.

$\operatorname{Gal}(\mathbf{Ł}/\mathbf{K})$ the Galois group of $\mathbf{Ł}$ over $\mathbf{K}$.

$\operatorname{Im}(\varphi)$ the image of φ.

$\operatorname{Ker}(\varphi)$ the kernel of φ.

$\operatorname{Coker}(\varphi)$ the cokernel of φ.

263

$\mathrm{Syz}(a_1,\ldots,a_n)$ the syzygy module of $(a_1,\ldots,a_n)$.

$\mathrm{GL}_s(\mathbf{R})$ the group of invertible matrices of size $s \times s$ with entries in $\mathbf{R}$.

$\mathrm{SL}_s(\mathbf{R})$ the subgroup of $\mathrm{GL}_s(\mathbf{R})$ formed by matrices of determinant 1.

$\mathrm{M}_{n,m}(\mathbf{R})$ the set of matrices of size $n \times m$ with entries in $\mathbf{R}$.

$\mathrm{M}_n(\mathbf{R})$ the set of matrices of size $n \times n$ with entries in $\mathbf{R}$.

$\mathrm{GL}(\mathbf{R})$ the group $\cup_{n \geq 1} \mathrm{GL}_n(\mathbf{R})$.

$\mathrm{E}_s(\mathbf{R})$ the subgroup of $\mathrm{SL}_s(\mathbf{R})$ generated by elementary matrices.

$\mathbf{R}^\times$ the group of units of $\mathbf{R}$.

$\mathrm{Rad}(\mathbf{R})$ the set of all $x \in \mathbf{R}$ such that $1 + x\mathbf{R} \subseteq \mathbf{R}^\times$.

$a^{\mathbb{N}}$ the monoid $\{a^n; n \in \mathbb{N}\}$.

$\mathscr{M}(a)$ the monoid $a^{\mathbb{N}}$.

$S^{-1}\mathbf{R}$ or $\mathbf{R}_S$ the localization of $\mathbf{R}$ at S.

$\mathrm{Spec}(\mathbf{R})$ the set of prime ideals of $\mathbf{R}$.

$\mathscr{M}(U)$ the monoid generated by U.

$\mathscr{M}(u_1,\ldots,u_\ell)$ the monoid $\mathscr{M}(\{u_1,\ldots,u_\ell\})$.

$\mathbf{R}_a$ the localization of the ring $\mathbf{R}$ at the monoid $a^{\mathbb{N}}$.

M_a the localization of the module M at the monoid $a^{\mathbb{N}}$.

$\mathrm{Res}(f,g)$ the resultant of f and g.

$\mathrm{Res}_X(f,g)$ the resultant of f and g with respect to X.

$\gcd(f,g)$ the greatest common divisor of f and g.

$\mathbf{R}\langle X \rangle$ the localization of $\mathbf{R}[X]$ at the monoid of monic polynomials.

$\mathbf{R}(X)$ the localization of $\mathbf{R}[X]$ at the monoid of primitive polynomials.

$\mathbf{R}_\mathfrak{p}$ the localization of $\mathbf{R}$ at the monoid $\mathbf{R} \setminus \mathfrak{p}$, where $\mathfrak{p}$ is a prime ideal.

$\mathrm{LC}(f)$ the leading coefficient of the polynomial f.

$\mathrm{LM}(f)$ the leading monomial of the polynomial f.

$\mathrm{LT}(f)$ the leading term of the polynomial f.

$\mathrm{LT}(I)$ the ideal $\langle \mathrm{LT}(f) : f \in I \rangle$.

$\mathrm{mdeg}(f)$ the multidegree of the polynomial f.

$\mathrm{tdeg}(f)$ the total degree of the polynomial f.

$E_{i,j}(a)$ the matrix with 1s on the diagonal, a on position (i,j) and 0s elsewhere.

$I : J$ the colon ideal or conductor of J in I, i.e., $\{x \in \mathbf{R} \mid xJ \subseteq I\}$, where I, J are ideals of the ring $\mathbf{R}$.

$I : a$ the conductor $I : \langle a \rangle$ of $\langle a \rangle$ in I.

$\mathrm{Ann}(x)$ the annihilator of x, i.e., $\langle 0 \rangle : \langle x \rangle$.

$\mathrm{Kdim}\mathbf{R}$ or $\dim \mathbf{R}$ the Krull dimension of $\mathbf{R}$.

$\mathrm{diag}(u_1,\ldots,u_n)$ the matrix in $\mathrm{M}_n(\mathbf{R})$ with u_i on position (i,i) for $1 \leq i \leq n$, and 0s elsewhere.

$\mathbb{M}_n^m$ the set of monomials in $\mathbf{R}[X_1,\ldots,X_n]^m$.

$\mathbb{M}_n$ the set of monomials in $\mathbf{R}[X_1,\ldots,X_n]$.

$Y^{\uparrow}$ the set $\{Z \in E \mid Z \geq Y\}$.

$\mathscr{M}_E^+(Y_1,\ldots,Y_m)$ the final subset of E of finite type $\cup_{i=1}^m Y_i^{\uparrow}$.

$\mathscr{F}(E)$ the set of final subsets of finite type of E, including the empty subset considered as generated by the empty family.

$\mathcal{M}_d$ the set $\mathcal{F}(\mathbb{N}^d) \setminus \{\mathbf{0}\}$.

$\bar{f}^F$ a remainder of f on division by F.

$S(f,g)$ the S-polynomial of f and g.

$S(f,f)$ the auto-S-polynomial of f.

$\mathbb{Z}_{p\mathbb{Z}}$ the localization $\{\frac{a}{b} \in \mathbb{Q} \mid a \in \mathbb{Z} \text{ and } b \in \mathbb{Z} \setminus p\mathbb{Z}\}$ of $\mathbb{Z}$.

$\mathbb{F}_q$ the field with q elements.

$\mathbf{R}_{a_1.a_2.\dots.a_n}$ the ring $\mathcal{M}(a_1,\dots,a_n)^{-1}\mathbf{R} = \mathbf{R}[\frac{1}{a_1\cdots a_n}]$.

$\mathrm{HS}_I(t)$ the Hilbert series of I.

K_C the canonical divisor of the curve C.

$\mathcal{I}(D)$ the $\bar{\mathbf{K}}$-vector space of all rational differential forms w over C such that $(w) \geq D \ i(D) := \dim_{\bar{\mathbf{K}}} \mathcal{I}(D)$.

$\mathcal{I}(D)$ the $\bar{\mathbf{K}}$-vector space of all rational differential forms w over C such that $(w) \geq D$.

$i(D) := \dim_{\bar{\mathbf{K}}} \mathcal{I}(D)$.

$\mathcal{I}(V) := \{f \in \mathbf{K}[X_1,\dots,X_n] \mid f(a_1,\dots,a_n) = 0 \ \forall\, (a_1,\dots,a_n) \in V\}$ the ideal of $\mathbf{K}[X_1,\dots,X_n]$ defined by the subset V of $\mathbb{A}^n(\mathbf{K})$.

$V(f_1,\dots,f_s) := \{(a_1,\dots,a_n) \in \mathbb{A}^n(\mathbf{K}) \mid f_i(a_1,\dots,a_n) = 0 \text{ for all } 1 \leq i \leq s\}$ the algebraic variety defined by $f_1,\dots,f_s$.

Index

Bibliography

[1] Abhyankar S.-S., Kravitz B. *Two counterexamples in normalization*. Proc. Amer. Math. Soc. **135** (2007) 3521–3523.

[2] Adams W.W., Loustaunau P. *An introduction to Gröbner bases*. Graduate Studies in Mathematics, vol. 3, American Mathematical Society, Providence, RI, 1994.

[3] Ai M., Jiang B., Li K. *Construction of sliced space-filling designs based on balanced sliced orthogonal arrays*. Statist. Sinica **24** (2014) 1685–1702.

[4] Arnold E. *Modular algorithms for computing Gröbner bases*. J. Symb. Comp. **35** (2003) 403–419.

[5] Atiyah M.F., Macdonald I.G. *Introduction to commutative algebra*, Addison-Wesley Publishing Co., Reading, Mass.-London-Don Mills, Ont., 1969.

[6] Bayer D., Stillman M. *On the Complexity of Computing Syzygies*. J. Symb. Comp. **6** (1988) 135–147.

[7] Brickenstein M., Dreyer A., Greuel G.-M., Wedler M., Wienand O. *New developments in the theory of groebner bases and applications to formal verification*. J. Pure Appl. Algebra **213** (2009) 1612–1635.

[8] Bourbaki N. *"Algèbre commutative", Chapitres 5–6*. Masson, Paris, 1985.

[9] Brunault F. *Courbes elliptiques*, ENS Lyon, 2008–2009.

[10] Buchberger B. *Ein Algorithmus zum Auffinden der Basiselemente des Restklassenringes nach einem nulldimensionalen polynomideal*. Ph.D. thesis, University of Innsbruck, Austria, 1965.

[11] Byrne E., Fitzpatrick P. *Gröbner bases over Galois rings with an application to decoding alternant codes*. J. Symbolic Comput. **31** (2001) 565–584.

[12] Cai Y., Kapur D. *An algorithm for computing a Gröbner basis of a polynomial ideal over a ring with zero divisors*. Math. Comput. Sci. **2** (2009) 601–634.

[13] Coste M., Lombardi H., Roy M.-F. *Dynamical method in algebra: Effective Nullstellensätze*. Annals of Pure and Applied Logic **111** (2001) 203–256.

[14] Cox D., Little J., O'Shea D. *Ideals, varieties and algorithms*, 2nd edition. New York, Springer-Verlag, 1997.

[15] Delaunay C. *Chapitre 1 - Courbes elliptiques*.

http://math.univ-lyon1.fr/~wagner/coursDelaunay.pdf

[16] Della Dora J., Dicrescenzo C., Duval D. *About a new method for computing in algebraic number fields.* In Caviness B.F. (Ed.) EUROCAL '85. Lecture Notes in Computer Science 204, 289–290. Springer (1985).

[17] Dubé T.W. (1990), *The structure of polynomial ideals and Gröbner bases.* SIAM J. Comput. **19** (1990) 750–773.

[18] Ebert G.L. *Some comments on the modular approach to Gröbner-bases.* ACM SIGSAM Bulletin **17** (1983) 28–32.

[19] Eisenbud D. *Commutative Algebra with a view toward Algebraic Geometry.* Springer Verlag, 1995.

[20] Ene V., Herzog J. *Gröbner bases in Commutative Algebra.* Graduate Studies in Mathematics 130, A.M.S. 2012.

[21] Faugère J.-C. *A new efficient algorithm for computing Gröbner bases without reduction to zero (F_5).* In Proc. ISSAC (2002).

[22] Faugère J.-C., Safey El Din M. *Chapitre d'introduction à la résolution des systèmes polynomiaux.* Ouvrage collectif chez Pearson Education.

[23] Fitchett S. *Bézout's Theorem: A taste of algebraic geometry.* Proceedings of the Thirty-Fourth Annual Meeting, Florida Section, The Mathematical Association of America.

[24] Fulton W. *Algebraic Curves - An Introduction to Algebraic Geometry.* Addison-Wesley, Redwood City CA, 1989.

[25] Gamanda M., Lombardi H., Neuwirth S., Yengui I. *The syzygy theorem for Bézout rings.* Math. Comp. **89** (2020) 941–964.

[26] Gamanda M., Yengui I. *Noether Normalization theorem and dynamical Gröbner bases over Bézout domains of Krull dimension 1.* J. Algebra **492** (2017) 52–56.

[27] Gamanda M., Yengui I. *Corrigendum to Noether Normalization theorem and dynamical Gröbner bases over Bézout domains of Krull dimension 1.* J. Algebra **492** (2017) 52–56. J. Algebra **533** (2019) 376–377.

[28] von zur Gathen J., Gerhard J. *Modern Computer Algebra.* Cambridge University Press, Cambridge (2003).

[29] Gräbe H. *On lucky primes.* J. Symb. Comp. **15** (1994) 199–209.

[30] Greuel G.M., Pfister G. *A Singular introduction to commutative algebra.* Springer Verlag Berlin, Heidelberg, New York, 2002.

[31] Greuel G.-M., Seelisch F., Wienand O. *The Gröbner Basis of the Ideal of Vanishing Polynomials.* J. Symb. Comp. **46** (2011) 561–570.

[32] Hadj Kacem A., Yengui I. *Dynamical Gröbner bases over Dedekind rings.* J. Algebra **324** (2010) 12–24.

[33] Hartshorne R. *Algebraic Geometry*, Springer, New York, 1977.

[34] Hochster M. *Noether normalization and Hilbert's Nullstellensatz*:

http://www.math.lsa.umich.edu/~hochster/615W10/supNoeth.pdf

[35] Hong H. *Gröbner bases under composition.* J. Symb. Comp. **25** (1998) 643–663.

[36] Kapur D., Narendran P. *An equational approach to theoretical proving in first-order predicate calculus*. IJCAI (1985) 1146–1153.

[37] Kemper G. *A course in commutative algebra*. Graduate Texts in Mathematics, 2011.

[38] Kiming I. *Reduction modulo p*. University of Copenhagen.

[39] Koblitz N. *Algebraic Aspects of Cryptography*. Algorithms and Computation in Mathematics, Springer, 2004.

[40] Krauss A. *Cours de cryptographie MM067 - 2012/13 - Chapitre VII - Courbes elliptiques*. Université Pierre et Marie-Curie.

[41] Kreuzer M., Robbiano L. *Computational Commutative Algebra 1*. Springer, 2000.

[42] Kreuzer M., Robbiano L. *Computational Commutative Algebra 2*. Springer, 2005.

[43] Lang, S. *Algebra*. Second edn., Addison-Wesley, Redwood City, 1984.

[44] Laszlo R. *Ein kombinatorischer Satz*. Acta Sci. Math. (Szeged) **7** (1934–1935) 39–43.

[45] Lenstra Jr. H.W. *Factoring integers with elliptic curves*. Ann. Math. **126** (1987) 649–673.

[46] Liu J., Li D., Liu W. *Some criteria for Gröbner bases and their applications*. J. Symb. Comp. **92** (2019) 15–21.

[47] Lombardi H., Perdry H. *The Buchberger Algorithm as a Tool for Ideal Theory of Polynomial Rings in Constructive Mathematics*. In: Gröbner Bases and Applications (Proc. of the Conference 33 Years of Gröbner Bases), Cambridge University Press, London Mathematical Society Lecture Notes Series **251** (1998) 393–407.

[48] Mayr E.W., Meyer A.R. *The complexity of the word problems for commutative semigroups and polynomial ideals*. Adv. Math. **46** (1982) 305–329.

[49] Monceur S., Yengui I. *On the leading terms ideals of polynomial ideals over a valuation ring*. J. Algebra **351** (2012) 382–389.

[50] Norton G.H., Salagean A. *Strong Gröbner bases and cyclic codes over a finite-chain ring*. Applicable algebra in engineering, communication and computing **10** (2000) 489–506.

[51] Norton G.H., Salagean A. *Strong Gröbner bases for polynomials over a principal ideal ring*. Bull. of the Australian Mathematical Soc. **64** (2001) 505–528.

[52] Norton G.H., Salagean A. *Gröbner bases and products of coefficient rings*. Bull. of the Australian Mathematical Soc. **65** (2002) 145–152.

[53] Norton G.H., Salagean A. *Cyclic codes and minimal strong Gröbner bases over a principal ideal ring*. Finite fields and their applications **9** (2003) 237–249.

[54] Pauer F. *On lucky ideals for Gröbner basis computations*, J. Symb. Comp. **14** (1992) 471–482.

[55] Sasaki T., Takeshima T. *A modular method for Gröbner-basis construction over $\mathbb{Q}$ and solving system of algebraic equations*. Journal of Information Processing **12** (1989) 371–379.

[56] Saux Picart Ph., Rannou E. *Cours de Calcul Formel. Corps finis - Systèmes polynomiaux - Applications*, Ellipses, 2002.

[57] Schreyer F.-O. *Die Berechnung von Syzygien mit dem verallgemeinerten Weierstrassschen Divisionssatz*, Diploma Thesis, University of Hamburg, Germany, 1980.

[58] Serre J.-P. *Faisceaux algébriques cohérents.* Ann. Math. **61** (1955) 191–278.

[59] Shafarevich I.R. *Basic Algebraic Geometry*, Springer, Berlin, New York, 1974.

[60] Shekhar N., Kalla P., Enescu F., Gopalakrishnan S. Equivalence verification of polynomial datapaths with fixed-size bit-vectors using finite ring algebra. In ICCAD 05: Proceedings of the 2005 IEEE/ACM International conference on Computer-aided design, pages 291–296, Washington, DC, USA, 2005. IEEE Computer Society.

[61] Silverman J.H. *The arithmetic of elliptic curves*, Graduate Texts in Mathematics, vol. 106, Springer-Verlag, New York, 1994, Corrected reprint of the 1986 original.

[62] Smith B. *Introduction to Elliptic Curves*, Talk at ECRYPT II Winter School, Lausanne, January 2009.

[63] Talovikova V. *Riemann-Roch Theorem*, 2009.

[64] Yengui I. *Dynamical Gröbner bases.* J. Algebra **301** (2006) 447–458.

[65] Yengui I. *Corrigendum to Dynamical Gröbner bases [J. Algebra 301 (2) (2006) 447–458] and to Dynamical Gröbner bases over Dedekind rings [J. Algebra 324 (1) (2010) 12–24].* J. Algebra **339** (2011) 370–35.

[66] Yengui I. *Constructive Commutative Algebra - Projective modules over polynomial rings and dynamical Gröbner bases.* Lecture Notes in Mathematics, no. 2138, Springer 2015.

[67] Wienand O. *Algorithms for Symbolic Computation and their Applications.* PhD thesis, Kaiserslautern (2011).

[68] Winkler F. *A p-adic approach to the computation of Gröbner bases.* J. Symb. Comp. **6** (1987) 287–304.

[69] Zariski O., Samuel P. *Commutative Algebra, vol. I*, Springer-Verlag, New-York, 1960.

Printed in the United States
by Baker & Taylor Publisher Services

Printed in the United States
by Baker & Taylor Publisher Services